Assured Cloud Computing

IEEE PRESS

IEEE Computer Society

About IEEE Computer Society

IEEE Computer Society is the world's leading computing membership organization and the trusted information and career-development source for a global workforce of technology leaders including: professors, researchers, software engineers, IT professionals, employers, and students. The unmatched source for technology information, inspiration, and collaboration, the IEEE Computer Society is the source that computing professionals trust to provide high-quality, state-of-the-art information on an on-demand basis. The Computer Society provides a wide range of forums for top minds to come together, including technical conferences, publications, and a comprehensive digital library, unique training webinars, professional training, and the TechLeader Training Partner Program to help organizations increase their staff's technical knowledge and expertise, as well as the personalized information tool myComputer. To find out more about the community for technology leaders, visit http://www.computer.org.

IEEE/Wiley Partnership

The IEEE Computer Society and Wiley partnership allows the CS Press authored book program to produce a number of exciting new titles in areas of computer science, computing, and networking with a special focus on software engineering. IEEE Computer Society members continue to receive a 15% discount on these titles when purchased through Wiley or at wiley.com/ieeecs.

To submit questions about the program or send proposals, please contact Mary Hatcher, Editor, Wiley-IEEE Press: Email: mhatcher@wiley.com, Telephone: 201-748-6903, John Wiley & Sons, Inc., 111 River Street, Hoboken, NJ 07030-5774.

Assured Cloud Computing

Edited by Roy H. Campbell, Charles A. Kamhoua,
and Kevin A. Kwiat

WILEY

Registered Office
John Wiley & Sons, Inc., 111 River Street, Hoboken, NJ 07030, USA

Editorial Office
111 River Street, Hoboken, NJ 07030, USA

For details of our global editorial offices, customer services, and more information about Wiley products visit us at www.wiley.com.

Wiley also publishes its books in a variety of electronic formats and by print-on-demand. Some content that appears in standard print versions of this book may not be available in other formats.

Library of Congress Cataloging-in-Publication Data

Names: Campbell, Roy Harold, editor. | Kamhoua, Charles A., editor. | Kwiat,
  Kevin A., editor.
Title: Assured cloud computing / edited by Roy H. Campbell, Charles A.
  Kamhoua, Kevin A. Kwiat.
Description: First edition. | Hoboken, NJ : IEEE Computer Society,
  Inc./Wiley, 2018. | Includes bibliographical references and index. |
  Identifiers: LCCN 2018025067 (print) | LCCN 2018026247 (ebook) | ISBN
  9781119428503 (Adobe PDF) | ISBN 9781119428480 (ePub) | ISBN 9781119428633
  (hardcover)
Subjects: LCSH: Cloud computing.
Classification: LCC QA76.585 (ebook) | LCC QA76.585 .A87 2018 (print) | DDC
  004.67/82–dc23
LC record available at https://lccn.loc.gov/2018025067

Cover image: Abstract gray polka dots pattern background - ©shuoshu/Getty Images; Abstract modern background - ©tmeks/iStockphoto; Abstract wave - ©Keo/Shutterstock

Cover design by Wiley

Set in 10/12 pt WarnockPro-Regular by Thomson Digital, Noida, India

Printed in the United States of America

V10003772_081618

Table of Contents

Preface

Starting around 2009, higher bandwidth networks, low-cost commoditized computers and storage, hardware virtualization, large user populations, service-oriented architectures, and autonomic and utility computing together provided the foundation for a dramatic change in the scale at which computation could be provisioned and managed. Popularly, the resulting phenomenon became known as *cloud computing*. The National Institute of Standards and Technology (NIST), tasked with addressing the phenomenon, defines it in the following way:

> "Cloud computing is a model for enabling ubiquitous, convenient, on-demand network access to a shared pool of configurable computing resources (e.g., networks, servers, storage, applications and services) that can be rapidly provisioned and released with minimal management effort or service provider interaction." [1]

In 2011, the U.S. Air Force, through the Air Force Research Laboratory (AFRL) and the Air Force Office of Scientific Research (AFOSR), established the Assured Cloud Computing Center of Excellence (ACC-UCoE) at the University of Illinois at Urbana-Champaign to explore how cloud computing could be used to better support the computing and communication needs of the Air Force. The Center then pursued a broad program of collaborative research and development to address the core technical obstacles to the achievement of assured cloud computing, including ones related to design, formal analysis, runtime configuration, and experimental evaluation of new and modified architectures, algorithms, and techniques. It eventually amassed a range of research contributions that together represent a comprehensive and robust response to the challenges presented by cloud computing. The team recognized that there would be significant value in making a suite of key selected ACC-UCoE findings readily available to the cloud computing community under one cover, pulled together with newly written connective material that explains how the individual research contributions relate to each other and to the big picture of assured cloud computing. Thus, we produced this book, which offers in one volume some of the most important and highly cited research findings of the Assured Cloud Computing Center.

Military computing requirements are complex and wide-ranging. Indeed, rapid technological advances and the advent of computer-based weapon systems have created the need for network-centric military superiority. However, network-centricity is stretched in the context of global networking requirements and the desire to use cloud computing. Furthermore, cloud computing is heavily based on the use of commercial off-the-shelf technology. Outsourcing operations on commercial, public, and hybrid clouds introduces the challenge of ensuring that a computation and its data are secure even as operations are performed remotely over networks over which the military does not have absolute control. Finally, nowadays, military superiority requires agility and mobility. This both increases the benefits of using cloud computing, because of its ubiquitous accessibility, and increases the difficulty of assuring access, availability, security, and robustness.

However, although military requirements are driving major research efforts in this area, the need for assured cloud computing is certainly not limited to the military. Cloud computing has also been widely adopted in industry, and the government has asked its agencies to adopt it as well. Cloud computing offers economic advantages by amortizing the cost of expensive computing infrastructure and resources over many client services. A survivable and distributed cloud-computing-based infrastructure can enable the configuration of any dynamic systems-of-systems that contain both trusted and partially trusted resources (such as data, sensors, networks, and computers) and services sourced from multiple organizations. To assure mission-critical computations and workflows that rely on such dynamically configured systems-of-systems, it is necessary to ensure that a given configuration does not violate any security or reliability requirements. Furthermore, it is necessary to model the trustworthiness of a workflow or computations' completion to gain high assurances.

The focus of this book is on providing solutions to the problems of cloud computing to ensure a robust, dependable computational and data cyberinfrastructure for operations and missions. While the research has been funded by the Air Force, its outcomes are relevant and applicable to cloud computing across all domains, not just to military activities. The Air Force acknowledges the value of this interdomain transfer as exemplified by the Air Force's having patented – with an intended goal of commercialization – some of the cloud computing innovation described in this book.

This material is based on research sponsored by the Air Force Research Laboratory (AFRL) and the Air Force Office of Scientific Research (AFOSR) under agreement number FA8750-11-2-0084, and we would like to thank AFRL and AFOSR for their financial support, collaboration, and guidance.[1] The U.S. Government is authorized to reproduce and distribute reprints for governmental purposes notwithstanding any copyright notation thereon. The work described in this book was also partially supported by the Boeing Company and by other sources acknowledged in individual chapters.

The editors would like to acknowledge the contributions of the following individuals (in alphabetical order): Cristina L. Abad, Gul Agha, Masooda N. Bashir, Rakesh B. Bobba, Chris X. Cai, Roy H. Campbell, Tej Chajed, Brian Cho, Domenico Cotroneo, Fei Deng, Carlo Di Giulio, Peter Dinges, Zachary J. Estrada, Jatin Ganhotra, Mainak Ghosh, Jon Grov, Indranil Gupta, Gopala-krishna Holla, Jingwei Huang, Jun Ho Huh, Ravishankar K. Iyer, Zbigniew Kalbarczyk, Charles A. Kamhoua, Manoj Kumar, Kevin A. Kwiat, Luke Kwiat, Luke M. Leslie, Tianwei Li, Philbert Lin, Si Liu, Yi Lu, Andrew Martin, José Meseguer, Priyesh Narayanan, Sivabalan Narayanan, Son Nguyen, David M. Nicol, Shadi A. Noghabi, Peter Csaba Ölveczky, Antonio Pecchia, Boyang Peng, Cuong Pham, Mayank Pundir, Muntasir Rahman, Nathan Roberts, Aashish Sharma, Reza Shiftehfar, Yosub Shin, Stephen Skeirik, Read Sprabery, Sriram Subramanian, Jian Tang, Gary Wang, Wenting Wang, Le Xu, Lok Yan, Mindi Yuan, and Mammad Zadeh. We would also like to thank Todd Cushman, Robert Herklotz, Tristan Nguyen, Laurent Njilla, Andrew Noga, James Perretta, Anna Weeks, and Stanley Wenndt. Finally, we would like to thank and acknowledge Jenny Applequist, who helped edit and collect the text into its final form, as well as Mary Hatcher, Vishnu Narayanan, Victoria Bradshaw, and Melissa Yanuzzi of Wiley and Vinod Pandita of Thomson Digital for their kind assistance in guiding this book through the publication process.

Reference

1 Mell, P. and Grance, T., The NIST Definition of Cloud Computing: Recommendations of the National Institute of Standards and Technology. Special Publication 800-145, National Institute of Standards and Technology, U.S. Department of Commerce, Sep. 2011. Available at http://dx.doi.org/10.6028/NIST.SP.800-145.

---

1 *Disclaimer:* The views and content expressed in this book are those of the authors and do not reflect the official policy or position of the Department of the Air Force, Department of Defense, or the U.S. Government.

Editors' Biographies

Roy H. Campbell is Associate Dean for Information Technology of the College of Engineering, the Sohaib and Sara Abbasi Professor in the Department of Computer Science, and Director of the NSA-designated Center for Academic Excellence in Information Assurance Education and Research at the University of Illinois at Urbana-Champaign (UIUC); previously, he was Director of the Air Force-funded Assured Cloud Computing Center in the Information Trust Institute at UIUC from 2011 to 2017. He received his Honors B.S. degree in Mathematics, with a Minor in Physics, from the University of Sussex in 1969 and his M.S. and Ph.D. degrees in Computer Science from the University of Newcastle upon Tyne in 1972 and 1976, respectively. Professor Campbell's research interests are the problems, engineering, and construction techniques of complex system software. Cloud computing, data analytics, big data, security, distributed systems, continuous media, and real-time control pose system challenges, especially to operating system designers. Past research includes path expressions as declarative specifications of process synchronization, real-time deadline recovery mechanisms, error recovery in asynchronous systems, streaming video for the Web, real-time Internet video distribution systems, object-oriented parallel processing operating systems, CORBA security architectures, and active spaces in ubiquitous and pervasive computing. He is a Fellow of the IEEE.

Charles A. Kamhoua is a researcher at the Network Security Branch of the U.S. Army Research Laboratory (ARL) in Adelphi, MD, where he is responsible for conducting and directing basic research in the area of game theory applied to cyber security. Prior to joining the Army Research Laboratory, he was a researcher at the U.S. Air Force Research Laboratory (AFRL), Rome, New York for 6 years and an educator in different academic institutions for more than 10 years. He has held visiting research positions at the University of Oxford and Harvard University. He has coauthored more than 100 peer-reviewed journal and conference papers. He has presented over 40 invited keynote and distinguished speeches and has co-organized over 10 conferences and workshops. He has mentored more than 50 young scholars, including students, postdocs, and AFRL Summer Faculty Fellowship scholars. He has been recognized for his scholarship and leadership with numerous prestigious awards, including the 2017 AFRL Information Directorate Basic Research Award "For Outstanding Achievements in Basic Research," the 2017 Fred I. Diamond Award for the best paper published at AFRL's Information Directorate, 40 Air Force Notable Achievement Awards, the 2016 FIU Charles E. Perry Young Alumni Visionary Award, the 2015 Black Engineer of the Year Award (BEYA), the 2015 NSBE Golden Torch Award – Pioneer of the Year, and selection to the 2015 Heidelberg Laureate Forum, to name but a few. He received a B.S. in electronics from the University of Douala (ENSET), Cameroon, in 1999, an M.S. in Telecommunication and Networking from Florida International University (FIU) in 2008, and a Ph.D. in Electrical Engineering from FIU in 2011. He is currently an advisor for the National Research Council, a member of the FIU alumni association and ACM, and a senior member of IEEE.

Kevin A. Kwiat retired in 2017 as Principal Computer Engineer with the U.S. Air Force Research Laboratory (AFRL) in Rome, New York after more than 34 years of federal service. During that time, he conducted research and development in a wide range of areas, including high-reliability microcircuit selection for military systems, testability, logic and fault simulation, rad-hard microprocessors, benchmarking of experimental computer architectures, distributed processing systems, assured communications, FPGA-based reconfigurable computing, fault tolerance, survivable systems, game theory, cyber-security, and cloud computing. He received a B.S. in Computer Science and a B.A. in Mathematics from Utica College of Syracuse University, and an M.S. in Computer Engineering and a Ph.D. in Computer Engineering from Syracuse University. He holds five patents. He is co-founder and co-leader of Haloed Sun TEK of Sarasota, Florida, which is an LLC specializing in technology transfer and has joined forces with the Commercial Applications for Early Stage Advanced Research (CAESAR) Group. He is also an adjunct professor of Computer Science at the State University of New York Polytechnic Institute, and a Research Associate Professor with the University at Buffalo.

List of Contributors

Cristina L. Abad
Escuela Superior Politecnica del
Litoral
ESPOL
Guayaquil
Ecuador

Gul Agha
Department of Computer Science
University of Illinois at Urbana-
Champaign
Urbana, IL
USA

Masooda Bashir
School of Information Sciences
University of Illinois at Urbana-
Champaign
Champaign, IL
USA

Rakesh Bobba
School of Electrical Engineering
and Computer Science
Oregon State University
Corvallis, OR
USA

Roy H. Campbell
Department of Computer Science
University of Illinois at Urbana-
Champaign
Urbana, IL
USA

Minas Charalambides
Department of Computer Science
University of Illinois at Urbana-
Champaign
Urbana, IL
USA

Domenico Cotroneo
Dipartimento di Ingegneria
Elettrica e delle Tecnologie
dell'Informazione
Università degli Studi di Napoli
Federico II
Naples
Italy

Fei Deng
Department of Electrical and
Computer Engineering
University of Illinois at Urbana-
Champaign
Urbana, IL
USA

Carlo Di Giulio
Information Trust Institute
University of Illinois at Urbana-
Champaign
Urbana, IL
USA

and

European Union Center
University of Illinois at Urbana-
Champaign
Champaign, IL
USA

Zachary Estrada
Department of Electrical and
Computer Engineering
Rose-Hulman Institute of
Technology
Terre Haute, IN
USA

and

Department of Electrical and
Computer Engineering
University of Illinois at Urbana-
Champaign
Urbana, IL
USA

Mainak Ghosh
Department of Computer Science
University of Illinois at Urbana-
Champaign
Urbana, IL
USA

Jon Grov
Gauge AS
Oslo
Norway

Indranil Gupta
Department of Computer Science
University of Illinois at Urbana-
Champaign
Urbana, IL
USA

Jingwei Huang
Department of Engineering
Management and Systems
Engineering
Old Dominion University
Norfolk, VA
USA

and

Information Trust Institute
University of Illinois at Urbana-
Champaign
Urbana, IL
USA

Jun Ho Huh
Samsung Research
Samsung Electronics
Seoul
South Korea

Ravishankar K. Iyer
Department of Electrical and
Computer Engineering and
Coordinated Science Laboratory
University of Illinois at Urbana-
Champaign
Urbana, IL
USA

Zbigniew Kalbarczyk
Department of Electrical and
Computer Engineering and
Coordinated Science Laboratory
University of Illinois at Urbana-
Champaign
Urbana, IL
USA

Charles A. Kamhoua
Network Security Branch
Network Sciences Division
U.S. Army Research Laboratory
Adelphi, MD
USA

Kevin A. Kwiat
Haloed Sun TEK
Sarasota, FL
USA

Luke Kwiat
Department of Industrial and
Systems Engineering
University of Florida
Gainesville, FL
USA

Si Liu
Department of Computer Science
University of Illinois at Urbana-
Champaign
Urbana, IL
USA

Kirill Mechitov
Department of Computer Science
University of Illinois at Urbana-
Champaign
Urbana, IL
USA

José Meseguer
Department of Computer Science
University of Illinois at Urbana-
Champaign
Urbana, IL
USA

David M. Nicol
Department of Electrical and
Computer Engineering and
Information Trust Institute
University of Illinois at Urbana-
Champaign
Urbana, IL
USA

Shadi A. Noghabi
Department of Computer Science
University of Illinois at Urbana-
Champaign
Urbana, IL
USA

Peter Csaba Ölveczky
Department of Computer Science
University of Illinois at Urbana-
Champaign
Urbana, IL
USA

and

Department of Informatics
University of Oslo
Oslo
Norway

Karl Palmskog
Department of Computer Science
University of Illinois at Urbana-
Champaign
Urbana, IL
USA

Antonio Pecchia
Dipartimento di Ingegneria
Elettrica e delle Tecnologie
dell'Informazione
Università degli Studi di Napoli
Federico II
Naples
Italy

Cuong Pham
Department of Electrical and
Computer Engineering
University of Illinois at Urbana-
Champaign
Urbana, IL
USA

Atul Sandur
Department of Computer Science
University of Illinois at Urbana-
Champaign
Urbana, IL
USA

Aashish Sharma
Lawrence Berkeley National Lab
Berkeley, CA
USA

Reza Shiftehfar
Department of Computer Science
University of Illinois at Urbana-
Champaign
Urbana, IL
USA

Stephen Skeirik
Department of Computer Science
University of Illinois at Urbana-
Champaign
Urbana, IL
USA

Jian Tang
Department of Electrical
Engineering and Computer Science
Syracuse University
Syracuse, NY
USA

Gary Wang
Department of Computer Science
University of Illinois at Urbana-
Champaign
Urbana, IL
USA

Le Xu
Department of Computer Science
University of Illinois at Urbana-
Champaign
Urbana, IL
USA

Lok Yan
Air Force Research Laboratory
Rome, NY
USA

1

Introduction

Roy H. Campbell

Department of Computer Science, University of Illinois at Urbana-Champaign, Urbana, IL, USA

Mission assurance for critical cloud applications is of growing importance to governments and military organizations, yet mission-critical cloud computing may face the challenge of needing to use hybrid (public, private, and/or heterogeneous) clouds and require the realization of "end-to-end" and "cross-layered" security, dependability, and timeliness. In this book, we consider cloud applications in which assigned tasks or duties are performed in accordance with an intended purpose or plan in order to accomplish an *assured* mission.

1.1 Introduction

Rapid technological advancements in global networking, commercial off-the-shelf technology, security, agility, scalability, reliability, and mobility created a window of opportunity in 2009 for reducing the costs of computation and led to the development of what is now known as *cloud computing* [1–3]. Later, in 2010, the Obama Administration [4] announced an

> "extensive adoption of cloud computing in the federal government to improve information technology (IT) efficiency, reduce costs, and provide a standard platform for delivering government services. In a cloud computing environment, IT resources—services, applications, storage devices and servers, for example—are pooled and managed centrally. These resources can be provisioned and made available on demand via the Internet. The cloud model strengthens the resiliency of mission-critical applications by removing dependency on underlying hardware. Applications can be easily moved from one system to another in the event of system failures or cyber attacks" [5].

In the same year, the Air Force signed an initial contract with IBM to build a mission-assured cloud computing capability [5].

Assured Cloud Computing, First Edition. Edited by Roy H. Campbell, Charles A. Kamhoua, and Kevin A. Kwiat.
© 2018 the IEEE Computer Society, Inc. Published 2018 by John Wiley & Sons, Inc.

Table 1.1 Model of cloud computing.

Service Models	Deployment Models
Software as a Service	**Private Cloud**
	Community Cloud
Platform as a Service	**Hybrid Cloud**
	Public Cloud
Infrastructure as a Service	
Essential Characteristics	Resource Pooling · Rapid Elasticity · Measured Service · Broad Network Access

Cloud computing was eventually defined by the National Institute of Standards and Technology (as finalized in 2011) as follows [6]: "Cloud computing is a model for enabling ubiquitous, convenient, on-demand network access to a shared pool of configurable computing resources (e.g., networks, servers, storage, applications, and services) that can be rapidly provisioned and released with minimal management effort or service provider interaction. This cloud model is composed of five essential characteristics, three service models, and four deployment models." That model of cloud computing is depicted in Table 1.1.

One of the economic reasons for the success of cloud computing has been the scalability of the computational resources that it provides to an organization. Instead of requiring users to size a planned computation exactly (e.g., in terms of the number of needed Web servers, file systems, databases, or compute engines), cloud computing allows the computation to scale easily in a time-dependent way. Thus, if a service has high demand, it can be replicated to make it more available. Instead of having two Web servers provide a mission-critical service, the system might allow five more Web servers to be added to the service to increase its availability. Likewise, if demand for a service drops, the resources it uses can be released, and thus be freed up to be used for other worthwhile computation. This flexible approach allows a cloud to economically support a number of organizations at the same time, thereby lowering the costs of cloud computation. In later chapters, we will discuss scaling performance and how to assure the correctness of a mission-oriented cloud computation as it changes in size, especially when the scaling occurs dynamically (i.e., is *elastic*).

1.1.1 Mission-Critical Cloud Solutions for the Military

As government organizations began to adopt cloud computing, security, availability, and robustness became growing concerns; there was a desire to use cloud computing even in mission-critical contexts, where a *mission-critical system* is one that is essential to the survival of an organization. In 2010, in response to military recognition of the inadequacy of the then state-of-the-art technologies, IBM was awarded an Air Force contract to build a secure cloud computing infrastructure capable of supporting defense and intelligence networks [5]. However, the need for cloud computing systems that could support missions involved more numerous major concerns than could easily be solved in a single, focused initiative and, in particular, raised the question of how to assure cloud support for mission-oriented computations—the subject of this book. Mission-critical cloud computing can stretch across private, community, hybrid, and public clouds, requiring the realization of "end-to-end" and "cross-layered" security, dependability, and timeliness. That is, cloud computations and computing systems should survive malicious attacks and accidental failures, should be secure, and should execute in a timely manner, despite the heterogeneous ownership and nature of the hardware components.

End-to-end implies that the properties should hold throughout the lifetime of individual events, for example, a packet transit or a session between two machines, and that they should be assured in a manner that is independent of the environment through which such events pass. Similarly, *cross-layer* encompasses multiple layers, from the end device through the network and up to the applications or computations in the cloud. A survivable and distributed cloud-computing-based infrastructure requires the configuration and management of dynamic systems-of-systems with both trusted and partially trusted resources (including data, sensors, networks, computers, etc.) and services sourced from multiple organizations. For mission-critical computations and workflows that rely on such dynamically configured systems-of-systems, we must ensure that a given configuration doesn't violate any security or reliability requirements. Furthermore, we should be able to model the trustworthiness of a workflow or computation's completion for a given configuration in order to specify the right configuration for high assurance.

Rapid technological advances and computer-based weapons systems have created the need for net-centric military superiority. Overseas commitments and operations stretch net-centricity with global networking requirements, use of government and commercial off-the-shelf technology, and the need for agility, mobility, and secure computing over a mixture of blue and gray networks. (*Blue* networks are military networks that are considered secure, while *gray* networks are those in private hands, or run by other nations, that may not be secure.) An important goal is to ensure the confidentiality and integrity of data and communications needed to get missions done, even amid cyberattacks and failures.

1.2 Overview of the Book

This book encompasses the topics of architecture, design, testing, and formal verification for assured cloud computing. The authors propose approaches for using formal methods to analyze, reason, prototype, and evaluate the architectures, designs, and performance of secure, timely, fault-tolerant, mission-oriented cloud computing. They examine a wide range of necessary assured cloud computing components and many urgent concerns of these systems.

The chapters of this book provide research overviews of (1) flexible and dynamic distributed cloud-computing-based architectures that are survivable; (2) novel security primitives, protocols, and mechanisms to secure and support assured computations; (3) algorithms and techniques to enhance end-to-end timeliness of computations; (4) algorithms that detect security policy or reliability requirement violations in a given configuration; (5) algorithms that dynamically configure resources for a given workflow based on security policy and reliability requirements; and (6) algorithms, models, and tools to estimate the probability of completion of a workflow for a given configuration. Further, we discuss how formal methods can be used to analyze designed architectures, algorithms, protocols, and techniques to verify the properties they enable. Prototypes and implementations may be built, formally verified against specifications, and tested as components in real systems, and their performance can be evaluated.

While our research has spanned most of the cloud computing phenomenon's lifetime to date, it has had, like all fast-moving technological advances, only a short history (starting 2011). Much work is still to be done as cloud computing evolves and "mission-critical" takes on new meanings within the modern world. Wherever possible, throughout the volume (and in the concluding chapter) we have offered reflections on the state of the art and commented on future directions.

- Chapter 2: **Survivability: Design, Formal Modeling, and Validation of Cloud Storage Systems Using Maude,** José Meseguer in collaboration with Rakesh Bobba, Jon Grov, Indranil Gupta, Si Liu, Peter Csaba Ölveczky, and Stephen Skeirik

  To deal with large amounts of data while offering high availability and throughput and low latency, cloud computing systems rely on distributed, partitioned, and replicated data stores. Such cloud storage systems are complex software artifacts that are very hard to design and analyze. We argue that formal specification and model checking analysis should significantly improve their design and validation. In particular, we propose rewriting logic and its accompanying Maude tools as a suitable framework for formally specifying and analyzing both the correctness and the performance of cloud storage systems. This chapter largely focuses on how we have used rewriting logic to model and analyze industrial cloud storage systems such as Google's Megastore, Apache Cassandra, Apache ZooKeeper, and RAMP. We also touch on the use of formal methods at Amazon Web Services. Cloud computing relies on software systems that store large amounts of data correctly and efficiently. These cloud systems are expected to achieve high performance (defined as high availability and throughput) and low latency. Such performance needs to be assured even in the presence of congestion in parts of the network, system or network faults, and scheduled hardware and software upgrades. To achieve this, the data must be replicated both across the servers within a site and across geo-distributed sites. To achieve the expected scalability and elasticity of cloud systems, the data may need to be partitioned. However, the CAP theorem states that it is impossible to have both high availability and strong consistency (correctness) in replicated data stores in today's Internet.

Different storage systems therefore offer different trade-offs between the levels of availability and consistency that they provide. For example, weak notions of consistency of multiple replicas, such as "eventual consistency," are acceptable for applications (such as social networks and search) for which availability and efficiency are key requirements, but for which it would be tolerable if different replicas stored somewhat different versions of the data. Other cloud applications, including online commerce and medical information systems, require stronger consistency guarantees.

The key challenge addressed in this chapter is that of how to design cloud storage systems with high assurance such that they satisfy desired correctness, performance, and quality of service requirements.

- Chapter 3: **Risks and Benefits: Game-Theoretical Analysis and Algorithm for Virtual Machine Security Management in the Cloud,** Luke A. Kwiat in collaboration with Charles A. Kamhoua, Kevin A. Kwiat, and Jian Tang

  Many organizations have been inspired to move to the cloud the services they depend upon and offer because of the potential for cost savings, ease of access, availability, scalability, and elasticity. However, moving services into a multitenancy environment raises many difficult problems. This chapter uses a game-theoretic approach to take a hard look at those problems. It contains a broad overview of the ways game theory can contribute to cloud computing. Then it turns to the more specific question of security and risk. Focusing on the virtual machine technology that supports many cloud implementations, the chapter delves into the security issues involved when one organization using a cloud may impact other organizations that are using that same cloud. The chapter provides an interesting insight that a cloud and its multiple tenants represent many different opportunities for attackers and asks some difficult questions: To what extent, independent of the technology used, does multitenancy create security problems, and to what extent, based on a "one among many" argument, does it help security? In general, what, mathematically, can one say about multitenancy clouds and security? It is interesting to note that it may be advantageous for cloud applications that have the same levels of security and risk to be clustered together on the same machines.

- Chapter 4: **Detection and Security: Achieving Resiliency by Dynamic and Passive System Monitoring and Smart Access Control**, Zbigniew Kalbarczyk in collaboration with Rakesh Bobba, Domenico Cotroneo, Fei Deng, Zachary Estrada, Jingwei Huang, Jun Ho Huh, Ravishankar K. Iyer, David M. Nicol, Cuong Pham, Antonio Pecchia, Aashish Sharma, Gary Wang, and Lok Yan

  System reliability and security is a well-researched topic that has implications for the difficult problem of cloud computing resiliency. Resiliency is described as an interdisciplinary effort involving monitoring, detection, security, recovery from failures, human factors, and availability. Factors of concern include design, assessment, delivery of critical services, and interdependence among systems. None of these are simple matters, even in a static system. However, cloud computing can be very dynamic (to manage elasticity concerns, for example), and this raises issues of situational awareness, active and passive monitoring, automated reasoning, coordination of monitoring and system activities (especially when there are accidental failures or malicious attacks), and use of access control to modify the attack surface. Because use of virtual machines is a significant aspect of reducing costs from shared resources, the chapter features virtualization resilience issues. One practical topic focused on is that of whether hook-based monitoring technology has a place in instrumenting virtual machines and hypervisors with probes to report anomalies and attacks. If one creates a strategy for hypervisor monitoring that takes into account the correct behavior of guest operating systems, then it is possible to construct a "return-to-user" attack detector and a process-based "key logger," for example. However, even with such monitoring in place, attacks can still occur by means of hypervisor introspection and cross-VM side-channels. A number of solutions from the literature, together with the hook-based approach, are reviewed, and partial solutions are offered.

  On the user factors side of attacks, a study of data on credential-stealing incidents at the National Center for Supercomputing Applications revealed that a threshold for correlated events related to intrusion can eliminate many false positives while still identifying compromised users. The authors pursue that approach by using Bayesian networks with event data to estimate the likelihood that there is a compromised user. In the example data evaluated, this approach proved to be very effective. Developing the notion that stronger and more precise access controls would allow for better incident analysis and fewer false positives, the researchers combine attribute-based access control (ABAC) and role-based access control (RBAC). The scheme describes a flexible RBAC model based on ABAC to allow more formal analysis of roles and policies.

- Chapter 5: **Scalability, Workloads, and Performance: Replication, Popularity, Modeling, and Geo-Distributed File Stores**, Roy H. Campbell in collaboration with Shadi A. Noghabi and Cristina L. Abad

  Scalability allows a cloud application to change in size, volume, or geographical distribution while meeting the needs of the cloud customer. A practical approach to scaling cloud applications is to improve the availability of the application by replicating the resources and files used, including creating multiple copies of the application across many nodes in the cloud. Replication improves availability through redundant resources, services, networks, file systems, and nodes but also creates problems with respect to whether clients observe consistency as they are served from the multiple copies. Variability in data sizes, volumes, and the homogeneity and performance of the cloud components (disks, memory, networks, and processors) can impact scalability. Evaluating scalability is difficult, especially when there is a large degree of variability. This leads one to estimate how applications will scale on clouds based on probabilistic estimates of job load and performance. Scaling can have many different dimensions and properties. The emergence of low-latency worldwide services and the desire to have higher fault tolerance and reliability have led to the design of geo-distributed storage with replicas in multiple locations. Scalability in terms of global information systems implemented on the cloud is also geo-distributed. We consider, as a case example, scalable geo-distributed storage.

- Chapter 6: **Resource Management: Performance Assuredness in Distributed Cloud Computing via Online Reconfigurations**, Indranil Gupta in collaboration with Mainak Ghosh and Le Xu

  Building systems that perform *predictably* in the cloud remains one of the biggest challenges today, both in mission-critical scenarios and in non-real-time scenarios. Many cloud infrastructures do not easily support, in an assured manner, reconfiguration operations such as changing of the shard key in a sharded storage/database system, or scaling up (or down) of the number of VMs being used in a stream or batch processing system. We discuss *online* reconfiguration operations whereby the system does not need to be shut down and the user/client-perceived behavior is indistinguishable regardless of whether a reconfiguration is occurring in the background, that is, the performance continues to be assured in spite of ongoing background reconfiguration. We describe ways to scale-out and scale-in (increase or decrease) the number of machines/VMs in cloud computing frameworks, such as distributed stream processing and distributed graph processing systems, again while offering assured performance to the customer in spite of the reconfigurations occurring in the background. The ultimate performance assuredness is the ability to support SLAs/SLOs (service-level agreements/objectives) such as deadlines. We present a new real-time scheduler that supports priorities and hard deadlines for Hadoop jobs.

  This chapter describes multiple contributions toward solution of key issues in this area. After a review of the literature, it provides an overview of five systems that were created in the Assured Cloud Computing Center that are oriented toward offering performance assuredness in cloud computing frameworks, even while the system is under change:

1) Morphus (based on MongoDB), which supports reconfigurations in sharded distributed NoSQL databases/storage systems.
2) Parqua (based on Cassandra), which supports reconfigurations in distributed ring-based key-value stores.
3) Stela (based on Storm), which supports scale-out/scale-in in distributed stream processing systems.
4) A system (based on LFGraph) to support scale-out/scale-in in distributed graph processing systems.
5) Natjam (based on Hadoop), which supports priorities and deadlines for jobs in batch processing systems.

We describe each system's motivations, design, and implementation, and present experimental results.

- Chapter 7: **Theoretical Considerations: Inferring and Enforcing Use Patterns for Mobile Cloud Assurance**, Gul Agha in collaboration with Minas Charalambides, Kirill Mechitov, Karl Palmskog, Atul Sandur, and Reza Shiftehfar

  The mobile cloud combines cloud computing, mobile computing, smart sensors, and wireless networks into well-integrated ecosystems. It offers unrestricted functionality, storage, and mobility to serve a multitude of mobile devices anywhere, anytime. This chapter shows how support for fine-grained mobility can improve mobile cloud security and trust while maintaining the benefits of efficiency. Specifically, we discuss an actor-based programming framework that can facilitate the development of mobile cloud systems and improve efficiency while enforcing security and privacy. There are two key ideas. First, by supporting fine-grained units of computation (actors), a mobile cloud can be agile in migrating components. Such migration is done in response to a system context (including dynamic variables such as available bandwidth, processing power, and energy) while respecting constraints on information containment boundaries. Second, through specification of constraints on interaction patterns, it is possible to observe information flow between actors and flag or prevent suspicious activity.

- Chapter 8: **Certifications Past and Future: A Future Model for Assigning Certifications that Incorporate Lessons Learned from Past Practices**, Masooda Bashir in collaboration with Carlo Di Giulio and Charles A. Kamhoua

  This chapter describes the evolution of three security standards used for cloud computing and the improvements made to them over time to cope with new threats. It also examines their adequacy and completeness by comparing them to each other. Understanding their evolution, resilience, and adequacy sheds light on their weaknesses and thus suggests improvements needed to keep pace with technological innovation. The three security certifications reviewed are as follows:

1) ISO/IEC 27001, produced by the International Organization for Standardization and the International Electrotechnical Commission to address the building and maintenance of information security management systems.

2) SOC 2, the Service Organization Control audits produced by the American Institute of Certified Public Accountants (AICPA), which has controls relevant to confidentiality, integrity, availability, security, and privacy within a service organization.

3) FedRAMP, the Federal Risk and Authorization Management Program, created in 2011 to meet the specific needs of the U.S. government in migrating its data on cloud environments.

References

1 "Cloud computing: Clash of the clouds," *The Economist*, October 15, 2009. Available at http://www.economist.com/node/14637206. (accessed November 3, 2009).

2 "Gartner says cloud computing will be as influential as e-business" (press release), Gartner, Inc., June 26, 2008. Available at http://www.gartner.com/newsroom/id/707508 (accessed August 22, 2010).

3 Knorr, E., and Gruman, G.,"What cloud computing really means," *ComputerWorld*, April 8, 2008. Available at https://www.computerworld.com.au/article/211423/what_cloud_computing_really_means/ (accessed June 2, 2009).

4 Obama, B.,"Executive order 13571: Streamlining service delivery and improving customer service," Office of the Press Secretary, the White House, April 27, 2011. Available at https://obamawhitehouse.archives.gov/the-press-office/2011/04/27/executive-order-13571-streamlining-service-delivery-and-improving-custom.

5 "U.S. Air Force selects IBM to design and demonstrate mission-oriented cloud architecture for cyber security" (press release), IBM, February 4, 2010. Available at https://www-03.ibm.com/press/us/en/pressrelease/29326.wss.

6 Mell, P. and Grance, T.,"The NIST definition of cloud computing: recommendations of the National Institute of Standards and Technology," Special Publication 800-145, National Institute of Standards and Technology, U.S. Department of Commerce, September 2011. Available at https://csrc.nist.gov/publications/detail/sp/800-145/final.

2

Survivability: Design, Formal Modeling, and Validation of Cloud Storage Systems Using Maude

Rakesh Bobba,[1] Jon Grov,[2] Indranil Gupta,[3] Si Liu,[3] José Meseguer,[3] Peter Csaba Ölveczky,[3,4] and Stephen Skeirik[3]

[1]*School of Electrical Engineering and Computer Science, Oregon State University, Corvallis, OR, USA*
[2]*Gauge AS, Oslo, Norway*
[3]*Department of Computer Science, University of Illinois at Urbana-Champaign, Urbana, IL, USA*
[4]*Department of Informatics, University of Oslo, Oslo, Norway*

To deal with large amounts of data while offering high availability, throughput, and low latency, cloud computing systems rely on distributed, partitioned, and replicated data stores. Such cloud storage systems are complex software artifacts that are very hard to design and analyze. We argue that formal specification and model checking analysis should significantly improve their design and validation. In particular, we propose rewriting logic and its accompanying Maude tools as a suitable framework for formally specifying and analyzing both the correctness and the performance of cloud storage systems. This chapter largely focuses on how we have used rewriting logic to model and analyze industrial cloud storage systems such as Google's Megastore, Apache Cassandra, Apache ZooKeeper, and RAMP. We also touch on the use of formal methods at Amazon Web Services.

2.1 Introduction

Cloud computing relies on software systems that store large amounts of data correctly and efficiently. These cloud systems are expected to achieve high performance, defined as high availability and throughput, and low latency. Such performance needs to be *assured* even in the presence of congestion in parts of the network, system or network faults, and scheduled hardware and software upgrades. To achieve this, the data must be *replicated* across both servers within a site, and across geo-distributed sites. To achieve the expected scalability and elasticity of cloud systems, the data may need to be *partitioned*. However, the *CAP theorem* [1] states that it is impossible to have both high availability and strong consistency (correctness) in replicated data stores in today's Internet. Different storage systems therefore offer different trade-offs between the levels of availability and consistency

Assured Cloud Computing, First Edition. Edited by Roy H. Campbell, Charles A. Kamhoua, and Kevin A. Kwiat.

that they provide. For example, weak notions of consistency of multiple replicas, such as "eventual consistency," are acceptable for applications like social networks and search, where availability and efficiency are key requirements, but where one can tolerate that different replicas store somewhat different versions of the data. Other cloud applications, including online commerce and medical information systems, require stronger consistency guarantees.

The following key challenge is addressed in this chapter:

> How can cloud storage systems be designed with high assurance that they satisfy desired correctness, performance, and quality-of-service requirements?

2.1.1 State of the Art

Standard system development and validation techniques are not well suited for addressing the above challenge. Designing cloud storage systems is hard, as the design must take into account wide-area asynchronous communication, concurrency, and fault tolerance. Experimentation with modifications and extensions of an existing system is often impeded by the lack of a precise description at a suitable level of abstraction and by the need to understand and modify large code bases (if available) to test the new design ideas. Furthermore, test-driven system development [2] – where a suite of tests for the planned features are written before development starts, and is used both to give the developer quick feedback during development and as a set of regression tests when new features are added—has traditionally been considered to be unfeasible for ensuring fault tolerance in complex distributed systems due to the lack of tool support for testing large numbers of different scenarios.

It is also very difficult or impossible to obtain high assurance that the cloud storage system satisfies given correctness and performance requirements using traditional validation methods. Real implementations are costly and error-prone to implement and modify for experimentation purposes. Simulation tool implementations require building an additional artifact that cannot be used for much else. Although system executions and simulations can give an idea of the performance of a design, they cannot give any (quantified) assurance on the performance measures. Furthermore, such implementations cannot verify consistency guarantees: Even if we execute the system and analyze the read/write operations log for consistency violations, this would only cover certain scenarios and cannot guarantee the absence of subtle bugs. In addition, nontrivial fault-tolerant storage systems are too complex for "hand proofs" of key properties based on an informal system description. Even if attempted, such proofs can be error-prone, informal, and usually rely on implicit assumptions.

The inadequacy of current design and verification methods for cloud storage systems in industry has also been pointed out by engineers at Amazon in [3] (see also Section 2.6). For example, they conclude that "the standard verification techniques in industry are necessary but not sufficient. We routinely use deep design reviews, code reviews, static code analysis, stress testing, and fault-injection testing but still find that subtle bugs can hide in complex concurrent fault-tolerant systems."

2.1.2 Vision: Formal Methods for Cloud Storage Systems

Our vision is to use *formal methods* to design cloud storage systems and to provide high levels of assurance that their designs satisfy given correctness and performance requirements. In a formally based system design and analysis methodology, a *mathematical model S* describes the system design at the appropriate level of abstraction. This system specification S should be complemented by a formal *property specification P* that describes mathematically (and therefore precisely) the requirements that the system S should satisfy. Being a mathematical object, the model S can be subjected to mathematical reasoning (preferably fully automated or at least machine-assisted) to guarantee that the design satisfies the properties P. If the mathematical description S is *executable*, then it can be immediately simulated; there is no need to generate an extra artifact for testing and verification. An executable model can also be subjected to various kinds of *model checking* analyses that automatically explore *all possible* system behaviors from a given initial system configuration. From a system developer's perspective, such model checking can be seen as a powerful debugging and testing method that can automatically find subtle "corner case" bugs and that automatically executes a comprehensive "test suite" for complex fault-tolerant systems. We advocate the use of formal methods throughout the design process to quickly and easily explore many design options and to validate designs as early as possible, since errors are increasingly costly the later in the development process they are discovered. Of course, one can also do a *postmortem* formal analysis of an existing system by defining a formal model of it in order to analyze the system formally; we show the usefulness of such *postmortem* analysis in Section 2.2.

Performance is as important as correctness for storage systems. Some formal frameworks provide probabilistic or statistical model checking that can give performance assurances with a given confidence level.

What properties should a formal framework have in order to be suitable for developing and analyzing cloud storage systems in an industrial setting? In Ref. [4], Chris Newcombe of Amazon Web Services, the world's largest cloud computing provider, who has used formal methods during the development of key components of Amazon's cloud computing infrastructure, lists key requirements for formal methods to be used in the development of such cloud computing systems in industry. These requirements can be summarized as follows:

1) *Expressive languages and powerful tools that can handle very large and complex distributed systems.* Complex distributed systems at different levels of abstraction must be expressible without tedious workarounds of key

concepts (e.g., time and different forms of communication). This requirement also includes the ability to express and verify complex liveness properties. In addition to automatic methods that help users diagnose bugs, it is also desirable to be able to machine-check proofs of the most critical parts.

2) *The method must be easy to learn, apply, and remember, and its tools must be easy to use.* The method should have clean simple syntax and semantics, should avoid esoteric concepts, and should use just a few simple language constructs. The author also recommends against distorting the language to make it more accessible, as the effect would be to obscure what is really going on.

3) *A single method should be effective for a wide range of problems, and should quickly give useful results with minimal training and reasonable effort.* A single method should be useful for many kinds of problems and systems, including data modeling and concurrent algorithms.

4) *Modeling and analyzing performance,* since performance is almost as important as correctness in industry.

2.1.3 The Rewriting Logic Framework

Satisfying the above requirements is a tall order. We suggest the use of *rewriting logic* [5] and its associated Maude tool [6], and their extensions, as a suitable framework for formally specifying and analyzing cloud storage systems.

In rewriting logic, data types are defined by *algebraic equational specifications.* That is, we declare sorts and function symbols; some function symbols are *constructors* used to define the *values* of the data type; the others denote *defined functions* – functions that are defined in a functional programming style using equations. Transitions are defined by *rewrite rules* of the form $t \rightarrow t'$ if *cond*, where t and t' are terms (possibly containing variables) representing *local state patterns*, and *cond* is a condition. Rewriting logic is particularly suitable for specifying distributed systems in an object-oriented way, in which case the states are multisets of objects and messages (traveling between the objects), and where an object o of class C with attributes att_i to att_n, having values val_1 to val_n, is represented by a term $\langle o : C \mid att_1 : val_1, \ldots, att_n : val_n \rangle$. A rewrite rule

```
rl [1] : m(O,w)
          < O : C | a1 : x, a2 : O', a3 : z >
          =>
          < O : C | a1 : x + w, a2 : O', a3 : z >
          m'(O',x) .
```

then defines a family of transitions in which a message m, with parameters O and w, is read and consumed by an object O of class C, the attribute a1 of the object O is changed to x + w, and a new message m'(O',x) is generated.

Maude [6] is a specification language and high-performance simulation and model checking tool for rewriting logic. *Simulations* – which simulate *single* runs of the system – provide a first quick initial feedback of the design. Maude *reachability analysis* – which checks whether a certain (un)desired state pattern can be reached from the initial state – and *linear temporal logic (LTL) model checking* – which checks whether all possible behaviors from the initial state satisfy a given LTL formula – can be used to analyze all possible behaviors from a given initial configuration.

The Maude tool ecosystem also includes Real-Time Maude [7], which extends Maude to *real-time systems*, and *probabilistic rewrite theories* [8], a specification formalism for specifying distributed systems with probabilistic features. A fully probabilistic subset of such theories can be subjected to statistical model checking analysis using the PVeStA tool [9]. Statistical model checking [10] performs randomized simulations until a probabilistic query can be answered (or the value of an expression be estimated) with the desired statistical confidence.

Rewriting logic and Maude address the above requirements as follows:

1) Rewriting logic is an expressive logic in which a wide range of complex concurrent systems, with different forms of communication and at various levels of abstractions, can be modeled in a natural way. In addition, its real-time extension supports the modeling of real-time systems. The Maude tools have been applied to a range of industrial and state-of-the-art academic systems [11,12]. Complex system requirements, including safety and liveness properties, can be specified in Maude using *linear temporal logic*, which seems to be the most intuitive and easy-to-understand advanced property specification language for system designers [13]. We can also define functions on states to express nontrivial reachability properties.

2) *Equations and rewrite rules:* These intuitive notions are all that have to be learned. In addition, object-oriented programming is a well-known programming paradigm, which means that Maude's simple model of concurrent objects should be attractive to designers. We have experienced in other projects that system developers find object-oriented Maude specifications easier to read and understand than their own use case descriptions [14], and that students with no previous formal methods background can easily model and analyze complex distributed systems in Maude [15]. The Maude tools provide automatic (push-button) reachability and temporal logic model checking analysis, and simulation for rapid prototyping.

3) As mentioned, this simple and intuitive formalism has been applied to a wide range of systems, and to all aspects of those systems. For example, data types are modeled as equational specification and dynamic behavior is modeled by rewrite rules. Maude simulations and model checking are easy to use and provide useful feedback automatically: Maude's search and LTL model checking provides a counterexample trace if the desired property does not hold.

4) We show in Ref. [16] that randomized Real-Time Maude simulations (of wireless sensor networks) can give performance estimates as good as those of domain-specific simulation tools. More importantly, we can analyze performance measures and provide performance estimations with given

confidence levels using probabilistic rewrite theories and statistical model checking; e.g., "I can claim with 90% confidence that at least 75% of the transactions satisfy the property *P*." For performance estimation for cloud storage systems, see Sections 2.2, 2.3, and 2.5.

To summarize, a formal executable specification in Maude or one of its extensions allows us to define *a single artifact* that is, simultaneously, a mathematically precise high-level description of the system design and an executable system model that can be used for rapid prototyping, extensive testing, correctness analysis, and performance estimation.

2.1.4 Summary: Using Formal Methods on Cloud Storage Systems

In this chapter, we summarize some of the work performed at the Assured Cloud Computing Center at the University of Illinois at Urbana-Champaign using Maude and its extensions to formally specify and analyze the correctness and performance of several important industrial cloud storage systems and a state-of-the-art academic one. In particular, we describe the following contributions:

i) *Apache Cassandra* [17] is a popular open-source industrial key-value data store that only guarantees *eventual consistency*. We were interested in (i) evaluating a proposed variation of Cassandra, and (ii) analyzing under what circumstances – and how often in practice – Cassandra also provides stronger consistency guarantees, such as read-your-writes or strong consistency. After studying Cassandra's 345,000 lines of code, we first developed a 1000-line Maude specification that captured the main design choices. Standard model checking allowed us to analyze under what conditions Cassandra guarantees strong consistency. By modifying a single function in our Maude model, we obtained a model of our proposed optimization. We subjected both of our models to statistical model checking using PVeStA; this analysis indicated that the proposed optimization did *not* improve Cassandra's performance. But how reliable are such formal performance estimates? To investigate this question, we modified the Cassandra code to obtain an implementation of the alternative design, and executed both the original Cassandra code and the new system on representative workloads. These experiments showed that PVeStA statistical model checking provides reliable performance estimates. To the best of our knowledge, this was the first time that for key-value stores, model checking results were checked against a real system deployment, especially on performance-related metrics.

ii) *Megastore* [18] is a key part of Google's celebrated cloud infrastructure. Megastore's trade-off between consistency and efficiency is to guarantee consistency only for transactions that access a single *entity group*. It is obviously interesting to study such a successful cloud storage system.

Furthermore, one of us had an idea on how to extend Megastore so that it would also guarantee strong consistency for certain transactions accessing *multiple* entity groups *without* sacrificing performance. The first challenge was to develop a detailed formal model of Megastore from the short high-level description in Ref. [18]. We used Maude simulation and model checking throughout the formalization of this complex system until we obtained a model that satisfied all desired properties. This model also provided the first reasonable detailed public description of Megastore. We then developed a formal model of our extension, and estimated the performance of both systems using randomized simulations in Real-Time Maude; these simulations indicated that Megastore and our extension had about the same performance. (Note that such *ad hoc* randomized simulations do not give a precise level of confidence in the performance estimates.)

iii) *RAMP* [19] is a state-of-the-art academic partitioned data store that provides efficient lightweight transactions that guarantee the simple "read atomicity" consistency property. Reference [19] gives hand proofs of correctness properties and proposes a number of variations of RAMP without giving details. We used Maude to (i) check whether RAMP indeed satisfies the guaranteed properties, and (ii) develop detailed specifications of the different variations of RAMP and check which properties they satisfy.

iv) *ZooKeeper* [20] is a fault-tolerant distributed key/value data store that provides reliable distributed coordination. In Ref. [21] we investigate whether a useful group key management service can be built using ZooKeeper. PVeStA statistical model checking showed that such a Zoo-Keeper-based service handles faults better than a traditional centralized group key management service, and that it scales to a large number of clients while maintaining low latencies.

To the best of our knowledge, the above-mentioned work at the Assured Cloud Computing Center represents the first published papers on the use of formal methods to model and analyze such a wide swathe of industrial cloud storage systems. Our results are encouraging, but the question arises: Is the use of formal methods feasible in an industrial setting? The recent paper [3] from Amazon tells a story very similar to ours, and formal methods are now a key ingredient in the system development process at Amazon. The Amazon experience is summarized in Section 2.6, which also discusses the formal framework used at Amazon.

The rest of this chapter is organized as follows: Sections 2.2–2.5 summarize our work on Cassandra, Megastore, RAMP, and ZooKeeper, respectively, while Section 2.6 gives an overview of the use of formal methods at Amazon. Section 2.7 discusses related work, and Section 2.8 gives some concluding remarks.

2.2 Apache Cassandra

Apache Cassandra [17] is a popular open-source key-value data store originally developed at Facebook.[1] According to the DB-Engines Ranking [22], Cassandra has advanced into the top 10 most popular database engines among 315 systems, and is currently used by, for example, Amadeus, Apple, IBM, Netflix, Facebook/ Instagram, GitHub, and Twitter.

Cassandra only guarantees *eventual consistency* (if no more writes happen, then eventually all reads will see the *last* value written). However, it might be possible that Cassandra offers stronger consistency guarantees in certain cases.

It is therefore interesting to analyze both the circumstances under which Cassandra offers stronger consistency guarantees, and how often stronger consistency properties hold in practice.

The task of accurately predicting when consistency properties hold is non-trivial. To begin with, building a large-scale distributed key-value store is a challenging task. A key-value store usually consists of a large number of components (e.g., membership management, consistent hashing, and so on), and each component is given by source code that embodies many complex design decisions. If a developer wishes to improve the performance of a system (e.g., to improve consistency guarantees, or reduce operation latency) by implementing an alternative design choice for a component, then the only option available is to make changes to huge source code bases (Apache Cassandra has about 345,000 lines of code). Not only does this require many man-months of effort, it also comes with a high risk of introducing new bugs, requires understanding a huge code base before making changes, and is not repeatable. Developers can only afford to explore very few design alternatives, which may in the end fail to lead to a better design.

To be able to reason about Cassandra, experiment with alternative design choices and understand their effects on the consistency guarantees and the performance of the system, we have developed in Maude both a formal non-deterministic model [23] and a formal probabilistic model [24] of Cassandra, as well as a model of an alternative Cassandra-like design [24]. To the best of our knowledge, these were the first formal models of Cassandra ever created. Our Maude models include main components of Cassandra such as data partitioning strategies, consistency levels, and timestamp policies for ordering multiple versions of data. Each Maude model consists of about 1000 lines of Maude code with 20 rewrite rules. We use the nondeterministic model to answer *qualitative* consistency queries about Cassandra (e.g., whether a key-value store read operation is strongly (respectively weakly) consistent); and we use the

---

1 A key-value store can be seen as a transactional data store where transactions are single read or write operations.

probabilistic model to answer *quantitative* questions like: how often are these stronger consistency properties satisfied in practice?

Apache Cassandra is a distributed, scalable, and highly available NoSQL database design. It is distributed over collaborative servers that appear as a single instance to the end client. Data items are dynamically assigned to several servers in the cluster (called the *ring*), and each server (called a *replica*) is responsible for different ranges of the data stored as key-value pairs. Each key-value pair is stored at multiple replicas to support fault tolerance. In Cassandra a client can perform *read* or *write* operations to query or update data. When a client requests a read/write operation to a cluster, the server connected to the client acts as a *coordinator* and forwards the request to all replicas that hold copies of the requested key. According to the specified *consistency level* in the operation, after collecting sufficient responses from replicas, the coordinator will reply to the client with a value. Cassandra supports tunable consistency levels, with ONE, QUORUM, and ALL being the three major ones, meaning that the coordinator will reply with the most recent value (namely, the value with the highest timestamp) to the client after hearing from *one* replica, a *majority* of the replicas, or *all* replicas, respectively.

We show below one rewrite rule to illustrate our specification style. This rewrite rule describes how the coordinator S reacts upon receiving a read reply message {T, S <- ReadReplySS(ID,KV,CL,A)} from a replica at global time T, with KV the returned key-value pair of the form (key,value, timestamp), ID and A the read operation's and the client's identifiers, respectively; and CL the read's consistency level. The coordinator S adds KV to its local buffer (which stores the replies from the replicas) by add(ID,KV, BF). If the coordinator S now has collected the required number of responses (according to the desired consistency level CL for the operation), which is determined by the function cl?, then the coordinator returns to A the highest timestamped value, determined by the function tb, by sending the message [D, A <- ReadReplyCS(ID,tb(BF'))] to A. This outgoing message is equipped with a message delay D nondeterministically selected from the delay set delays, where DS describes the other delays in the set. If the coordinator has not yet received the required number of responses, then no message is sent. (Below, none denotes the empty multiset of objects and messages).

```
crl [on-rec-rrep-coord-nondet]    :
    {T, S <- ReadReplySS(ID,KV,CL,A)}
    < S : Server | buffer: BF, delays:   (D,DS), AS >
 =>
    < S : Server | buffer: BF', delays:   (D,DS), AS >
    (if cl?(CL,BF') then
       [D, A <- ReadReplyCS(ID,tb(BF'))]
     else none fi)
    if BF'   := add(ID,KV,BF)    .
```

We analyze *strong consistency* (where each read returns the value of the last write that occurred before that read) and *eventual consistency* using experimental scenarios with *one* or *multiple* clients. The purpose of our experiments is to answer the following question: Does strong/eventual consistency depend on the following:

- With one client, the combination of consistency levels of consecutive requests?
- With multiple clients, also on the latency between consecutive requests?

Our model checking results show that strong consistency holds in some scenarios, and that eventual consistency holds in all scenarios regardless of the consistency level combinations. Although Cassandra is expected to violate strong consistency under certain conditions, previously there was no formal way of discovering under which conditions such violations could occur.

Table 2.1 shows the results of model checking strong/eventual consistency with one client, with × denoting a violation. Three out of nine combinations of consistency levels violate strong consistency, where at least one of the read and the write operations has a consistency level of ONE.

The results of model checking strong consistency with *two* clients are shown in Table 2.2. We experiment with different relations between the set of possible message delays for the coordinator ($D1$ and $D2$, with $D1 < D2$) and the latencies ($L1, L2,$ and $L3$, with $L1 < L2 < L3$) between consecutive requests. We observe that satisfaction of strong consistency depends on the time between requests. For eventual consistency, no violation occurs, regardless of the consistency level and time between requests.

We then wanted to experiment with a possible optimization of the Cassandra design in which the *values* of the keys are considered instead of their *timestamps*

Table 2.1 Results of checking strong consistency (*top*) and eventual consistency (*bottom*) with one client.

Write₁\Read₂	ONE	QUORUM	ALL
ONE	×	×	✓
QUORUM	×	✓	✓
ALL	✓	✓	✓
Write₁\Write₂	ONE	QUORUM	ALL
ONE	✓	✓	✓
QUORUM	✓	✓	✓
ALL	✓	✓	✓

Table 2.2 Results of checking strong consistency (*top*) and eventual consistency (*bottom*) with two clients; the delay set for the coordinator to nondeterministically select a message delay from is {D1, D2} with D1 < D2.

	Latency\Consistency Lv.	ONE	QUORUM	ALL
Strong	L1 (L1<D1)	✗	✗	✗
	L2 (D1<L2<D2)	✗	✗	✗
	L3 (D2<L3)	✓	✓	✓
	Latency\Consistency Lv.	ONE	QUORUM	ALL
Eventual	L1 (L1<D1)	✓	✓	✓
	L2 (D1<L2<D2)	✓	✓	✓
	L3 (D2<L3)	✓	✓	✓

when deciding which value should be returned upon a read operation. Our goal was to see whether that would provide better consistency or at least lower operation latency. Having a formal specification of Cassandra, we were able to easily specify this possible optimization by just modifying the function tb.

To estimate how often Cassandra satisfies stronger consistency *in practice*, and to compare the original Cassandra design with our proposed optimization, we transformed our nondeterministic specification into a *fully probabilistic* rewrite theory that can be subjected to statistical model checking using the PVeStA tool [9]. The main idea behind turning a nondeterministic model into a fully probabilistic model is to let the message delays be sampled from a dense probabilistic distribution. Density implies that the probability that two messages arrive at the same time, and hence that two events happen at the same time, is zero.

To illustrate this transformation from a nondeterministic model to a probabilistic one, we show below the probabilistic rewrite rule obtained by transforming the above rewrite rule. In the transformed rule, the message delay D of the generated ReadReply message is now probabilistically chosen according to the parameterized probability distribution function distr(...), which in our model was selected to be a lognormal distribution:

```
crl [on-rec-rrep-coord-prob]   :
    {T, S <- ReadReplySS(ID,KV,CL,A) }
    < S : Server | buffer: BF, AS >
=>
    < S : Server | buffer: BF', AS >
```

```
(if cl?(CL,BF') then
  [D,  A <- ReadReplyCS(ID,tb(BF'))]
  else none fi)
if BF'  := add(ID,KV,BF)
with probability D := distr(...)   .
```

We quantitatively analyzed the following five consistency guarantees in Cassandra and in our alternative design:

- *Strong consistency* (SC) ensures that each read returns the value of the last write that occurred before that read.
- *Read your writes* (RYW) guarantees that the effects of all writes performed by a client are visible to that client's subsequent reads.
- *Monotonic reads* (MR) ensure that a client observes a key-value store increasingly up-to-date over time.
- *Consistent prefix* (CP) guarantees that a client will observe an ordered sequence of writes starting with the first write to the system.
- *Causal consistency* (CC) guarantees that effects are observed only after their causes: Reads will not see a write unless its dependencies are also seen.

Our PVeStA analysis indicated that Cassandra frequently achieves much higher consistency (up to strong consistency) than the promised eventual consistency, especially with QUORUM and ALL reads, in which cases the probability of achieving strong consistency starts to approach 1 even with fairly short latencies (see Figure 2.1a).

Another interesting observation comes from the PVeStA analysis of consistent prefix. By fixing the consistency level of writes (see Figure 2.2a, c, and e for ONE/QUORUM/ALL write), we can see how the consistency level of reads affect consistent prefix over issuing latency. Surprisingly, it appears that lower reads can achieve higher consistent prefix consistency.

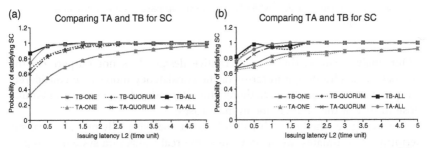

Figure 2.1 PVeStA-based estimation (a) and actual measures running Cassandra (b) of the probability of satisfying strong consistency (SC) as a function of the time between a read request and the latest previous write request. TA stands for *time-agnostic strategy*, our proposed alternative design, while TB stands for *timestamp-based strategy*, the original Cassandra design.

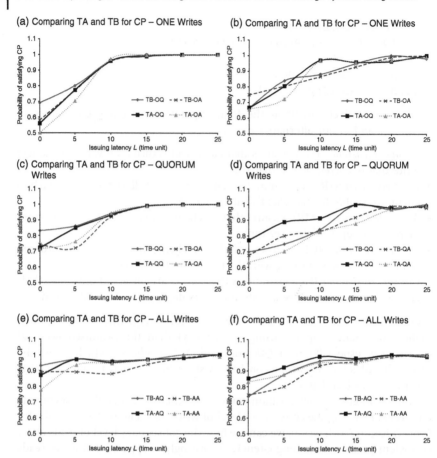

Figure 2.2 Probabilities of satisfying consistent prefix (CP) by statistical model checking (a, c, and e) and by real deployment run (b, d, and f). TA stands for time-agnostic strategy, our proposed alternative design, while TB for timestamp-based strategy, the original Cassandra design.

Our analysis shows that our alternative design does not outperform the original Cassandra design in terms of the consistency models we considered, except for *consistent prefix*, even though the alternative design's behavior is quite close to that of the original one in most cases.

To investigate whether such PVeStA-based analysis really provides realistic performance estimates, we also executed the real Cassandra system on representative workloads (see Figures 2.1a and 2.2b, d, and f). Both the model predictions and the implementation-based evaluations reached the same conclusion. We could show that results derived from model checking agree reasonably well with experimental reality. We showed that this agreement holds true for various consistency models, even if changes are introduced on how timestamps are used in responding to queries. For more details on our quantitative analysis of Cassandra and the alternative design we considered, we refer the reader to Ref. [25], which substantially extends our conference paper [24].

2.3 Formalizing, Analyzing, and Extending Google's Megastore

Megastore [18] is a distributed data store developed and widely applied at Google. It is a key component of Google's celebrated cloud computing infrastructure, and is used internally at Google for Gmail, Google+, Android Market, and Google AppEngine. It is one of a few industrial replicated data stores that provide both data replication, fault tolerance, and support for transactions with some data consistency guarantees. Megastore has handled 3 billion write and 20 billion read transactions per day [18].

In Megastore, data are divided into different *entity groups* (e.g., "Peter's email" or "books on rewriting logic"). Google Megastore's trade-off among high availability, concurrency, and consistency is that data consistency is only guaranteed for transactions that access a *single* entity group. There are no guarantees if a transaction reads *multiple* entity groups.

One of us, Jon Grov, a researcher on transactional systems with no background in formal methods, had some idea about how to add consistency also for certain transactions that access multiple entity groups, *without* significant performance penalty. Grov was therefore interested in experimenting with his ideas to extend Megastore, to analyze the correctness of his extension, and to compare its performance with that of Megastore.

There was, however, a problem: There was no detailed description of Megastore, or publicly available code for this industrial system that could be used for experimentation. The only publicly available description of Megastore was a brief overview paper [18]. To fully understand Megastore (more precisely, the Megastore algorithm/approach), two of us, Grov and Ölveczky, first had to develop our own sufficiently detailed executable specification of Megastore from the description in Ref. [18]. This specification could then be used to estimate the performance of Megastore, which could then be compared with the estimated performance of our extension. We employed Real-Time Maude simulations and, in particular, LTL model checking throughout our development effort.

2.3.1 Specifying Megastore

Our specification of Megastore [26] is the first publicly available formalization and reasonably detailed description of Megastore. It contains 56 rewrite rules, of which 37 deal with fault tolerance features. An important point is that even *if* we had access to Megastore's code base, understanding and extending it would likely have been much more time-consuming than developing our own 56-rule description and simulation model in Real-Time Maude.

To show an example of the specification style, the following shows one (of the smallest) of the 56 rewrite rules:

```
rl [bufferWriteOperation]  :
  < SID : Site | localTransactions : LOCALTRANS
  < TID : Transaction | operations : write(KEY, VAL) :: OL, writes :
WRITEOPS,
     status : idle > >
  =>
  < SID : Site | localTransactions : LOCALTRANS
  < TID : Transaction | operations : OL, writes : WRITEOPS :: write
(KEY, VAL) > >.
```

In this rule, the Site object SID has a number of transaction objects to execute in its localTransactions attribute; one of them is the transaction TID (the variable LOCALTRANS ranges over multisets of objects, and therefore captures all the other objects in the site's transaction set). The next operation that the transaction TID should execute is the write operation write(KEY, VAL), which represents writing the value VAL to the key KEY. The effect of applying this rewrite rule is that the write operation is removed from the transaction's list of operations to perform and is added to its *write set* writes.

The second rewrite rule we show concerns the validation phase. In Megastore, a replicated transaction log is maintained for each entity group. When a transaction *t* is ready to commit, its coordinating site *s* prepares a new log entry for each entity group written by *t*, does some leader election part of Paxos, and (if the leader election phase is successful) multicasts the new proposed log entry to the other replicating sites. Each recipient of this message must then verify that it has not already granted an accept for the new log position. If so, the recipient replies with an accept message to the originating site.

The following rule shows the part when the replicating site THIS receives the multicasted message acceptAllReq with the proposed new log entry (TID′ LP SID OL) for entity group EG from the coordinator THIS. The site THIS verifies that it has not already granted an accept for that log position. (Since messages could be delayed for a long time, it checks both the transaction log and received proposals). If there are no such conflicts, the site responds with an accept message acceptAllRsp(...) and stores its accept in its proposals attribute. The record (TID′ LP SID OL) represents the candidate log entry, which contains the transaction identifier TID′, the log position LP, the proposed leader site SID, and the list of update operations OL:

```
crl [rcvAcceptAllReq]   :
   (msg acceptAllReq(TID, EG,   (TID' LP SID OL), PROPNUM)
from SENDER to THIS)
   < THIS : Site |
   entityGroups : EGROUPS
    < EG : EntityGroup |
 proposals : PROPSET, transactionLog : LEL > >
   =>
   < THIS : Site |
   entityGroups :   EGROUPS
   < EG : EntityGroup |
   proposals : accepted(SENDER,   (TID' LP SID OL), PROPNUM);
removeProposal(LP, PROPSET) > >
    dly(msg acceptAllRsp(TID, EG, LP, PROPNUM)
 from THIS to SENDER, T)
    if not (containsLPos(LP, LEL) or hasAcceptedForPosition
(LP, PROPSET)) /\ T;  TS := possibleMessageDelays(THIS, SENDER)  .
```

Our model assumes that `possibleMessageDelays`(s_1, s_2) gives the set $d_1; d_2; \dots; d_n$ of possible messaging delays between sites s_1 and s_2, where the union operator `_;_` is associative and commutative. The matching equation

```
T;  TS := possibleMessageDelays(THIS, SENDER)
```

then nondeterministically assigns to the variable T (of sort `Time`) *any* delay d_i from the set `possibleMessageDelays(THIS, SENDER)` of possible delays, and uses this value as the messaging delay when sending the message

```
dly(msg acceptAllRsp(TID, EG, LP, PROPNUM)
 from THIS to SENDER, T).
```

2.3.2   Analyzing Megastore

We wanted to analyze both the correctness and the performance of our model of Megastore throughout its development, to catch functional errors and performance bottlenecks quickly. For our analysis, we generated the following two additional models from the main model already described:

1) Our model of Megastore is a real-time model. However, this means that any exhaustive model checking analysis of our model only analyzes those behaviors that are possible *within the given timing parameters* (for messaging delays, etc.). To exhaustively analyze all possible system behaviors *irrespective* of particular timing parameters, we generated an *untimed* model from our real-time model by just ignoring messaging delays in our

specification. For example, in this specification, the above rewrite rule becomes (where the parts represented by ". . . "are as before):

```
crl [rcvAcceptAllReq]  :
    (msg acceptAllReq(...) from SENDER to THIS)
    < THIS : Site | entityGroups : EGROUPS < EG : EntityGroup | ... > >
    =>
    < THIS : Site | entityGroups :   EGROUPS < EG : EntityGroup | ... > >
    (msg acceptAllRsp(TID, EG, LP, PROPNUM) from THIS to SENDER)
    if not (containsLPos(LP, LEL) or hasAcceptedForPosition(LP, PROPSET)) .
```

2) For *performance estimation* purposes, we also defined a real-time model in which certain parameters, such as the messaging delays between two nodes, are selected probabilistically according to a given probability distribution. For example, we used the following probability distribution for the network delays (in milliseconds):

	30%	30%	30%	10%
London ↔ Paris	10	15	20	50
London ↔ New York	30	35	40	100
Paris ↔ New York	30	35	40	100

An important difference between our Megastore work and the Cassandra effort described in Section 2.2 is that in the Cassandra work we used (fully) probabilistic rewrite theories and the dedicated statistical model checker PVeStA for the probabilistic analysis. In contrast, our Megastore work stayed within Real-Time Maude and simulated the selection of a value from a distribution by using an ever-changing seed and Maude's built-in function random. In this "lightweight" probabilistic real-time model, we maintain the value n of the seed in an object < seed : Seed | value : n > in the state, and the above rewrite rule is transformed to the following:

```
crl [rcvAcceptAllReq]  :
      (msg acceptAllReq(...)  from SENDER to THIS)
      < THIS : Site | entityGroups : EGROUPS < EG : EntityGroup | ... > >
      < seed : Seed / value : N > =>
    < THIS : Site | entityGroups :   EGROUPS < EG : EntityGroup | ... > >
      < seed : Seed / value : N + 1 >
      dly(msg acceptAllRsp(...) from THIS to SENDER,
      selectFromDistribution(delayDistribution(THIS, SENDER), random(N)))
      if not (containsLPos(LP, LEL) or hasAcceptedForPosition(LP, PROPSET)) .
```

where selectFromDistribution "picks" the appropriate delay value from the distribution delayDistribution(THIS, SENDER) using the random number random(N). While such lightweight probabilistic performance analysis using Real-Time Maude has previously been shown to give performance estimates on par with those of dedicated domain-specific simulation tools, for example, for wireless sensor networks [16], such analysis cannot provide a (quantified) level of confidence in the estimates; for that we need statistical model checking.

Performance Estimation

For performance estimation, we also added a transaction generator that generates transaction requests at random times, with an adjustable average rate measured in *transactions per second,* and used the above probability distribution for the network delays.

The key performance metrics to analyze are the average transaction latency and the number of committed/aborted transactions. We also added a *fault injector* that randomly injects short outages in the sites, with mean time to failure 10 s and mean time to repair 2 s for each site. The results from the randomized simulations of our probabilistic real-time model for 200 s, with an average of 2.5 transactions generated per second, in this fairly challenging setting, are given in the following table[2]:

Site	Average latency (ms)	Commits	Aborts
London	218	109	38
New York	336	129	16
Paris	331	116	21

These latency figures are consistent with Megastore itself [18]: "Most users see average write latencies of 100–400 milliseconds, depending on the distance between datacenters, the size of the data being written, and the number of full replicas."

Model Checking Analysis

For model checking analysis – which exhaustively analyzes all possible system behaviors from a given initial system configuration – we added to the state a *serialization graph,* which is updated whenever a transaction commits. The combination of properties we analyzed were as follows: (i) The serialization graph never contains cycles. (ii) Eventually all transactions have finished executing. (iii) All entity groups and all transaction logs are equal (or invalid).

---

2 Because of site failures, not all the generated 500 transactions were committed or aborted; however, our analysis shows that all 500 transactions are validated when there are no site failures. See Ref. [26] for details.

Model checking these properties in combination can be done by giving the following Real-Time Maude command:

```
Maude> (mc initMegastore /=u
([] isSerializable)
/\  (<> [] (allTransFinished /\ entityGroupsEqualOrInvalid
                /\ transLogsEqualOrInvalid)).)
```

This command returns true if the temporal logic formula (the last three lines in the command) holds from the initial state initMegastore, and a counterexample showing a behavior that does satisfy the formula in case the formula does not hold for all behaviors.

We used model checking throughout the development of our model and discovered many unexpected corner cases. To give an idea of the size of the configurations that can be model checked, we summarize below the execution time of the above model checking command for different system parameters, where $\{n_1, \ldots, n_k\}$ means that the corresponding value was selected non-deterministically from the set. All the model checking commands that finished executing returned true. *DNF* means that the execution was aborted after 4 h.

Message delay	No. of transactions	Transaction start time	No. of failures	Failure time	Run (s)
{20,100}	4	{19,80} and {50,200}	0	—	1367
{20,100}	3	{10, 50, 200}	1	60	1164
{20,40}	3	20, 30, and {10, 50}	2	{40,80}	872
{20,40}	4	20, 20, 60, and 110	2	70 and {10,130}	241
{20,40}	4	20, 20, 60, and 110	2	{30,80}	DNF
{10, 30, 80}, and {30, 60,120}	3	20, 30, 40	1	{30,80}	DNF
{10, 30, 80}, and {30, 60,120}	3	20, 30, 40	1	60	DNF

As mentioned, we also model checked an untimed model (of the nonfault-tolerant part of Megastore) that covers all possible behaviors from an initial system configuration irrespective of timing parameters. The disadvantage of such untimed model checking is that we can only analyze smaller systems: model checking the untimed system proved unfeasible even for four transactions. Furthermore, failure detection and fault tolerance features rely heavily on timing, which means that the untimed model is not suitable for modeling and analyzing the fault-tolerant version of Megastore.

2.3.2.1 Megastore-CGC

As already mentioned, Jon Grov had some idea on how to extend Megastore to also provide consistency for transactions that access *multiple* entity groups while maintaining Megastore's performance and strong fault tolerance features. He observed that, in Megastore, a site replicating a set of entity groups participates in all updates of these entity groups and should therefore be able to maintain an ordering on those updates. The idea behind our extension, called Megastore-CGC, is that by making this ordering explicit, such an "ordering site" can validate transactions [27].

Since Megastore-CGC exploits the implicit ordering of updates during Megastore commits, it *piggybacks* ordering and validation onto Megastore's commit protocol. Megastore-CGC therefore does not require additional messages for validation and commit and should maintain Megastore's performance and strong fault tolerance. A *failover* protocol deals with failures of the ordering sites.

We again used both simulations (to discover performance bottlenecks) and Maude model checking extensively during the development of Megastore-CGC, whose formalization contains 72 rewrite rules. The following table compares the performance of Megastore and Megastore-CGC in a setting without failures, where sites London and New York also handle transactions accessing multiple entity groups. Notice that Megastore-CGC will abort some transactions that access multiple entity groups (validation aborts) that Megastore does not care about.

	Megastore			Megastore-CGC			
	Commits	Aborts	Average latency	Commits	Aborts	Validation aborts	Average latency
Paris	652	152	126	660	144	0	123
London	704	100	118	674	115	15	118
New York	640	172	151	631	171	10	150

Designing and validating a sophisticated protocol such as Megastore-CGC is very challenging. Maude's intuitive and expressive formalism allowed a domain expert to define both a precise, formal description and an executable prototype in a single artifact. We found that anticipating all possible behaviors of

Megastore-CGC is impossible. A similar observation was made by Google's Megastore team, which implemented a pseudorandom test framework and stated that *the tests have found many surprising problems* [18]. Compared to such a testing framework, Real-Time Maude model checking analyzes not only a set of pseudorandom behaviors but also all possible behaviors from an initial system configuration. Furthermore, we believe that Maude provides a more effective and low overhead approach to testing than that of a real testing environment.

In a test-driven development method, a suite of tests for the planned features are written before development starts. This set of tests is then used both to give the developer quick feedback during development and as a set of regression tests when new features are added. However, test-driven development has traditionally been considered to be unfeasible when targeting fault tolerance in complex concurrent systems, due to the lack of tool support for testing a large number of different scenarios. Our experience with Megastore-CGC showed that with Maude, a test-driven approach is possible also in such systems, since many complex scenarios can be quickly tested by model checking.

Simulating and model checking this prototype automatically provided quick feedback about both the performance and the correctness of different design choices, even for very complex scenarios. Model checking was especially helpful, both to verify properties and to find subtle "corner case" design errors that were not found during extensive simulations.

2.4 RAMP Transaction Systems

Read-Atomic Multipartition (RAMP) transactions were proposed by Bailis *et al.* [19] to offer lightweight multipartition transactions that guarantee one of the fundamental consistency levels, namely, *read atomicity*: Either all updates or no updates of a transaction are visible to other transactions.

Reference [19] presents a pseudocode description of the RAMP algorithms and "hand proofs" of key properties. It also mentions some optimizations and variations of RAMP, but without providing any details or correctness arguments about these variations. All this means that there was a clear need for formally specifying and analyzing RAMP and its proposed extensions. Providing correctness of the protocol and its extensions is a critical step toward making RAMP a production-capable system.

We therefore formalized and formally analyzed RAMP and its variations in Maude, including the sketched variations with fast commit and one-phase writes. We also experimented with decoupling one of the key building blocks of RAMP, the two-phase commit protocol (2PC), which is used to ensure that either all partitions commit a transaction or that none do.

To show an example of the specification, the following rewrite rule illustrates what happens during the two-phase commit part when a client O receives a `prepared` message for the write `ID` from a partition O′, meaning that partition O′ can commit the write `ID` (since there is no data replication in RAMP, each write operation in a transaction must only be accepted and performed by a single partition). The client O can then remove `ID` from its set `NS` of pending `prepared` messages. If all write operations can commit, that is, the resulting set `NS′` of pending replies is empty, the client starts committing the transaction using the function `startCommit`, which generates a `commit` message for each write:

```
crl [receive-prepared]   :
    msg prepared(ID) from O′ to O
    < O : Client | pendingPrep : NS, pendingOps : OI, sqn : SQN >
=>
    < O : Client | pendingPrep : NS′ >
    (if NS′ == empty then startCommit(OI,SQN,O) else none fi) if NS′ :=
delete(ID,NS)   .
```

We used reachability analysis to analyze whether the different variants of RAMP satisfy all the following properties (from Ref. [19]) that RAMP transactions should satisfy:

- *Read atomic isolation:* Either all updates or no update of a transaction is visible to other transactions.
- *Companions present:* If a version is committed, then each of the version's sibling versions is present on their respective partitions.
- *Synchronization independence*: Each transaction will eventually commit or abort.
- *Read your writes:* A client's writes are visible to her subsequent reads.

We analyzed these properties for our seven versions of RAMP, for all initial configurations with four operations and two clients, as well as for a number of configurations with six operations. Our analysis results agree with the theorems and conjectures in Ref. [19]: All versions satisfy the above properties, except that

- RAMP without 2PC only satisfies synchronization independence; and
- RAMP with one-phase writes does not satisfy read-your-writes.

2.5 Group Key Management via ZooKeeper

Group key management is the management of cryptographic keys so that multiple authorized entities can securely communicate with each other. A central group key controller can fulfill this need by (a) authenticating/admitting authorized users into the group, and (b) generating a *group key* and distributing it to authorized group members [28]. The group key needs to be updated periodically to ensure the secrecy of the group key, and whenever a new member joins or leaves the group to preserve *backward secrecy* and *forward secrecy*, respectively. In particular, the group has some secure mechanism to designate a *key encrypting key* (KEK) which the group controller then uses to securely distribute group key updates. In settings with a centralized group controller, its failure can impact both group dynamics and periodic key updates, leaving the group key vulnerable. If the failure occurs during a key update, then the group might be left in an inconsistent state where both the updated and old keys are still in use. This is especially significant when designing a cloud-based group key management service, since such a service will likely manage many groups.[3]

In Ref. [21] we investigated whether a fault-tolerant cloud-based group key management service could be built by leveraging existing coordination services commonly available in cloud infrastructures and if so, how to design such a system. In particular, we: (a) designed a group key management service built using Zookeeper [20], a reliable distributed coordination service supporting Internet-scale distributed applications; (b) developed a rewriting logic model of our design in Maude [6], based on Ref. [30], where key generation is handled by a centralized key management server and key distribution is offloaded to a ZooKeeper cluster and where the group controller stores its state in ZooKeeper to enable quick recovery from failure; and (c) analyzed our model using the PVeStA [9] statistical model checking tool. The analysis centered on two key questions: (i) Can a ZooKeeper-based group key management service handle faults more reliably than a traditional centralized group key manager. (ii) Can it scale to a large number of concurrent clients with a low-enough latency to be useful?

2.5.1 Zookeeper Background

ZooKeeper is used by many distributed applications, such as Apache Hadoop MapReduce and Apache HBase, and by many companies, including Yahoo and Zynga in their infrastructures. From a bird's eye view, the ZooKeeper system consists of two kinds of entities: servers and clients. All of the servers together form a distributed and fault-tolerant key/value store that the clients may read data from or write data to. In ZooKeeper terminology, each key/value pair is called a *znode*, and these znodes are then organized into a tree structure similar to that of a standard filesystem. In order to achieve fault tolerance, ZooKeeper requires that a majority of servers acknowledge and log each write to disk before it is committed. Since ZooKeeper requires a majority of servers to be alive and aware of each other in order to operate, updates will not be lost. A detailed

---

3 Please refer to Ref. [29] for a survey of group key management schemes.

description of ZooKeeper design and features can be found in Ref. [20] and in ZooKeeper documentation.[4]

ZooKeeper provides a set of guarantees that are useful for key management applications. Updates to ZooKeeper state are atomic, and once committed they will persist until they are overwritten. Thus, a client will have a consistent view of the system regardless of which server it connects to and is guaranteed to be up-to-date within a certain time-bound. For our purposes, this means updates, once committed, will not be lost. Furthermore, ZooKeeper provides an event-driven notification mechanism through *watches.* A client sets a *watch* on a znode that will be notified whenever the znode is changed or deleted. Watches can enable us, for example, to easily propagate key change events to interested clients.

2.5.2 System Design

In our design, if a user wishes to join/leave a group, she will contact the group controller who will: (a) perform the necessary authentication and authorization checks; (b) if the user is authorized, add/remove the user to/from the group; (c) update the group membership state in the ZooKeeper service; (d) generate and update the group key in the ZooKeeper service; and (e) provide any new users with the updated group key and necessary information to connect to ZooKeeper and obtain future key updates.

The system state is stored in ZooKeeper as follows: Each secure group and authorized user is assigned a unique znode storing an encrypted key. Whenever a user joins a group, the znode corresponding to that user is added as a child of the znode representing the group. The znode corresponding to the group also stores the current group key encrypted by the KEK. During periodic group key updates, the old group key is overwritten by the new group key (encrypted by the same KEK). However, when an authorized user joins or leaves a group, the group controller generates a *new* KEK and distributes it using the client's pairwise keys. Specifically, the group controller updates the value stored at each user's assigned znode with the new KEK encrypted by the client's pairwise key. Since each group member sets a watch on the znode corresponding to itself and its group, it will be notified whenever the group key or the KEK changes.

Thus, the group controller uses the ZooKeeper service to maintain group information and to distribute and update cryptographic keys. Since the group controller's operational state is already saved within the ZooKeeper service (e.g., group member znodes it has updated, last key update time, etc.), if the controller were to fail a backup controller could take over with minimal downtime.

---

4 http://zookeeper.apache.org/doc/trunk/.

2.5.3 Maude Model

The distributed system state in our model is a multiset structure, populated by objects, messages, and the scheduler. Messages are state fragments passed from one object to another marked with a timestamp, an address, and a payload. The scheduler's purpose is to deliver messages to an object at the time indicated by the message's timestamp and to provide a total ordering of messages in case some have identical timestamps. ZooKeeper system state is modeled by three classes of objects: the ZooKeeper *service, servers,* and *clients.* The group key management system also consists of three different classes of objects: the group key *service, managers,* and *clients.* All these objects are designed to operate according to the system design discussed above, but we must be careful to clarify a few details.

To simplify the model, we abstracted away many complex – but externally *invisible* – details of ZooKeeper. We also say a few words about failure modes. We assume that both ZooKeeper servers and group key managers may fail. When a ZooKeeper server fails, it will no longer respond to any messages. After a variable repair timeout, the server will come back online. When any connected clients discover the failure, they will attempt to migrate to another randomly chosen server. Similarly, when a key manager fails, any client waiting on that key manager will time out and try again. A key manager buffers clients' requests and answers them in order. Before answering each request, it saves information about the requested operation to ZooKeeper. If a manager dies, the succeeding manager loads the stored state from ZooKeeper and completes any pending operations before responding to new requests. In order that our model would accurately reflect the performance of a real ZooKeeper cluster, we chose parameters that agree with the data gathered in Ref. [20]. In particular, we set the total latency of a read or write request to ZooKeeper to under 2*ms* on average (which corresponds to a ZooKeeper system under *full* load as in Ref. [20]). Note that this average time only occurs in practice if there is no server failure; our model permits servers to fail. We also assume that all of the ZooKeeper servers are located in the same data center, which is the recommended way to deploy a ZooKeeper cluster.

Here we show an example Maude rewrite rule from our specification:

```
crl [ZKSERVER-PROCESS]    :
      < A : ZKServer | leader: L, live: true, txid: Z, store: S,
          clients: C, requests: R, updates: U, AS >
      {T, A <- process}
   =>
      < A : ZKServer | leader: L, live: true, txid: Z', store: S',
          clients: C, requests: null, updates: null, AS >
        M M'
      if (Z',S',M)  := commit(C,U,S) /\ M'  := process(L,R,S')   .
```

This rule, specified in Maude, illustrates how a live ZooKeeper server object processes a batch of updates (service internal messages) and requests from clients during a processing cycle. The condition of this rule invokes two auxiliary functions: `commit` and `process`, which repeatedly commit key updates and process requests from connected clients. When this rule completes, the server will have a new transaction ID and key/value store, an empty update and request list, and a set of messages M M' to be sent, where M are notifications to interested clients that keys were updated and M' are client requests forwarded onto the ZooKeeper leader.

2.5.4 Analysis and Discussion

Our analysis consisted of two experiments. Both were run hundreds of times via PVeStA and average results were collected. The first experiment was designed to test whether saving snapshots of the group key manager's state in the ZooKeeper store could increase the overall reliability of the system. In the experiment, we set the failure rate for the key manager to 50% per experiment, the time to failover from one server to another to 5 s and the experiment duration to 50 s. We then compared the average key manager availability (i.e., the time it is available to distribute keys to clients) between a single key manager and two key managers where they share a common state saved in the ZooKeeper store. In former case, the average availability was 32.3 seconds, whereas in the latter case it was 42.31 s. This represents an availability improvement from 65 to 85%. Of course, we expect a system with replicated state to be more reliable as there is no longer a single point of failure.

Our second experiment was designed to examine whether using ZooKeeper to distribute shared keys is efficient and scalable enough for real-world use. The experiment measured the variations in: (a) the percentage of keys successfully received by group members, and (b) the key distribution latency, as increasing numbers of clients joined a group per second. The percentage of keys successfully received is defined as the total number of keys expected to be received by any client over its lifetime versus the amount actually received; and distribution latency is the average time measured from when a key is generated by the key manager until that key is received by a client. We sampled these parameters at client join rates of once every 4, 2, 1, 0.5, 0.25, and 0.125 s. We kept the same duration as in the first experiment, but picked failure probabilities such that system will have 99.99% availability and specified link latency between the ZooKeeper service and ZooKeeper clients to vary according to a uniform distribution on the interval from 0.05 to 0.25 s.

We present our results in Figures 2.3 and 2.4, where the black curve corresponds to our initial experiments while the gray curve corresponds to a slightly modified model where we added a 2 second wait time between key updates from the key manager. While our initial experiments show that naively

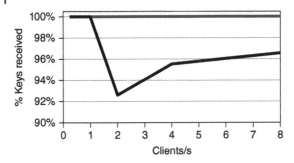

Figure 2.3 % Keys versus join rate.

using ZooKeeper as a key distribution agent works well, at high client join rates, the key reception rate seems to level out around 96%. This occurs because ZooKeeper can apply key updates internally more quickly than clients can download them; after all the ZooKeeper servers enjoy a high-speed intra-cloud connection. By adding extra latency between key updates, the ZooKeeper servers are forced to wait enough time for the correct keys to propagate to clients. As shown in Figure 2.3, this change achieves a 99% key reception in all cases. On the other hand, key distribution latency remained relatively constant, at around half a second, regardless of the join rate because ZooKeeper can distribute keys at a much higher rate than a key manager can update them [20]. Of course, the artificial latency added in the second round of experiments has a cost; it increases the time required for a client to join or leave the group by the additional wait time.

In essence, our analysis confirmed that a scalable and fault-tolerant key management service can indeed be built using ZooKeeper, settling various doubts raised about the effectiveness of ZooKeeper for key management by an earlier, but considerably less-detailed, model and analysis [31]. This result is not

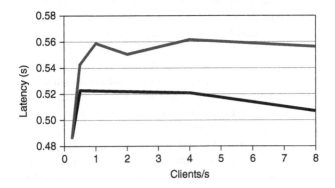

Figure 2.4 Latency versus join rate.

particularly surprising, especially considering that many man-hours would be needed to optimize an actual system. More interestingly, the analysis also showed that system designs may suffer from performance bottlenecks not readily apparent in the original description – highlighting the power of formal modeling and analysis as a method to explore the design space.

2.6 How Amazon Web Services Uses Formal Methods

The previous sections have made the case for the use of formal methods in general, and rewriting logic and Maude in particular, during the design and development of cloud computing storage systems. The reported work was conducted in academic settings. How about industry?

In 2015, engineers at Amazon Web Services (AWS) published a paper entitled "How Amazon Web Services Uses Formal Methods" [3]. AWS is probably the world's largest provider of cloud computing services, with more than a million customers and almost $10 billion in revenue in 2015, and is now more profitable than Amazon's North American retail business [32]. Key components of its cloud computing infrastructure include the DynamoDB highly available replicated database and the Simple Storage System (S3), which stores more than two trillion objects and handles more than 1.1 million requests per second [3].

The developers at AWS have used formal specification and model checking extensively since 2011. This section summarizes their use of formal methods and their reported experiences.

2.6.1 Use of Formal Methods

The AWS developers used the approach that we advocate in this chapter: the use of an intuitive, expressive, and executable (sort of) specification language together with model checking for automatic push-button exploration of all possible behaviors.

More precisely, they used Leslie Lamport's specification formalism TLA+ [33] and its associated model checker TLC. The language is based on set theory and has the usual logic and temporal logic operators. In TLA+, a transition T is defined as a logical axiom relating the "current value" of a variable x with the *next* value x' of the variable. For example, a transition T that increases the value of x by 1 and the value of *sum* with x can be defined as $T = x' = x + 1 \wedge sum' = sum + x$. The model checker TLC can analyze invariants and generate random behaviors.

Formal methods were applied on different components of S3, DynamoDB, and other components, and a number of subtle bugs were found.

2.6.2 Outcomes and Experiences

The experience of the AWS engineers was remarkably similar to that generally advocated by the formal methods community. The quotations below are all from Ref. [3].

Model checking finds "corner case" bugs that would be hard to find with standard industrial methods:

- "We have found that standard verification techniques in industry are necessary but not sufficient. We routinely use deep design reviews, static code analysis, stress testing, and fault-injection testing but still find that subtle bugs can hide in complex fault-tolerant systems."
- "T.R. learned TLA+ and wrote a detailed specification of [components of DynamoDB] in a couple of weeks. [. . .] the model checker found a bug that could lead to losing data if a particular sequence of failures and recovery steps would be interleaved with other processing. This was a very subtle bug; the shortest error trace exhibiting the bug included 35 high-level steps. [. . .] The bug had passed unnoticed through extensive design reviews, code reviews, and testing."

A formal specification is a valuable precise description of an algorithm:

- "There are at least two major benefits to writing precise design: the author is forced to think more clearly, helping eliminating 'hand waving,' and tools can be applied to check for errors in the design, even while it is being written. In contrast, conventional design documents consist of prose, static diagrams, and perhaps pseudo-code in an ad hoc untestable language."
- "Talk and design documents can be ambiguous or incomplete, and the executable code is much too large to absorb quickly and might not precisely reflect the intended design. In contrast, a formal specification is precise, short, and can be explored and experimented on with tools."
- "We had been able to capture the essence of a design in a few hundred lines of precise description."

Formal methods are surprisingly feasible for mainstream software development and give good return on investment:

- "In industry, formal methods have a reputation for requiring a huge amount of training and effort to verify a tiny piece of relatively straightforward code, so the return on investment is justified only in safety-critical domains (such as medical systems and avionics). Our experience with TLA+ shows this perception to be wrong. [. . .] Amazon engineers have used TLA+ on 10 large complex real-world systems. In each, TLA+ has added significant value. [. . .] Amazon now has seven teams using TLA+, with encouragement from senior management and technical leadership. Engineers from entry level to principal have been able to learn TLA+ from scratch and get useful results in two to three weeks."

- "Using TLA+ in place of traditional proof writing would thus likely have improved time to market, in addition to achieving greater confidence in the system's correctness."

Quick and easy to experiment with different design choices:

- "We have been able to make innovative performance optimizations [. . .] we would not have dared to do without having model checked those changes. A precise, testable description of a system becomes a what-if tool for designs."

The paper's conclusions include the following:

> "Formal methods are a big success at AWS, helping us to prevent subtle but serious bugs from reaching production, bugs we would not have found using other techniques. They have helped us devise aggressive optimizations to complex algorithms without sacrificing quality. [. . .] seven Amazon teams have used TLA+, all finding value in doing so [. . .] Using TLA+ will improve both time-to-market and quality of our systems. Executive management actively encourages teams to write TLA+ specs for new features and other significant design changes. In annual planning, managers now allocate engineering time to TLA+."

2.6.3 Limitations

The authors point out that there are two main classes of problems with large distributed systems: bugs and performance degradation when some components slow down, leading to unacceptable response times from a user's perspective. While TLA+ was effective to find bugs, it had one significant limitation in that it was not, or could not be, used to analyze performance degradation.

Why TLA + ? We have advocated using the Maude framework for the specification and analysis of cloud computing storage systems. What are the differences between Maude and its toolset and TLA+ and its model checker? And why did the Amazon engineers use TLA+ instead of, for example, Maude?

On the specification side, Maude supports hierarchical system states, object-oriented specifications, with dynamic object creation deletion, subclasses, and so on, as well as the ability to specify any computable data type as an equational specification. These features do not seem to be supported by TLA+. Maude also has a clear separation between the system specification and the property specification, whereas TLA+ uses the same language for both parts. Hence, the fact that a system S satisfies its property P can be written in TLA+ as the logical implication $S \rightarrow P$.

Perhaps the most important difference is that real-time systems can be modeled and analyzed in Real-Time Maude, and that probabilistic rewrite theories can be statistically model checked using PVeStA, whereas TLA+ seems to lack support for the specification and analysis of real-time and probabilistic systems. (Lamport argues that special treatment of real-time systems is not needed: just add a system variable *clock* that denotes the current time [34].) The lack of support for real-time and probabilistic analysis probably explains why the TLA+ engineers could only use TLA+ and TLC for correctness analysis but *not* for performance analysis, whereas we have shown that the Maude framework can be used for both aspects.

So, why did the Amazon engineers choose TLA+? The main reason seems to be that TLA+ was developed by one of the most prominent researcher in distributed systems, Leslie Lamport, whose algorithms (like the many versions of Paxos) are key components in today's cloud computing systems:

"C.N. eventually stumbled on a language [. . .] when he found a TLA+ specification in the appendix of a paper on a canonical algorithm in our problem domain—the Paxos consensus algorithm. The fact that TLA+ was created by the designer of such a widely used algorithm gave us confidence that TLA+ would work for real-world systems."

Indeed, it seems that they did not explore too many formal frameworks:

"When we found [that] TLA+ met those requirements, we stopped evaluating methods."

2.7 Related Work

Regarding related work on Cassandra, on the model-based performance estimation side, Osman and Piazzola use queueing Petri nets (an extension of colored stochastic Petri nets) to study the performance of a single Cassandra node on realistic workloads. They also compare their model-based predictions with actual running times of Cassandra. The main difference between our work and this and other work on model-based performance estimation [35,36] is that we do both functional correctness analysis and model-based performance estimation.

We discuss the use of formal methods at Amazon [3,4] in Section 6.6 and note that their approach is very similar to ours, except that they do not use their models also for performance estimation. In the same vein, the designers of the TAPIR transaction protocol targeting large-scale distributed storage systems have specified and model checked correctness properties of their system design using TLA+ [37,39].

Instead of developing and analyzing high-level formal models to quickly analyze different design choices and finding bugs early, other approaches [39,40] use distributed model checkers to model check the *implementation* of cloud systems such as Cassandra and ZooKeeper, as well as the BerkeleyDB database and a replication protocol implementation. This method can discover implementation bugs as well as protocol-level bugs.

Verifying both protocols and code is the goal of the IronFleet framework at Microsoft Research [41]. Their methodology combines TLA+ analysis to reason about protocol-level concurrency (while ignoring implementation complexities) and Floyd-Hoare-style imperative verification to reason about implementation complexities (while ignoring concurrency). Their verification methodology includes a wide range of methods and tools, including SMT solving, and requires "considerable assistance from the developer" to perform the proofs, as well as writing the code in a new verification-friendly language instead of a standard implementation language. To illustrate their method's applicability, they built and proved the correctness of a Paxos-based replicated-state-machine library and a sharded key-value store.

Whereas most of the work discussed in this chapter employs model checking, the Verdi framework [42] focuses on specifying, implementing, and verifying distributed systems using the higher order theorem prover Coq. The Verdi framework provides libraries and a toolchain for writing distributed systems (in Coq) and verifying them. The Verdi methodology has been used to mechanically check correctness proofs for the Raft consensus algorithm [43], a back-up replication system, and a key-value data store. Much like our approach, their executable Coq "implementations" can be seen as executable high-level formal models/specifications, an approach that eliminates the "formality gap between the model and the implementation." The key difference is that Verdi relies on theorem proving, which requires nontrivial user interaction, whereas model checking is automatic. Model checking and theorem proving are somewhat orthogonal analysis methods. On the one hand, model checking only verifies the system for single initial configurations, whereas theorem proving can prove the model correct for all initial configurations. On the other hand, model checking can find subtle bugs automatically, whereas theorem proving is not a good method for finding bugs. (Failure of a proof attempt does not imply that there is a bug in the system, only that the particular proof attempt did not work.)

Coq is also used in the Chapar framework, which aims to verify the causal consistency property for key-value store implementations [44].

Finally, in Ref. [45], Bouajjani *et al.* study the problem of verifying the *eventual consistency* property (guaranteed by, e.g., Cassandra) for optimistic replication systems (ORSs) in general. They show that verifying eventual consistency for such systems can be reduced to an LTL model checking problem. For this purpose, they formalize both the eventual consistency property and, in particular, ORSs, and characterize the classes of ORSs for which the problem of verifying eventual consistency is decidable.

2.8 Concluding Remarks

We have proposed rewriting logic, with its extensions and tools, as a suitable framework for formally specifying and analyzing both the correctness and the performance of cloud storage systems. Rewriting logic is a simple and intuitive yet expressive formalism for specifying distributed systems in an object-oriented way. The Maude tool supports both simulation for rapid prototyping and automatic "push-button" model checking exploration of all possible behaviors from a given initial system configuration. Such model checking can be seen as an exhaustive search for "corner case" bugs or as a way to automatically execute a more comprehensive "test suite" than is possible in standard test-driven system development. Furthermore, PVeStA-based statistical model checking can provide assurance about quantitative properties measuring various performance and quality of service behavior of a design with a given confidence level, and Real-Time Maude supports model checking analysis of real-time distributed systems.

We have used Maude and Real-Time Maude to develop quite detailed formal models of a range of industrial cloud storage systems (Apache Cassandra, Megastore, and Zookeeper) and an academic one (RAMP), and have also designed and formalized significant extensions of these systems (a variant of Cassandra, Megastore-CGC, a key management system on top of ZooKeeper, and variations of RAMP) and have provided assurance that they satisfy desired correctness properties; we have also analyzed their performance. Furthermore, in the case of Cassandra, we compared the performance estimates provided by PVeStA analysis with the performance actually observed when running the real Cassandra code on representative workloads; they differed only by 10–15%.

We have also summarized the experience of developers at Amazon Web Services, who have successfully used formal specification and model checking with TLA+ and TLC during the design of critical cloud computing services such as the DynamoDB database and the Simple Storage System. Their experience showed that formal methods: (i) can feasibly be mastered quickly by engineers, (ii) reduce time-to-market and improve product quality, (iii) are cost-effective, and (iv) discover bugs not discovered during traditional system development. Their main complaint was that they could not use formal methods to analyze the performance of a design (since TLA+ lacks the support for analyzing real-time and probabilistic systems).

We believe that the Maude formalism should be at least as easy to master as TLA+ (as our experience with Megastore indicates), and might be even more convenient; furthermore, as shown in this chapter, the Maude tools can also be used to analyze the performance of a system.

2.8.1 The Future

Much work remains ahead. To begin with, one current limitation is that explicit state model checking explores the behaviors starting from a *single* initial system configuration, and hence – since this can only be done for a finite number of states – it cannot be used to verify that an algorithm is correct for all possible initial system configurations, which can be infinite.

A *first* approach to obtain a more complete coverage is illustrated by Refs [26,46], where we have extended coverage by model checking all initial system configurations up to n operations, m replicas, and so on. However, naïvely generating initial states can yield a large number of symmetric initial states; greater coverage and symmetry reduction methods should therefore be explored. A *second*, more powerful approach is to use *symbolic model checking* techniques, where a possibly infinite number of initial states is described symbolically by formulas in a theory whose satisfiability is decidable by an SMT solver. In the Maude context, this form of symbolic model checking is supported by *rewriting modulo SMT* [47], which has already been applied to various distributed real-time systems, and by *narrowing-based symbolic model checking* [48,49]. An obvious next step is to verify properties of cloud storage systems for possibly infinite sets of initial states using this kind of symbolic model checking. A *third* approach is to take to heart the complementary nature of model checking and theorem proving, so that these methods do help each other when used *in combination*, as explained below.

In the approach we have presented, *design exploration* based on formal executable specification comes first. Once a promising design has been identified, fairly exhaustive *model checking* debugging and verification of such a design can be carried out using: (i) LTL explicit-state model checking, (ii) statistical model checking of quantitative properties, and (iii) symbolic model checking from possibly infinite sets of initial configurations for greater assurance. Since all these methods are *automatic,* this is relatively easy to do based on the system's formal executable specification with rewrite rules. Only after we are fairly sure that we have obtained a good design and have eliminated many subtle errors by these automatic methods does it become cost-effective to attempt a more labor-intensive *deductive verification* through theorem proving of properties not yet fully verified. Indeed, some properties may already have been *fully verified* in an automatic way by symbolic model checking, so that only some additional properties not expressible within the symbolic model checking framework need to be verified deductively. Two compelling reasons for such deductive verification are (i) the *safety-critical* nature of a system or a system component, and (ii) the *high reusability* of a component or algorithm, for example, Paxos or Raft, so that the deductive verification effort becomes amortized over many uses.

In the context of rewriting logic verification, a simple logic to deductively verify reachability properties of a distributed system specified as a rewrite theory has been recently developed, namely, *constructor-based reachability logic* [50], which itself extends and applies to rewrite theories of the original reachability logic in Ref. [51]. In fact, a number of distributed algorithms have already been proved correct in Ref. [50]; and we plan to soon extend and apply the Maude reachability logic tool to verify specific properties of cloud storage systems.

Of course, high reusability of designs is not just a good thing for amortizing verification efforts: It is a good thing in all respects. By breaking distributed system designs into modular, well-understood, and highly reusable components that come with strong correctness guarantees, as advocated by the notion of *formal pattern* in Ref. [52], unprecedented levels of assurance and of software quality could be achieved for cloud storage systems in the near future. It is for this reason that, from the beginning of our research on the application of formal methods to gain high assurance for cloud-based systems, we have tried to understand cloud system designs as suitable compositions of more basic components providing key functionality in the form of generic distributed algorithms. This is still work in progress. Our medium-term goal is to develop a library of generic components formally specified and verified in Maude as formal patterns in the above sense.

Out of these generic components, different system designs for key-value stores and for cloud storage systems supporting transactions could then be naturally and easily obtained as suitable compositions.

Acknowledgments

We thank Jatin Ganhotra, Muntasir Raihan Rahman, and Son Nguyen for their contributions to the work on Cassandra reported in this chapter. This work is based on research sponsored by the Air Force Research Laboratory and the Air Force Office of Scientific Research, under agreement number FA8750-11-2-0084. The U.S. Government is authorized to reproduce and distribute reprints for Government purposes notwithstanding any copyright notation thereon. This work is also based on research supported by the National Science Foundation under Grant Nos. NSF CNS 1409416 and NSF CNS 1319527.

References

1 Brewer, E.A. (2000) Towards robust distributed systems (abstract), in Proceedings of the PODC'00, ACM.

2 Munir, H., Moayyed, M., and Petersen, K. (2014) Considering rigor and relevance when evaluating test driven development: a systematic review. *Inform. Softw. Technol.*, **56** (4), 375–394.

3 Newcombe, C., Rath, T., Zhang, F., Munteanu, B., Brooker, M., and Deardeuff, M. (2015) How Amazon Web Services uses formal methods. *Commun. ACM*, **58** (4), 66–73.

4 Newcombe, C. (2014) Why Amazon chose TLA+, in Abstract State Machines, Alloy, B, TLA, VDM, and Z (ABZ'14), Lecture Notes in Computer Science, vol. 8477, Springer.

5 Meseguer, J. (1992) Conditional rewriting logic as a unified model of concurrency. *Theor. Comp. Sci.*, **96**, 73–155.

6 Clavel, M. *et al.* (2007) All About Maude, LNCS, vol. 4350, Springer.

7 Ölveczky, P.C. and Meseguer, J. (2007) Semantics and pragmatics of Real-Time Maude. *Higher-Order Symbol. Comput.*, **20** (1–2) 161–196.

8 Agha, G.A., Meseguer, J., and Sen, K. (2006) PMaude: rewrite-based specification language for probabilistic object systems. *Electr. Notes Theor. Comput. Sci.*, **153** (2), 213–239.

9 AlTurki, M. and Meseguer, J. (2011) PVeStA: a parallel statistical model-checking and quantitative analysis tool, in Proceedings of the CALCO 2011, LNCS, vol. 6859, Springer.

10 Sen, K., Viswanathan, M., and Agha, G. (2005) On statistical model checking of stochastic systems, in Proceedings of the CAV'05, LNCS, vol. 3576, Springer.

11 Meseguer, J. (2012) Twenty years of rewriting logic. *J. Logic Algebr. Program.*, **81** (7–8), 721–781.

12 Ölveczky, P.C. (2014) Real-Time Maude and its applications, in Proceedings of the WRLA'14, LNCS, vol. 8663, Springer.

13 Vardi, M.Y. (2001) Branching vs. linear time: final showdown, in Proceedings of the TACAS'01, LNCS, vol. 2031, Springer.

14 Ölveczky, P.C., Meseguer, J., and Talcott, C.L. (2006) Specification and analysis of the AER/NCA active network protocol suite in Real-Time Maude. *Formal Meth. Syst. Des.*, **29** (3), 253–293.

15 Lien, E. and Ölveczky, P.C. (2009) Formal modeling and analysis of an IETF multicast protocol, in Proceedings of the SEFM'09, IEEE Computer Society.

16 Ölveczky, P.C. and Thorvaldsen, S. (2009) Formal modeling, performance estimation, and model checking of wireless sensor network algorithms in Real-Time Maude. *Theor. Comp. Sci.*, **410** (2–3) 254–280.

17 Hewitt, E. (2010) *Cassandra: The Definitive Guide*, O'Reilly Media.

18 Baker, J. et al. (2011) Megastore: providing scalable, highly available storage for interactive services, in CIDR'11. Available at www.cidrdb.org.

19 Bailis, P., Fekete, A., Hellerstein, J.M., Ghodsi, A., and Stoica, I. (2014) Scalable atomic visibility with RAMP transactions, in Proceedings of the SIGMOD'14, ACM.

20 Hunt, P., Konar, M., Junqueira, F., and Reed, B. (2010) Zookeeper: wait-free coordination for Internet-scale systems, in: USENIX ATC, vol. 10.

21 Skeirik, S., Bobba, R.B., and Meseguer, J. (2013) Formal analysis of fault-tolerant group key management using ZooKeeper, in 13th IEEE/ACM International Symposium on Cluster, Cloud, and Grid Computing (CCGrid'13), IEEE Computer Society.

22 DB-Engines (2016) http://db-engines.com/en/ranking.

23 Liu, S., Rahman, M.R., Skeirik, S., Gupta, I., and Meseguer, J. (2014) Formal modeling and analysis of Cassandra in Maude, in Proceedings of the ICFEM'14, LNCS, vol. 8829, Springer.

24 Liu, S., Nguyen, S., Ganhotra, J., Rahman, M.R., Gupta, I., and Meseguer, J. (2015) Quantitative analysis of consistency in NoSQL key-value stores, in Proceedings of the QEST'15, LNCS, vol. 9259, Springer.

25 Liu, S., Ganhotra, J., Rahman, M.R., Nguyen, S., Gupta, I., and Meseguer, J. (2016) Quantitative analysis of consistency in NoSQL key-value stores. *LITES*, **3** (2), 02:1–02:26.

26 Grov, J. and Ölveczky, P.C. (2014) Formal modeling and analysis of Google's Megastore in real-time Maude, in Specification, Algebra, and Software, LNCS, vol. 8373, Springer.

27 Grov, J. and Ölveczky, P.C. (2014) Increasing consistency in multi-site data stores: Megastore-CGC and its formal analysis, in Proceedings of the SEFM'14, LNCS, vol. 8702, Springer.

28 Wong, C.K., Gouda, M.G., and Lam, S.S. (2000) Secure group communications using key graphs. *IEEE/ACM Trans. Netw.*, **8** (1), 16–30.

29 Rafaeli, S. and Hutchison, D. (2003) A survey of key management for secure group communication. *ACM Comput. Surv.*, **35** (3), 309–329.

30 Gupta, J. (2011) Available group key management for NASPInet, Master's thesis, University of Illinois at Champaign-Urbana.

31 Eckhart, J. (2012) Security analysis in cloud computing using rewriting logic. Master's thesis, Ludwig-Maximilans-Universität München.

32 Statt, N. Amazon's earnings soar as its hardware takes the spotlight, in The Verge, April 28, 2016. Available at http://www.theverge.com/2016/4/28/11530336/amazon-q1-first-quarter-2016-earnings (accessed May 29, 2016).

33 Lamport, L. (2002) *Specifying Systems: The TLA+ Language and Tools for Hardware and Software Engineers*, Addison-Wesley.

34 Abadi, M. and Lamport, L. (1994) An old-fashioned recipe for real time. *ACM Trans. Program. Lang. Syst.*, **16** (5), 1543–1571.

35 Barbierato, E., Gribaudo, M., and Iacono, M. (2014) Performance evaluation of NoSQL bigdata applications using multi-formalism models. *Future Generation Comp. Syst.*, **37**, 345–353.

36 Gandini, A., Gribaudo, M., Knottenbelt, W.J., Osman, R., and Piazzolla, P. (2014) Performance evaluation of NoSQL databases, in EPEW.

37 Zhang, I., Sharma, N.K., Szekeres, A., Krishnamurthy, A., and Ports, D.R.K. (2014) Building consistent transactions with inconsistent replication. Technical Report UW-CSE-2014-12-01 v2, University of Washington. Available at https://syslab.cs.washington.edu/papers/tapir-tr14.pdf.

38 Zhang, I., Sharma, N.K., Szekeres, A., Krishnamurthy, A., and Ports, D.R.K. (2015) Building consistent transactions with inconsistent replication, in Proceedings of the Symposium on Operating Systems Principles, (SOSP'15), ACM.

39 Leesatapornwongsa, T., Hao, M., Joshi, P., Lukman, J.F., and Gunawi, H.S. (2014) SAMC: semantic-aware model checking for fast discovery of deep bugs in cloud systems, in 11th USENIX Symposium on Operating Systems Design and Implementation (OSDI'14), USENIX Association.

40 Yang, J., Chen, T., Wu, M., Xu, Z., Liu, X., Lin, H., Yang, M., Long, F., Zhang, L., and Zhou, L. (2009) MODIST: transparent model checking of unmodified distributed systems, in Proceedings of the 6th USENIX Symposium on Networked Systems Design and Implementation (NSDI'09), USENIX Association, pp. 213–228.

41 Hawblitzel, C., Howell, J., Kapritsos, M., Lorch, J.R., Parno, B., Roberts, M.L., Setty, S.T.V., and Zill, B. (2015) IronFleet: proving practical distributed systems correct, in Proceedings of the 25th Symposium on Operating Systems Principles (SOSP'15), ACM.

42 Wilcox, J.R., Woos, D., Panchekha, P., Tatlock, Z., Wang, X., Ernst, M.D., and Anderson, T.E. (2015) Verdi: a framework for implementing and formally verifying distributed systems, in Proceedings of the 36th ACM SIGPLAN Conference on Programming Language Design and Implementation (PLD'15), ACM.

43 Ongaro, D. and Ousterhout, J.K. (2014) In search of an understandable consensus algorithm, in 2014 USENIX Annual Technical Conference, (USENIX ATC '14), Philadelphia, PA, June 19–20, 2014, USENIX Association, pp. 305–319.

44 Lesani, M., Bell, C.J., and Chlipala, A. (2016) Chapar: certified causally consistent distributed key-value stores, in Proceedings of the 43rd Annual ACM SIGPLAN-SIGACT Symposium on Principles of Programming Languages (POPL'16), ACM.

45 Bouajjani, A., Enea, C., and Hamza, J. (2014) Verifying eventual consistency of optimistic replication systems, in Proceedings of the 41st Annual ACM SIGPLAN-SIGACT Symposium on Principles of Programming Languages (POPL'14), ACM.

46 Liu, S., Ölveczky, P.C., Rahman, M.R., Ganhotra, J., Gupta, I., and Meseguer, J. (2016) Formal modeling and analysis of RAMP transaction systems, in Proceedings of the SAC'16, ACM.

47 Rocha, C., Meseguer, J., and Muñoz, C.A. (2017) Rewriting modulo SMT and open system analysis. *J. Logic Algebr. Meth. Program.*, **86** (1), 269–297.

48 Bae, K. and Meseguer, J. (2013) Abstract logical model checking of infinite-state systems using narrowing, in Rewriting Techniques and Applications (RTA'13), LIPIcs, vol. 21, pp. 81–96 (Schloss Dagstuhl–Leibniz-Zentrum fuer Informatik).

49 Bae, K. and Meseguer, J. (2014) Infinite-state model checking of LTLR formulas using narrowing, in WRLA'14, LNCS, vol. 8663, Springer.

50 Skeirik, S., Stefanescu, A., and Meseguer, J. (2017) A constructor-based reachability logic for rewrite theories. Technical Report,, University of Illinois Computer Science Department (March). Available at http://hdl.handle.net/2142/95770.

51 Stefanescu, A., Ciobâcă S., Mereuta, R., Moore, B.M., Serbanuta, T., and Rosu G., (2014) All-path reachability logic, in Proceedings of the RTA-TLCA, LNCS, vol. 8560, Springer.

52 Meseguer, J. (2014) Taming distributed system complexity through formal patterns. *Sci. Comp. Program.*, **83**, 3–34.

3

Risks and Benefits: Game-Theoretical Analysis and Algorithm for Virtual Machine Security Management in the Cloud

Luke Kwiat,[1] Charles A. Kamhoua,[2] Kevin A. Kwiat,[3] and Jian Tang[4]

[1]*Department of Industrial and Systems Engineering, University of Florida, Gainesville, FL, USA*
[2]*Network Security Branch, Network Sciences Division, U.S. Army Research Laboratory, Adelphi, MD, USA*
[3]*Haloed Sun TEK, Sarasota, FL, USA*
[4]*Department of Electrical Engineering and Computer Science, Syracuse University, Syracuse, NY, USA*

The growth of cloud computing has spurred many entities, both small and large, to use cloud services in order to achieve cost savings. Public cloud computing has allowed for quick, dynamic scalability without much overhead or long-term commitments. However, there are some disincentives to using cloud services, and one of the biggest is the inherent and unknown danger stemming from a shared platform – namely, the hypervisor. An attacker who compromises a virtual machine (VM) and then goes on to compromise the hypervisor can readily compromise all virtual machines on that hypervisor. That brings into play the game-theoretic problem of negative externalities, in which the security of one player affects the security of another. Using game theory to model and solve these externalities, we show that there are multiple Nash equilibriums. Furthermore, we demonstrate that the VM allocation type can adversely affect the role that externality plays in the cloud security game. Finally, we propose an allocation method based on a Nash equilibrium such that the negative externality imposed on other players can be significantly lowered compared to that found with other common VM allocation methods.

3.1 Introduction

Cloud computing is rapidly becoming a standard resource in the IT field. It has been predicted that "by 2017, nearly two-thirds of all workloads will be processed in the cloud," and that cloud traffic would grow by 370% from 2012 to 2017 [1]. Given such astounding growth, it is becoming exceedingly challenging for security to keep pace, especially since demand for security is often overlooked by clients in favor of performance and reliability. Security is difficult to quantify and offers less tangible obvious benefits than performance or

Assured Cloud Computing, First Edition. Edited by Roy H. Campbell, Charles A. Kamhoua, and Kevin A. Kwiat.
© 2018 the IEEE Computer Society, Inc. Published 2018 by John Wiley & Sons, Inc.

other cost-cutting measures, but even so, there are very apparent issues with cloud security. Security-sensitive companies are fearful of the possible theft of confidential or other sensitive data, and that fear poses a "significant barrier to the adoption of cloud services" [2].

The argument that current security measures in the cloud are insufficient has significant merit. Many cloud providers are unaware of the extent of their own security vulnerabilities, and this issue is compounded by the unique structure of the cloud. Many different users run virtual machines (VMs) on the same physical hardware, and that can allow an attacker to distribute his or her attack throughout the hypervisor and onto all the VMs running on that hypervisor. Under those conditions, one user's VM may be indirectly attacked when a direct attack is launched on a different user's VM on the same hypervisor. That is possible because the hypervisor may have unknown security vulnerabilities that, once compromised, can allow an attacker to permeate every VM on the targeted hypervisor [3]. This phenomenon is not common in traditional networks, in which an attacker must use a multihop method to attack victims indirectly. Thus, the problem of interdependency between users on the same hypervisor presents cloud computing providers with a unique challenge: How can virtual machines be allocated for optimum security?

In the cloud, an attacker can launch an indirect attack on User *j* by first compromising the VMs of User *i*, successfully attacking the hypervisor and then compromising the VMs of User *j*. That possibility creates a risk phenomenon wherein the security of User *j* is dependent on that of User *i* and vice versa. Further, the problem acquires another dimension if there is a large difference between the potential total losses that could be realized by Users *i* and *j*. As a result, big-name users could be discouraged from using the cloud by their much greater potential for loss. With only 56% of cloud users saying they trust cloud providers with sensitive information, and an even smaller 30% stating that they know what procedures cloud providers use for security, the problem is still prevalent [4]. Because of the interacting and interdependent nature of cloud users, we may use game theory to find solutions by modeling those users' choices, since game theory offers "mathematical models of conflict and cooperation between intelligent rational decision-makers" [5].

This chapter reviews and summarizes several contributions. Primarily, its focus is on modeling the choices of several rational players in the cloud within a game-theoretical framework. We will describe the externality problem discussed in Ref. [6] and propose how it can be solved through effective VM allocation management by the cloud provider to ensure delivery of maximum security and minimal loss potential for all cloud users. We aim to use the knowledge gained from this model to influence cloud providers to make optimal security-based virtual machine allocation decisions.

Section 3.2 discusses the potential role of cloud technologies, including those detailed in this chapter, in military missions. Section 3.3 describes related prior work and how the work discussed in the remainder of this chapter is different. Section 3.4 elaborates on the cloud architecture that is used in our game model and that is common to many cloud structures. Section 3.5 sets up the game model and diagrams the game in normal (matrix) form. After analyzing the game result for the simplest case in Section 3.6, we will extend our model in Section 3.7. Section 3.8 will present numerical results and compare our model's negative externality to that of other, more commonly used allocation methods. Section 3.9 concludes the chapter with thoughts on future developments.

3.2 Vision: Using Cloud Technology in Missions

In the term *cloud computing*, the modifier *cloud* is simply a metaphor for the Internet. Indeed, graphic illustrations of the Internet almost always depict it as an iconic cloud with white semicircles gathered together to form a large, puffy shape. It is a fitting symbol for the Internet because like a cloud, the Internet has the qualities of expansiveness and a seemingly continuously changing conformation. Cloud computing, or Internet-based computing, embraces those qualities. Its expansiveness inheres in its broad network access and abundantly available computing resources for on-demand service, and its continuously changing conformation is evident in its broad resource pooling and the rapid elasticity of those resources. The derived benefits of cloud computing – or simply "the cloud" – include faster deployment, infrastructure flexibility, and freedom from upfront investment. Although it appears amorphous, the cloud offers computing services that are monitored, controlled, and reported.

Unfortunately, another visual image that can be associated with cloud computing is that of a dark cloud that imparts a sense of foreboding. It is applicable because the aforementioned benefits of cloud computing are overshadowed by security risks that stem from cloud computing's openness and sharing of resources. In spite of the presence of monitoring, controls, and reporting, the inherent security risks create justifiable apprehension. Sound security practice dictates that the risks be continually assessed and appropriately mitigated.

In 2007, Air Force General T. Michael Moseley stated that the Air Force's mission is to "deliver sovereign options for the defense of the United States of America and its global interests – to fly and fight in Air, Space, and Cyberspace" [7]. To accomplish its mission, the Air Force takes the fight into the clouds, by which we now mean two types of clouds: not just the atmospheric phenomena but now the computational cloud as well. The forerunners of today's Air Force were the World War I flyers, for whom the air was a new arena for military combat. They realized that their missions should not be restricted to the open air of clear skies; instead, they would have to venture into the clouds (of the first type). This was a risky endeavor not just because of the turbulence they encountered in their rickety, canvas-skinned aircraft but also because the clouds could conceal something harmful within. Eventually techniques were developed to overcome this limitation of airpower. Now, the Air Force has leapt ahead to venture into contemporary clouds (of the second type). Therefore, like the early pioneers of military flight, we must address the attendant cyber risks so that pursuit of the Air Force's mission can continue to reach "into the clouds."

The cloud metaphor applies to multiple mission environments (i.e., both air and cyberspace), but cloud computing performs a role in the national information infrastructure that creates distinct mission considerations. Historically, a nation's commercial infrastructure can be found at the forefront of its military operations. For example, at the time of the Civil War, railways of the United States had become a key part of the nation's transportation infrastructure. The railways were viewed as so strategically critical for the expeditious movement of troops that in January of 1862, an act was passed authorizing President Lincoln to have the government take possession of important railway lines [8]. Assuming the total control of a public computing cloud in such a manner would be drastic; however, in the military acronym *C4I*, which stands for "command, control, communications, computers, and intelligence," the fourth "C" indicates that computation is viewed by the military authorities as critical to meeting strategic requirements. Given this criticality, it is conceivable that in times of urgency, the military might look to commercial resources such as the public cloud; cloud computing's openness has ushered in the prospect of garnering needed military computing resources directly from the public cloud. (Note that the term "public cloud" does not imply government-owned resources; instead, like the railways, it is a privately owned resource that is made available to the public.) The hypothetical commandeering by the military of the public cloud would likely not be done on a permanent basis; instead, following the mission, the resources would be returned to their original public state. At the end of the Civil War, an Executive Order of August 8, 1865 restored the government-controlled railways to their original owners [8]; similarly, when the military's needs in the cloud have been fulfilled, any computing resources it has taken over can be readily released in accordance with the cloud's property of elasticity. The seizure of the railways was done to ensure that they could be used with militarily precise efficiency and regularity. Likewise, complete possession of the cloud would be a significant risk-reducer: If cloud resources are not being shared, the possibility of negative externalities can be eliminated. However, resorting to the heavy-handed action of assuming complete public cloud possession would defeat almost all of the economic advantages offered by cloud computing. Indeed, the public cloud would no longer be "public"; instead, the computing needs of nonmilitary users would be disregarded. The situation would be analogous to one in which military-controlled railways halted all commercial passenger and freight traffic, placing extreme hardship upon civilians. A loss of public cloud services would have a similarly negative impact on the civilian sector. The magnitude of the negative impact could be especially great during postconflict nation-building or humanitarian missions following natural disasters. In such scenarios, the military's exclusive use of the (perhaps fledgling) public cloud could disastrously interfere with the computational needs of the citizens whose country is being rebuilt. Therefore, our goal for the computational cloud should be, as part of a national information infrastructure, to allow unfettered access to both the public and those performing military missions. That is a core tenet of our game-theoretical analysis and algorithm for virtual machine security management in the cloud: It admits the coexistence of nonmilitary users together with users engaged in military missions. Permitting coexistence, however, requires caution, because it can present unacceptable risks to a military mission's virtual machines. Consequently, our virtual machine security management scheme's ability to balance security risks buttresses the public cloud's ability to accommodate military and nonmilitary users simultaneously.

It is not enough that our game-theoretical algorithm for virtual machine security management embraces the principles of coexistence and balanced risk; on top of that, it is of paramount importance that the algorithm's implementation be conducive to public cloud adoption. Our implementation avoids a major pitfall by not taking the form of a military specification (*mil-spec*). In a previous work [9], we discussed at length the tension between proponents and detractors of mil-specs. In brief, the latter group argues that mil-specs are overburdened with inspection and screening requirements that slow the building of systems to a glacial pace. Use of mil-specs, they contend, has led not to leading-edge, but to trailing-edge designs. We avoid this controversy by not having our algorithm impose specific implementation details on the underlying cloud infrastructure, such as hardware or software applied to individual virtual machines in the form of guards, partitions, or other isolation enforcers. Our algorithm assumes a much greater generality in the cloud's baseline design. That approach makes the public cloud amenable for military use while leaving undisturbed the cloud's inherent properties of openness, broad resource pooling, and rapid elasticity of resources. Preservation of the public cloud's properties will promote its continued development by commercial entities, and those developments can be leveraged to support completion of missions. We view our game-theoretical algorithm for virtual machine security management as one such potential development. From a commercialization perspective, the algorithm provides a lightweight means to lower the cost of allocating cloud computing resources such that the risk associated with so-called "bad neighbors" is ameliorated. One of its advantages is that it offers underprivileged prospective cloud users access that is on par with that of cloud users with more generous security budgets. Increasing the involvement of users from underrepresented classes who seek more affordable computing will expand the overall user base of the cloud, and with an expanding user base, the public cloud will thrive.

If the public cloud is to be a supporting element of C4I, then our game-theoretical algorithm for virtual machine security must keep us on track toward use of cloud technology in missions. We have argued that a public cloud can be a valuable resource for military missions but only if the mission use is done judiciously. Our vision for the game-theoretical algorithm for virtual machine security management is that it will cultivate public cloud growth for military and nonmilitary users alike.

3.3 State of the Art

A major issue in cloud computing is how the cloud provider will allocate the virtual machine instances created by a constant stream of clients. Previous work has looked at many ways to approach the problem. Many of the solutions offered have involved allocation of virtual machines based on load-balancing techniques, energy consumption, or both [10–13].

Wei *et al.* [11] use game theory in order to maximize the efficiency of resources within a cloud computing environment. Their solution was split into two parts: an independent solution for nonmultiplexing of resource assignments, and a solution that includes multiplexing. The main idea of the problem model is to maximize the utility that is subject to an allocation vector. For the first part of the problem, it was assumed that each task (user) was assigned a specific resource in the cloud and that the subtasks are dependent and communicate with each other. For the independent optimization, only initial strategies (of allocation) are used in computation, and thus an evolutionary game is used for multiple instances. In evolutionary computations, the initial allocation is given, and later ones are calculated using the algorithms given in Ref. [11].

Beloglazov *et al.* [13] offered several algorithmic solutions rooted in heuristics. A lower utilization threshold was first established, ensuring that if CPU usage below this threshold was encountered, then all the VMs hosted on this server would be migrated elsewhere. For the upper utilization threshold, several algorithms were studied. They included the minimization of migration (MM) policy, which, as stated in its name, served to minimize the amount of VM migrations used to keep the CPU below the upper utilization threshold. The highest potential growth (HPG) policy migrates the VMs that have the lowest usage of the CPU relative to the CPU capacity in order to minimize the potential increase of the host's utilization and prevent a service-level agreement (SLA) violation. Finally, the random choice (RC) policy randomly selects VMs to migrate in order to decrease CPU utilization. After carrying out 10 different tests at different lower utilization thresholds, the authors found that as the lower utilization threshold increased, the SLA violations increased and energy consumption decreased. That gave an optimum lower utilization threshold for each algorithm. Furthermore, out of the three options, it was found that the MM policy was the best policy overall in terms of energy consumption and SLA violation.

Xu and Fortes [10] introduced several algorithms to achieve allocative efficiency such as that described by Beloglazov *et al.* They used two different research areas to solve the optimization problem: combinatorial (discrete) optimization and multiobjective (fuzzy) optimization. To determine the optimal level of resource management, they suggested a two-level control system, which involves mapping of application workloads and a mapping from virtual resources to physical resources. That gives rise to a local and a global controller. The focus of their paper is on the global controller and its resource mapping problem, which aims to optimize three main variables: the minimization of power usage, the minimization of thermal dissipation, and the minimization of maximum temperature. For the evaluation, six different types of algorithms were pitted against each other, including four offline bin-packing heuristics and two single-objective GGAs (grouping genetic algorithms). The results were measured by performance, robustness, and scalability.

Jalaparti *et al.* [12] attempted to solve the issue of cloud resource allocation with game theory, much like Wei *et al.* They used game theory to model the intricate client-to-client and client-to-provider interactions. Thus, they simultaneously solved the cloud provider's problem with allocative efficiency (since allocative procedures today are fairly simplistic and nonoptimal) and allowed the provider to charge near-optimal prices for cloud usage. (Pricing policies are also usually simplistic, with the norm being a fixed pricing model.) The solution is what they called a *CRAG* (cloud resource allocation game), which was always found to have a Nash equilibrium. They documented the worst case of bounds for the solution (bounds were needed as a result of the price of *anarchy*, a concept in economics and game theory that measures how the efficiency of a system degrades because of the selfish behavior of its players) and provided several heuristic algorithms that gave near-optimal allocation and pricing policies.

Game theory has also been applied to certain specific aspects of cloud computing, including infrastructure resilience [14], cloud usage pricing [15], and virtual machine allocation [11,12,16].

Han *et al.* [16] demonstrated a new method of infiltration that can be exploited through the cloud and not through traditional computing: infiltration through side-channels. That threat gives rise to new risks, as attackers can exploit the sharing by users of the hardware used to create VMs. By starting a VM in the same server where a target user has a VM, an attacker can siphon sensitive information from the victim, including Web traffic and even encryption keys. Han *et al.*'s study uses game theory to find the best method for mitigating such attacks, in which attackers have been shown to have a 40% success rate in colocating VMs with the target user's VM [17]. The setup was a two-player game with an attacker and a defender (the cloud administrator). The defender's utility function was to include weights for power distribution, security, and workload balance of the servers. It was determined that the optimal strategy for each player is a mixed strategy, with each strategy governing a different probability of using a common VM allocation setup used by cloud providers. The relative probabilities of the different VM allocations being played were dependent on several factors, including the lag time used by an attacker to launch a VM.

Rao *et al.* [14] studied the ability of a cloud computing entity, within the framework of game theory, to provide a given capacity C at a certain probability P given a physical or cyberattack on its infrastructure. A simple game was set up in which a cloud provider with a certain number of servers S was to defend against an attacker attacking these servers. One of the possible ways to mitigate an attack was for the provider to use reinforcement strategies that decreased attacker utility. The formulas for utility included the number of servers, the number of sites, the number of attacked servers, the attacked site, and the distribution of servers across the attacked site. It was concluded through several tests that the survivability of an attack (the ability of the system to operate at capacity C under probability P despite the attack) is heavily influenced by the cost of defense and the cost of attack. If the cost of defense is too high, the provider will choose not to defend the site, and thus the survivability will be 0. Furthermore, if the number of servers that are spread over given sites is changed and the number or the distribution are not made public, then the attacker's utility is diminished, since a linear attack method will not be as effective against a nonlinear setup of servers.

Künsemöller and Karl [15] examined the economics of cloud computing and the viability of its use for a given business. Game theory was specifically used to model a few circumstances in which it would be economically beneficial to use cloud services. In particular, the authors used an extensive-form game that first branched out a decision tree from the cloud provider (based on the different prices the provider would charge for cloud services) and then reached a decision from the potential customer based on that price. The decision for the customer was binary, as he could either use cloud services or build his own data center. The payoff for the provider is clearly maximized if it charges the highest price at which there is a break-even cost-benefit point for the client in using cloud services. Another issue that Künsemöller and Karl addressed was a cloud–data hybrid, in which a client uses a private data center for a base load of information and cloud services during peak demand/hours. A $price_{low}$ was used to denote the price at which a complete cloud solution was the best option, and a $price_{high}$ represented the price at which using cloud services in conjunction with a private data center became unviable. Above that highest price for cloud usage, a wholly owned private data center would be the best option for maximum utility. In practice, given the oligopolistic nature of cloud providers, cloud services can be provided with large economies of scale, which can keep the price of cloud services around $price_{low}$. Thus, there are few reasons left for not outsourcing to the cloud.

Outside the areas of game theory and VM allocation, another important effort was that of Ristenpart *et al.* [17], who pursued discovery of new vulnerabilities in the cloud structure. They looked at the idea of coresident attacks on virtual machines within the cloud network. Unlike any predecessors, they used empirical evidence and actual data from experiments run on the Amazon EC2 cloud. They began by running all five instance types that EC2 offers across three available placement zones within the cloud. From that, they determined that IP assignment was very uniform, in that IP addresses were partitioned by availability zones (most likely in order to make it easy to manage separate network connectivity for these zones). Using that information, the authors developed a test to determine coresidence on network-based technology (and also showed that that type of technology need not be used). Eventually, they showed that an efficient attacker can achieve coresidence in up to 40% of attacks. Once coresidence has been achieved, several methods can be used to extract information from or damage the victim, including measurement of cache usage, launch of denial-of-service or keystroke-timing attacks, theft of cryptographic keys, and estimation of traffic rates. Ristenpart *et al.* concluded that to ensure unconditional security against cross-VM attacks, it is currently necessary to avoid coresidence.

Kamhoua *et al.* [6] viewed attacks on the hypervisor and compromise of virtual machines from a game-theoretical perspective. One of the larger issues they presented in Ref. [6], interdependency, is also addressed in this chapter. However, they addressed interdependency mainly as it related to the compromise of one user's security integrity through the lack of investment of another user who is on the same hypervisor, since an attack on the hypervisor may propagate from one user to other users. We touch on that specific scenario in the present chapter, but to a much lesser extent; it is not our main focus here, although we will discuss interdependency in general, and how the choices of some players reflect the relevant payoffs of other players as well. In this chapter, we resolve the negative externality that one player imposes on another as described in Ref. [6]. Our solution is through effective VM allocation management by the cloud provider to ensure delivery of maximum security for all cloud users. That is one of the prominent points of this chapter.

3.4 System Model

Figure 3.1 is the given system model. A public cloud infrastructure that is running on hypervisor H_1 has n users, denoted by user $U_{11}, U_{12}, \ldots, U_{1n}$, each of whom runs a virtual machine, as denoted by $VM_{11}, VM_{12}, \ldots, VM_{1n}$. Note that the first subscript character is the number of the hypervisor on which the user is located and the second is the user number. For example, the first user operating on hypervisor 3 is U_{31}, and the second user on hypervisor 3 is U_{32}. In addition, we do not make a distinction between the cloud tenant and user. For practical purposes, the tenant is the true entity that manages the VMs as a liaison, while the *user* in cloud terminology is the end user who hires the tenant and benefits from the cloud (and also the one to realize any ill effects of asset loss). We assume that the cloud tenant will act in good faith (i.e., do what is most secure or best for the end user); thus, the remainder of this chapter will refer to the end user only.

Operating systems that manage the VMs on the cloud are designated (on the first hypervisor) as operating systems $OS_{11}, \ldots, OS_{1n}$. A single operating system

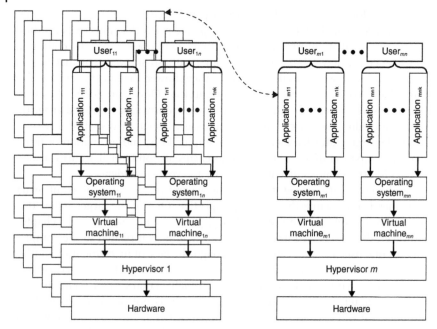

Figure 3.1 System model.

may run multiple applications, which are referred to as $Application_{111}, \ldots,$ $Application_{mnk}$. Each application needs three indices to be represented. The first index shows the hypervisor in which the application is located, the second represents the *OS* or *VM*, and the third differentiates the applications run by the same operating system. This setup holds true for all hypervisors, and thus the mth hypervisor will host users $U_{m1}, U_{m2}, \ldots, U_{mn}$ running virtual machines $VM_{m1}, VM_{m2}, \ldots, VM_{mn}$, and so forth. Each user may run multiple virtual machine instances (and multiple operating systems), but for simplification purposes it will be assumed that each user runs one VM, as it will be shown later that multiple VMs run by one user may be mathematically combined into one VM. The number of applications run by a user will also not impact the model. Thus, the architecture of Figure 3.1 is an accurate simplification and a common structure to the public cloud. Although the physical infrastructures that clouds use will vary (such as different hypervisors, like Xen, VMware, and KVM), the underlying principle of a shared platform in which users are exposed to collateral damage holds true.

Once that model is examined, it becomes evident that there are several issues with the cloud infrastructure. Users who run on the same hypervisor are susceptible to a "bad neighbor" effect, in which an attacker who has compromised one user's VMs may transverse across the hypervisor to launch an attack on another user's VMs on the same hypervisor. That is the problem of *interdependency*. We hold that if the hypervisor is compromised, then all users located on that hypervisor will be compromised (and suffer the consequences). The reason is that once an attacker has compromised the hypervisor, all VMs hosted on that hypervisor can be freely compromised by the attacker. However, a user who does not have VMs on the particular hypervisor that is being targeted will not suffer the consequences. This observation will play an important role later in this chapter. Next, we will explain and set up the problem in the context of game theory.

3.5　Game Model

We consider four players—an attacker and three users—acting across two hypervisors. We assume that the four players are rational and that they all have an understanding of the system in which the game is played. Furthermore, it is expected that each player can calculate and maximize his or her payoff (i.e., utility). In Section 3.7, we extend the problem to n players and m hypervisors.

Along with those commonly applied game-theoretic assumptions of rationality and common knowledge of the game's space, we further assume that the attacker has three strategies, which involve launching an attack on User 1 (a strategy that we will call A_1), on User 2 (A_2), or on User 3 (A_3). The attacker may directly attack only one user at a time. The strategy used in launching an attack may include steps such as collecting information, compromising credentials, executing an attack payload, establishing a backdoor, and scanning. The strategy for User 2 is binary, since that user's only choices are *to invest* (I) in security or *not to invest* (N) in security. In the instance of choosing to invest in security, the user will be allocated to hypervisor 2 (H_2), while no investment in security will result in User 2's being allocated to hypervisor 1 (H_1). The user who chooses to invest may take multiple courses of action, including updating software, buying new antivirus software, and applying stricter system monitoring. Thus, H_2 is the more secure platform for security-conscious cloud users. It is important to note that throughout the remainder of this chapter, we shall refer to users who *invest (do not invest)* in cloud security and users who are allocated to H_2 (H_1) interchangeably.

Furthermore, User 1 will automatically be allocated onto H_1 (*no investment in security*), and User 3 will be allocated to H_2 (*investment in security*). In other words, the only user making a choice on whether to invest in security will be User 2. Since User 2 will have two strategies (I or N) and the attacker has three strategies (A_1, A_2, or A_3), there are a total of six possible permutations in the normal-form game, as laid out in Table 3.1.

The reason for the automatic allocations of User 1 and User 3 is as follows. It is assumed that the relative importance of each user is determined by the total maximum loss that can be suffered by the user if compromised. Those potential total maximum losses are denoted by L_1, L_2, and L_3, where the subscripts correspond to the user numbers; for example, if User 1's virtual machines were

Table 3.1 Game model in normal form.

		User 2	
		N (H_1)	**I (H_2)**
Attacker	A_1	$q_N L_1 + q_N \pi L_2$	$q_N L_1$
		$R - q_N \pi L_2$	$R - e$
	A_2	$q_N L_2 + q_N \pi L_1$	$q_I L_2 + q_I \pi L_3$
		$R - q_N L_2$	$R - e - q_I L_2$
	A_3	$q_I L_3$	$q_I L_3 + q_I \pi L_2$
		R	$R - e - q_I \pi L_2$

compromised, then User 1 would suffer a loss totaling L_1. Further, we assume that

$$L_1 < L_2 < L_3. \tag{3.1}$$

This implies that User 3 will suffer the most costly damage (e.g., through loss of information, trade secrets, or client information) if its virtual machines are compromised, and User 1 faces the least amount of damage. Since User 3 faces the biggest potential for loss and User 1 the least, it is logical that User 3 would invest in security (and be allocated to H_2) and that User 1 would not invest in security (and be allocated to H_1). Thus, the only cloud user making an investment choice in this game will be User 2. The sufficient conditions under which Users 1 and 3 will always be allocated to H_1 and H_2, respectively, will be shown in the model extension. The strategy profile notation used throughout this chapter will be shown as (*attacker strategy, User 2 strategy*). For example, the profile of an attacker who attacks User 1 while User 2 invests in security is given as (A_1, I).

The probability of a successful attack on an individual user who has invested in security is given as q_I, where the user has paid cost e for his or her investment. If a user has not invested in security, then the probability of compromise is q_N. It is assumed that

$$0 < q_I < q_N < 1. \tag{3.2}$$

The reason is that if $q_I > q_N$, then no logical user would choose to invest in security, since it would not lower his chance of being compromised.

The probability of a successful attack on a hypervisor after one of its VMs has been compromised is given as π, where we assume

$$0 < \pi < 1. \tag{3.3}$$

It would be too bold to assume that there is no chance of a successful attack on a hypervisor ($\pi = 0$), especially since the current hypervisor security situation is very unclear [3]. On the other hand, not all attacks on the hypervisor are certain to result in a compromise ($\pi = 1$). Thus, Equation 3.3 results.

The reward of using cloud services (either security-invested, I, or not security-invested, N) is given as R, which could include monetary savings from the outsourcing of IT or on-demand resources that can dynamically change depending on need.

We turn again to the normal-form game outlined in Table 3.1. The attacker's strategies are represented by rows (and the top mathematical expression of the six game possibilities gives the attacker's payoff). User 2's strategies are shown in the columns (and the bottom six mathematical expressions give the user's payoff). Thus, a game profile of (A_1, I) would give the attacker a payoff of $q_N L_1$ and User 2 a payoff of $R - e$.

The payoffs are calculated as follows, taking strategy profile (A_1, I) as an initial example. User 2 will receive reward R from using the cloud services (which is true for all six game possibilities) and will pay expense e, the cost of extra security. Since User 2 is *not* being attacked directly and is located on a *different* hypervisor from the one being targeted, the user's expected loss from a successful attack is 0. Thus, the payoff for User 2 is $R - e$. For the attack targeting User 1, since User 1 has not invested in security, User 1's chance of being compromised is q_N. To find the probabilistic loss of User 1, we must multiply the chance of compromise by User 1's total possible loss (L_1), which gives an expected loss of $q_N L_1$. Since User 1 is the only user located on the first hypervisor, the total gain for the attacker who targets User 1 is $q_N L_1$.

Taking the strategy profile (A_1, N) as another example, we can see that the payoff for User 2 is the reward R minus $q_N \pi L_2$. The quantity $q_N L_2$ is the expected loss from a successful compromise of User 2. However, we must multiply that quantity by π, since User 2 is not a *direct target*, and in order to compromise User 2, the attacker must first compromise the hypervisor. If User 2 were the main target of the attacker, as seen in strategy profile (A_2, N), we can see that User 2's payoff would be the reward R minus the expected loss $q_N L_2$, that is, without the value π, since the attacker need not go through the hypervisor in order to compromise the virtual machines of User 2, because User 2 is a direct target.

When viewing the strategy profile (A_2, I) from the attacker's perspective, we can see that his reward is twofold. His payoff from attacking User 2 is $q_I L_2$ (User 2's expected loss). In addition, the quantity of $q_I \pi L_3$ is added to the attacker's payoff, since User 3 lies on the same hypervisor (H_2) as User 2, even though User 3 is not being directly targeted by the attacker. Since User 3 is not a direct target, the attacker must propagate his attack through the hypervisor before he can compromise User 3. As a result, User 3's expected loss is multiplied by π (giving $q_I \pi L_3$), and thus the total payoff for the attacker is $q_I L_2 + q_I \pi L_3$. Similar methods are used to derive the payoffs for all of the other profiles.

3.6 Game Analysis

In this analysis, we seek to find all the Nash equilibria from the game model. In game theory, when the Nash equilibrium profile is reached, no player can improve his or her utility by unilaterally deviating from the result. At that point, no player wants to change his or her strategy, since not changing is the best response given all the other players' actions. Thus, Nash equilibria can predict the behavior of rational agents.

We will begin by analyzing the strategy profiles that are *not* pure Nash equilibria.

Case 1: User 1 attacked, User 2 does not invest (A_1, N) In comparing the attacker's utility functions in (A_2, N) to (A_1, N), we notice that in order for the attacker to change strategies from $(A_1, N) \rightarrow (A_2, N)$, the following must hold:

$$q_N L_2 + q_N \pi L_1 > q_N L_1 + q_N \pi L_2. \tag{3.4}$$

Dividing both sides by q_N, we notice that

$$L_1 + \pi L_2 < L_2 + \pi L_1.$$

Bringing both L_1 terms to their opposite sides yields

$$\pi(L_2 - L_1) < L_2 - L_1,$$

which gives

$$\pi < 1,$$

which holds true, since the probability of the hypervisor's being compromised is given by $0 < \pi < 1$ (3.3).

The result shows that strategy (A_2, N) is always preferable to strategy (A_1, N) from the attacker's perspective, and thus (A_1, N) cannot be a Nash equilibrium, since a unilateral deviation by the attacker can improve his utility.

Case 2: User 2 attacked, User 2 invests (A_2, I) In comparing the attacker's utility functions in (A_3, I) to (A_2, I), we notice that in order for an attacker to change strategies from $(A_2, I) \rightarrow (A_3, I)$ the following must hold:

$$q_I L_3 + q_I \pi L_2 > q_I L_2 + q_I \pi L_3. \tag{3.5}$$

Dividing both sides by q_I, we notice that

$$L_2 + \pi L_3 < L_3 + \pi L_2.$$

Bringing both L_2 terms to their opposite sides yields

$$\pi(L_3 - L_2) < L_3 - L_2,$$

which gives

$$\pi < 1.$$

That holds true since the probability of the hypervisors being compromised is given by $0 < \pi < 1$ (3). The result shows that strategy (A_3, I) is always preferable to strategy (A_2, I) from the attacker's perspective, and thus (A_2, I) cannot be a Nash equilibrium, since a unilateral deviation by the attacker can improve his utility.

Case 3: User 3 attacked, User 2 invests In comparing User 2's utility functions in (A_3, N) to (A_3, I), we notice that in order for User 2 to change strategies from $(A_3, I) \rightarrow (A_3, N)$, the following must hold:

$$R > R - e - q_I \pi L_2 \tag{3.6}$$

That clearly holds for all $R > 0$. Thus, strategy profile (A_3, I) cannot be a pure Nash equilibrium, since User 2 can improve his or her utility by a unilateral deviation.

From those three cases, we will now prove that (A_3, N) is one possible Nash equilibrium.

Theorem 3.1 If

$$L_3 > \frac{q_N}{q_I}(L_2 + \pi L_1) \tag{3.7}$$

is true, then the strategy profile (A_3, N) is a Nash equilibrium of the game in Table 3.1.

Proof: For the specific strategy (A_3, N), it was already shown in Case 3 that User 2 will not deviate. Furthermore, it was also shown in Case 1 that the attacker will not choose to attack User 1 if User 2 has not invested in security (A_1, N). Therefore, if User 2 has not invested in security, that leaves the attacker with a choice of whether to attack User 2 or User 3. In order for an attacker to choose always to attack User 3 rather than User 2 in this situation, his utility function from (A_3, N) would need to be greater than (A_2, N). This translates to

$$q_I L_3 > q_N L_2 + q_N \pi L_1.$$

Solving for L_3, we factor our q_N on the left and divide both sides by q_I, leaving

$$L_3 > \frac{q_N}{q_I}(L_2 + \pi L_1),$$

which is (3.7), as stated in Theorem 3.1.

Remark This Nash equilibrium is probable if L_3 is very large compared to L_2 and L_1.

Theorem 3.2 If

$$L_3 < \frac{q_N}{q_I}(L_2 + \pi L_1) \tag{3.8}$$

and

$$e > (q_N - q_I)L_2, \tag{3.9}$$

then the strategy profile (A_2, N) is a Nash equilibrium.

Proof: In order for a Nash equilibrium to remain true in this instance, it must be the case that neither the attacker nor User 2 can improve their utility through a unilateral deviation from their selected strategy. It was already shown that Equation 3.7 must be true in order for (A_3, N) to remain strictly preferred to (A_2, N). For the opposite to be true, we must flip the inequality sign, thus yielding

$$L_3 < \frac{q_N}{q_I}(L_2 + \pi L_1).$$

With respect to User 2, in order to prevent a change of strategy from $(A_2, N) \rightarrow (A_2, I)$, the utility from (A_2, N) must be greater than (A_2, I) and as follows:

$$R - q_N L_2 > R - e - q_I L_2.$$

Rearranging the equation yields

$$e > (q_N - q_I)L_2.$$

This, along with the flipped inequality of Equation 3.7, forms the basis for the equations in Theorem 3.2.

Remark This Nash equilibrium is possible if e is so big that User 2 absorbs the loss involved in a cyberattack instead of investing to change the hypervisor (see Equation 3.9).

Theorem 3.3 If

$$L_1 > \frac{q_I}{q_N}(L_3 + \pi L_2) \tag{3.10}$$

and

$$e < q_N \pi L_2, \tag{3.11}$$

then the strategy profile (A_1, I) is a Nash equilibrium.

Proof: The same principle applies to the proof of this theorem as to those for the two previous theorems: In order for this strategy profile to be a Nash equilibrium, it must be the case that neither the attacker nor User 2 can improve his or her utility through a unilateral deviation. To start, in order for the attacker not to switch from $(A_1, I) \rightarrow (A_3, I)$ (note that we do not use (A_2, I), since it was proved in Case 2 that the strategy profile (A_3, I) is strictly preferred over it by the attacker), the following relationship must hold:

$$q_N L_1 > q_I L_3 + q_I \pi L_2.$$

Dividing by q_N and factoring out q_I on the right side leaves us with

$$L_1 > \frac{q_I}{q_N}(L_3 + \pi L_2),$$

which is Equation 3.10 in Theorem 3.3.

Now, to prevent User 2 from deviating from $(A_1, I) \to (A_1, N)$, the following relationship between utilities must hold:

$$R - e > R - q_N \pi L_2. \tag{3.12}$$

Cancelling out the Rs and negating both sides leave us with the simple inequality

$$e < q_N \pi L_2,$$

which is the second part to the conditions that satisfy Theorem 3.3.

Remark This Nash equilibrium is possible only when q_I is very small (i.e., the VMs on hypervisor 2 are very secure), or

$$L_1 \approx L_2 \approx L_3.$$

If none of the conditions of Theorems 3.1, 3.2, or 3.3 for a pure Nash equilibrium are fully met, then the problem admits a mixed Nash equilibrium.

Mixed Nash Equilibrium

A mixed Nash equilibrium is different from pure Nash in the sense that the players do not play a single strategy with an absolute certainty, but rather with a probabilistic strategy for each choice. For example, User 2 might play N with probability $3/4$ and I with probability $1/4$. The conditions and formulas for the equations of mixed Nash will be shown.

Theorem 3.4 If the following three conditions hold

$$L_3 < \frac{q_N}{q_I}(L_2 + \pi L_1), \tag{3.13}$$

$$e < (q_N - q_I)L_2, \tag{3.14}$$

$$L_1 < \frac{q_I}{q_N}(L_3 + \pi L_2), \tag{3.15}$$

then the game admits a mixed-strategy Nash equilibrium.

Proof: We can see that if (3.15) holds, the strategy A_1 is never an optimum, and thus it is discarded from further calculations for the equations of Nash equilibrium. Thus, we will use only strategy profiles A_2 and A_3 for the attacker. The reasoning is as follows. If the defender (User 2) chooses not to invest, the attacker will prefer to play A_2, because (3.13) holds. When the attacker plays A_2, the defender prefers to switch from not investing to investing to avoid the attacker, since Equation 3.14 holds. If User 2 invests and moves from H_1 to H_2, the attacker will prefer to switch his strategy from A_2 to A_3, as shown in Case 2. Given this outcome, since the attacker is now using strategy A_3, the defender will want to switch from investing to not investing in order to avoid an indirect compromise of virtual machines. As a result, the game will circulate among the four strategy profiles indefinitely with no final stoppage point. This shows that there is no pure Nash equilibrium (because there is no strategy in which both players are certain to remain) but rather a strategy profile of each player, who plays a given strategy probabilistically.

Remark Note that this circulation of strategies does not include A_1 as a strategy for the attacker, showing that this strategy will *never* be an optimum for the attacker and thus will never be played. The remaining four choices (A_2, N), (A_2, I) (A_3, N), and (A_3, I)) that result will be the payoffs that are used to calculate a mixed Nash equilibrium.

At the mixed Nash equilibrium of Theorem 3.4, User 2 must randomize in such a way that the *attacker* does not care which strategy is chosen, that is, comes to the realization that, in mathematical terms, $U_a(A_2) = U_a(A_3)$. This induces indifference in the attacker's targeted hypervisors, and attacker indifference can make a significant difference in cloud security! The intended outcome is that the attacker is kept off balance due to his inability to identify the high-value targets (i.e., hypervisors). Let α be the probability that User 2 will choose N. That gives

$$U_a(A_2) = \alpha(q_N L_2 + q_N \pi L_1) + (1 - \alpha)(q_I L_2 + q_I \pi L_3) \tag{3.16}$$

and

$$U_a(A_3) = \alpha(q_I L_3) + (1 - \alpha)(q_I L_3 + q_I \pi L_2). \tag{3.17}$$

In the context of User 2's randomizing, in order to find α we must equate (3.16) and (3.17) and solve them. This gives

$$\alpha = \frac{q_I[(L_3 + \pi L_2) - (L_2 + \pi L_3)]}{q_N(L_2 + \pi L_1) - q_I(L_2 + \pi L_3) + q_I \pi L_2}, \tag{3.18}$$

which means that User 2 will play strategy N with probability α and strategy I with probability $(1 - \alpha)$.

We will now examine the attacker's mixed Nash equilibrium by letting $\beta(A_2) + (1 - \beta)(A_3)$ be the strategy of the attacker. Given that, we can see that

$$U_d(N) = \beta(R - q_N L_2) + (1 - \beta)(R) \tag{3.19}$$

and

$$U_d(I) = \beta(R - e - q_I L_2) + (1 - \beta)(R - e - q_I \pi L_2). \tag{3.20}$$

At the mixed Nash equilibrium, the attacker must randomize in such a way that the *defender* (User 2) does not care which strategy is chosen. Equating (3.19) and (3.20) and solving for β gives

$$\beta = \frac{(e + q_I \pi L_2)}{q_N L_2 - q_I L_2(1 - \pi)}. \tag{3.21}$$

Next, verification that in all instances $0 < \alpha < 1$ and $0 < \beta < 1$ is straightforward, so there is no case in which any values produce an inappropriate value for α or β.

Theorem 3.5 If the following three conditions hold

$$L_1 > \frac{q_I}{q_N}(L_3 + \pi L_2), \tag{3.22}$$

$$e < (q_N - q_I)L_2, \tag{3.23}$$

and

$$e > q_N \pi L_2, \tag{3.24}$$

then the game admits a mixed-strategy Nash equilibrium.

Proof: With logic similar to that applied for Theorem 3.4, we can see that if (3.22) holds, the strategy A_3 is never an optimum and is not needed for calculations for the equations of Nash equilibria. Thus, only strategy profiles A_1 and A_2 will be used for the attack. If the defender (User 2) chooses not to invest, the attacker will prefer to play A_2, because $L_1 + \pi L_2 < L_2 + \pi L_1$. When the attacker plays A_2, the defender, because of (3.23), will prefer to switch from not investing to investing to avoid the attacker. Thus, if User 2 invests and moves from H_1 to H_2, the attacker will prefer to switch his strategy from A_2 to A_1 because of (3.22). Last, since the attacker is now playing A_1, the defender will want to switch from investing to not investing, because (3.24) means that an indirect compromise of virtual machines is less costly than the security investment. These four strategy profiles (A_2, N), (A_2, I), (A_1, N), (A_1, I) are not static; instead, they are used in circulation.

Applying similar logic to calculating λ and μ as in the calculation of α and β, we find that under the conditions for Theorem 3.5,

$$\lambda = \frac{q_N L_1 - q_I[L_2 + \pi L_3]}{q_N(L_2 + \pi L_1 - \pi L_2) - q_I(L_2 - \pi L_3)} \tag{3.25}$$

and

$$\mu = \frac{e + L_2(q_I - q_N)}{L_2(q_I + q_N \pi - q_N)}. \tag{3.26}$$

It can be readily verified that $0 < \lambda < 1$ and $0 < \mu < 1$.

3.7 Model Extension and Discussion

For our model extension, all our previously stated assumptions remain in place, except that the number of users is now increased to n. The number of hypervisors remains at two. In practice, there is a one-to-one mapping of hypervisors to physical servers, so with m hypervisors, cloud clients can pay for increasingly structured levels of security, with H_1 being the least secure and H_m being the most secure. The reasoning we applied for two hypervisors is the same for m hypervisors.

A significant issue in Ref. [6] is that an externality (either positive or negative) by one user is imposed upon another. As can be seen in the game analysis in Section 3.6, the mere presence of another player on the same hypervisor allows for collateral damage via attack propagation. In two of the three pure Nash equilibria, User 2 was allocated to the hypervisor that was not being attacked. (In the third Nash conditions stemming from the game analysis in Theorem 3.3, User 2, as the attack's direct target, could not avoid the attack.) These results have very important implications for how we should conduct user allocation in order to attain a maximum level of cloud security.

One of the main results that we can draw from n users is that there is a resultant family of Nash equilibria. We observe the following cases of Nash equilibria, which are similar to the setup of the game model in Section 3.5 but extended to n users:

Case 1: If $L_n > \frac{q_N}{q_I}(L_{n-1} + \pi L_{n-2} + \cdots + \pi L_2 + \pi L_1)$, then the strategy profile (A_n, U_{n-1}) is a Nash equilibrium.

Case 2: If $L_n < \frac{q_N}{q_I}(L_{n-1} + \pi L_{n-2} + \cdots + \pi L_2 + \pi L_1)$ and $e > (q_N - q_I)L_{n-1}$, then the strategy profile (A_{n-1}, U_{n-1}) is a Nash equilibrium.

Subcase 2_1: If $q_I L_n + q_I \pi L_{n-1} < q_N L_{n-2} + q_N \pi L_{n-3} + \cdots + q_N \pi L_1$, $e > (q_N - q_I)L_{n-1}$, and $e < q_N \pi L_{n-1}$, then the strategy profile (A_{n-2}, U_{n-2}) is a Nash equilibrium.

Subcase 2_2: If $q_I L_n + q_I \pi L_{n-1} + q_I \pi L_{n-2} < q_N L_{n-3} + q_N \pi L_{n-4} + \cdots + q_N \pi L_1$, $e > (q_N - q_I)L_{n-2}$, and $e < q_N \pi L_{n-2}$, then the strategy profile (A_{n-3}, U_{n-3}) is a Nash equilibrium.

Subcase 2_3: If $q_I L_n + q_I \pi L_{n-1} + q_I \pi L_{n-2} + q_I \pi L_{n-3} < q_N L_{n-4} + q_N \pi L_{n-5} + \cdots + q_N \pi L_1$, $e > (q_N - q_I)L_{n-2}$, and $e < q_N \pi L_{n-3}$, then the strategy profile (A_{n-4}, U_{n-4}) is a Nash equilibrium.

$\vdots$

Subcase 2_{n-1}: If $q_I L_n + q_I \pi L_{n-1} + q_I L_{n-2} + q_I \pi L_{n-3} + \cdots + q_I \pi L_3 < q_N L_2 + q_N \pi L_1$, $e > (q_N - q_I)L_2$, and $e < q_N \pi L_3$, then the strategy profile (A_2, U_2) is a Nash equilibrium.

Subcase 2_n: If $q_I L_n + q_I \pi L_{n-1} + q_I L_{n-2} + q_I \pi L_{n-3} + \cdots + q_I \pi L_2 < q_N L_1$, $e > (q_N - q_I)L_1$, and $e < q_N \pi L_2$, then the strategy profile (A_1, U_1) is a Nash equilibrium.

Case 3: If none of the conditions in Case 1, Case 2, or Case 2 subcases are met, then the game admits a mixed-strategy Nash equilibrium. The goal is to find the user that, when placed on a hypervisor, balances out the total loss on each hypervisor such that the attacker is indifferent as to which hypervisor to target.

Case 1 maps to Theorem 3.1 of the game analysis in Section 3.5. The basic Case 2 is analogous to Theorem 3.2. The final subcase of Case 2 maps to

Theorem 3.3, and Case 3 maps to Theorems 3.4 and 3.5. Those results are not unexpected, as the addition of n more users to the original three results in n more pure Nash equilibria, wherein each player has certain conditions that result in that user being targeted by the attacker. In the case of mixed Nash equilibria, we will also see a similar scenario in which each user has the potential to be of particular interest.

For mixed Nash equilibria, among the n users there exists a certain user who alone will make a decision on the hypervisor to which it will allocate itself, that is, all other users will remain static in their allocation choice, regardless of the number of players. That is why User 1 and User 3 were statically placed in H_1 and H_2, respectively. This placement affirms the observation that there exists only one user who will sit on the threshold of choosing between investing in security and not investing in security, because all other users' expected loss magnitudes balance out. That is, that user is unique: unlike all the other users, this one is unable to choose between invest or not-invest in a binary sense – the reason is that any hypervisor to which the user chooses to allocate itself will be attacked, because the user will "tip" the payoff for the attacker in that direction. Thus, this unique player has a mixed Nash equilibrium, whereas all other players have pure, and static, Nash equilibria.

Since the attacker will always attack the "largest" player in the targeted hypervisor, if the unique user were allocated to hypervisor H_1, the unique user would then be the direct victim of an attack, since by default it would be the largest player on H_1. The reason is that all players will be grouped by loss potential on the hypervisors, since a small loss gap between players will minimize the externality they impose on each other and thus maximize security. This is a well-studied situation in game theory and is known as *Hotelling's law* [18]. In accordance with Hotelling's law, players are self-grouping by potential maximum loss. Thus, a situation in which the largest and smallest potential loss players are grouped on one hypervisor, and all of the in-between potential loss players are on another hypervisor, is not observed; rather, players will be allocated with other users with similar total loss potentials.

For a game having n players and with critical User l $(1 < l < n)$, Users $1, 2, \ldots,$ $l-1$ will be located on H_1 and Users $l+1, l+2, \ldots, n-1, n$ will be located on H_2. If User l chooses to allocate himself to H_1, then he will be the direct target of the attack. If he chooses to allocate himself to H_2, then User n will be the direct target, and User l merely becomes an indirect target by factor π. Thus, in order for a user to be the pivotal user, the following two equations must be satisfied:

$$q_N \pi L_1 + \cdots + q_N \pi L_{i-1} + q_N L_i > q_l \pi L_{i+1} + \cdots + q_l \pi L_{n-1} + q_l L_n$$

$$(3.27)$$

and

$$q_N \pi L_1 + \cdots + q_N L_{i-1} < q_l \pi L_i + q_l \pi L_{i+1} + \cdots + q_l \pi L_{n-1} + q_l L_n. \quad (3.28)$$

As can be seen, the pivotal user shifts the inequality and thus the place where the attacker will focus his attack. Note that these equations will hold for $n \geq 3$. It is possible that User $l + 1 = n$ or that User $l - 1 = 1$.

For an example in which $n = 5$ and all the other variables are defined such that the critical player is $n = 3$, the two equations will look like

$$q_N \pi L_1 + q_N \pi L_2 + q_N L_3 > q_I \pi L_4 + q_I L_5$$

and

$$q_N \pi L_1 + q_N L_2 < q_I \pi L_3 + q_I \pi L_4 + q_I L_5.$$

VM allocation for User 3 becomes essentially a cat-and-mouse game in which the attacker follows and attacks the hypervisor wherever User 3 goes, since User 3 acts as a pivotal player that tips the equilibrium point just enough that switching the attack methods will result in an increased payoff. Thus, the best strategy for User 3 is not a discrete choice, but rather a mixed strategy, because there is no solution in which both User 3 and the attacker will remain binary in their choices. This phenomenon is not limited to User 3 as a pivotal user. Depending on the various values of q_N, q_I, $L_1, L_2, \ldots, L_n$, any user from 2 to $(n - 1)$ may be the pivotal user. In Section 3.8.5, in the course of the numerical analysis, we will look at specific values and calculate the pivotal user.

As previously stated, in our game model, it is assumed that Users 1 and 3 will always be allocated to H_1 and H_2, respectively. We will now prove the sufficient conditions that will guarantee that the biggest and smallest users will be allocated to opposite hypervisors.

To begin, in order for User 1 always to be allocated to H_1 (*not investing*), its utility from choosing N would always have to be strictly better than that from choosing I. That is shown by the inequality

$$R - q_N L_1 > R - e - q_I \pi L_1, \tag{3.29}$$

where the left side of the inequality is User 1's utility from an allocation to H_1, and the right side is User 1's utility from an allocation to H_2. For User 1, the two utility functions from choosing N or I will always be the same regardless of the attacker's strategy. The reason is that if the end user is making the decision of whether or not to invest (where User 1 is the pivotal user), then all the larger users have already allocated themselves to H_2 by investing. It follows that if User 1 were allocated to H_2, then H_1 would be completely empty of VMs. Thus, the expected loss of User 1 is $q_N L_1$ since there is no larger user on the first hypervisor to attack, as it is the only user. However, if User 1 chooses to invest and is allocated to H_2, then its expected loss includes π, since it will no longer be the largest user located on that hypervisor. (Recall that the attacker will always attack the largest user on its targeted hypervisor, since it is attempting to maximize its utility.) So for User 1 always to be allocated to H_1, its utility must always be higher than the utility it would gain from being allocated to H_2.

Continuing to solve the inequality, we show that

$$e > L_1(q_N - q_I\pi).$$

Thus, the cost for investment must reach a given threshold to show that User 1 will never invest in security. That conclusion is logical, since if the investment e is high enough, it will dissuade a user from investing. Note that the inequality is only sufficient to prove that User 1 will *always* choose not to invest.

If we apply similar reasoning to find the utility of User n, we find that it will always be true that User n will be allocated to H_2 if the following holds:

$$R - e - q_I L_n > R - q_n L_n. \tag{3.30}$$

Solving for e we discover that

$$e < L_n(q_N - q_I).$$

That too is logical, since the investment e needs to be low enough such that it encourages the user with the highest possible potential loss always to invest. This investment subsequently gives a total range e as

$$L_1(q_N - q_I\pi) < e < L_n(q_N - q_I) \tag{3.31}$$

to show that Users 1 and n will always choose not to invest and to invest, respectively. Since it is assumed in our game model that User 1 and User 3 ($n = 3$) will always be allocated to their respective hypervisors, we will use condition (3.31) in calculating the allowable range for e in our numerical results.

3.8 Numerical Results and Analysis

The following game analysis provides an in-depth explanation of the pure and mixed Nash equilibria. Our key variables in the analysis are $R, q_N, q_I, L_1, L_2, L_3, \pi$, and e. We show how User 2's payoff changes with respect to a change in some selected variables.

3.8.1   Changes in User 2's Payoff with Respect to L_2

In this section we provide specific values for all of the aforementioned constants except L_2 and graph the results. For our first example, we will take $R = 1.5$, $q_I = 0.1$, $q_N = 0.4$, $\pi = 0.1$, $L_1 = 1$, $L_3 = 100$, and $e = 0.4$. Applying (3.26), we see that $0.39 < e < 30$, which verifies that e is an appropriate value. Figure 3.2 shows how the payoff of User 2 changes as L_2 increases.

The first thing we can immediately see is that there is a strategy change from (A_3, N) to a mixed Nash equilibrium at $L_2 = 24.9$. The reason is that $1 < L_2 < 24.9$, the condition for a pure Nash equilibrium (3.7) is satisfied up until 24.9.

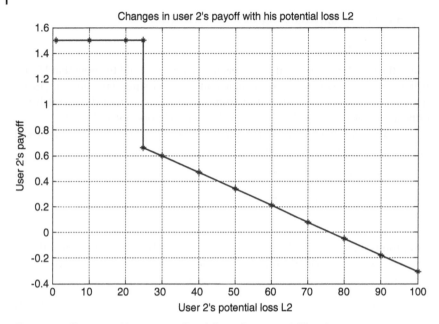

Figure 3.2 Changes in User 2's payoff with his or her potential loss L_2.

After that, (7) is falsified; all three conditions of mixed Nash equilibria are fulfilled; and, thus, a strategy change occurs. Notice that as L_2 approaches the value of L_3, there is a diminishing payoff from using the cloud, and indeed the payoff turns negative at $L_2 \approx 77$ for the given value of R. The reason is that as L_2 increases, α and β decrease and User 2 invests more often. As a result, the increased frequency of direct attacks on User 2 causes User 2's payoff to decrease. The implication of a negative payoff is that after reaching a negative payoff, User 2 will completely opt out of using cloud services.

3.8.2 Changes in User 2's Payoff with Respect to e

Moving on to Figure 3.3, we hold all of the same values as before, except we set $L_2 = 10$ and $L_3 = 20$, and graph the applicable value of e with respect to the changing payoff. The reason for the change in variables (L_3) is to show the mixed Nash equilibrium and how it changes to pure Nash with increasing e.

Too high a value for L_3 will always result in profile (A_3, N), so the potential loss values will be more closely clustered in this example. As shown before, the range for e is given in Equation 3.31, which results in $0.39 < e < 6$. That will be our range for the x-axis. We can see a strategy change from a mixed Nash equilibrium to pure Nash at $e = 3$. One can understand that change intuitively: After that threshold, the choice of investing becomes too expensive and

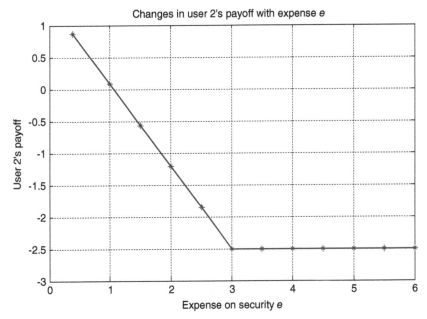

Figure 3.3 Changes in User 2's payoff with expense *e*.

infeasible for User 2, and thus the pure Nash equilibrium (A_2, N) results. Note that User 2, despite his or her awareness of being a direct target of the attacker, will not invest in cloud security at this Nash equilibrium. Furthermore, for the selected value of R, User 2 will not use cloud services at all unless there is a low value of e (specifically, $e = 1.213$).

3.8.3 Changes in User 2's Payoff with Respect to π

Figure 3.4 shows the changing payoff of User 2 as we change the probability of compromising the hypervisor, π. We use the same values as set in the Section 3.8.1 analysis, change π from a constant to a variable, and set $e = 0.4$.

It is apparent that there is no shift in the Nash equilibrium across all values of $0 < \pi < 1$. Those results are not surprising, as an increasing π only slightly increases the externality imposed onto User 2 by other users. The increasing externality problem imposed by an increasing π does not present a significant difference in terms of changing any strategies of the player. That is a very significant discovery since in Ref. [6] there was a Nash equilibrium shift if π reached a certain threshold, but in this analysis that is not the case, so we have achieved one of the main aims of this research: to reduce the externality imposed onto one user by another. However, it is possible that π may shift the Nash equilibrium in exceptional cases in which the conditions for two different Nash

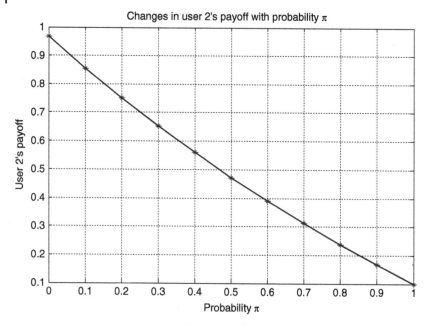

Figure 3.4 Changes in User 2's payoff with probability π.

equilibria are very close to being met. One instance is that if $L_3 \approx \frac{q_u}{q_i}(L_2 + \pi L_1)$, then π may shift the inequality either way and thus change the Nash equilibrium. In most cases, the Nash equilibrium will not change from the initial conditions, and that is a very positive sign that this security-based allocation will have the effect of mitigating the externality problem.

It is apparent that a direct attack is very important in deciding where to allocate users. As we will see in the next variable analysis, the q_I/q_N ratio is also very important in determining the prevalent Nash equilibrium.

3.8.4 Changes in User 2's Payoff with Respect to q_I

For this section, we will take $R = 1.5$, $q_N = 0.5$, $\pi = 0.1$, $L_1 = 1$, $L_2 = 10$, $L_3 = 20$, and $e = 0.4$ and note that q_N was increased to give a greater range of q_I (using q_I as a variable here). We take $0 < q_I < 0.5$ (since $q_I < q_N$); the results can be seen in Figure 3.5. For small values of q_I, the pure Nash profile (A_1, I) exists. That makes sense, since if the probability of compromising a VM on the secure hypervisor were so low as to discourage any type of attack, then the attacker would gain a higher payoff by targeting the users who chose not to invest.

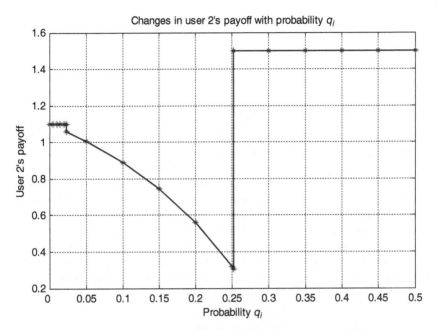

Figure 3.5 Changes in User 2's payoff with probability q_I.

At $q_I \approx 0.0238$, the Nash equilibrium changes to a mixed strategy and then changes again to the pure Nash equilibrium (A_3, N) at $q_I \approx 0.2525$. The second switch of Nash equilibria is feasible since as the q_I/q_N ratio becomes closer to 1, the L_3/L_2 ratio becomes a more dominant factor in the calculations; at the second threshold, the disparity becomes so large that $L_3 \gg L_2$ and the switch to strategy profile (A_3, N) occurs.

3.8.5   Model Extension to $n = 10$ Users

In this section, we will continue our model extension and not limit our discussion to just three players. We will take $R = 1.5$, $q_N = 0.4$, $q_I = 0.1$, $\pi = 0.1$, $e = 0.4$ as before and, per Section 3.7, set the number of users as $n = 10$. For all of our potential loss values, we will use $L_1 = 1$, $L_2 = 2$, $L_3 = 3$, $L_4 = 4$, $L_5 = 5$, $L_6 = 6$, $L_7 = 7$, $L_8 = 8$, $L_9 = 9$, $L_{10} = 10$.

Our next step is to find out which of the 10 users is the pivotal user. Thus, we must find the user for which Equations 3.27 and 3.28 are true.

By calculating individual potential losses, we find that User 4 is the pivotal user. This means that

$$(q_N \pi L_1 + q_N \pi L_2 + q_N \pi L_3 + q_N L_4) = 1.84 >$$
$$\left(q_I \pi L_5 + q_I \pi L_6 + q_I \pi L_7 + q_I \pi L_8 + q_I \pi L_9 + q_I L_{10} \right) = 1.35$$

and

$$(q_N \pi L_1 + q_N \pi L_2 + q_N L_3) = 1.32 <$$
$$(q_I \pi L_4 + q_I \pi L_5 + q_I \pi L_6 + q_I \pi L_7 + q_I \pi L_8 + q_I \pi L_9 + q_I L_{10}) = 1.39,$$

which are both true.

Using Equation 3.31, we find that $0.39 < e < 3$, which shows that our selected value of e is within the restricted range. It is important to note that the value of e will have a strong influence on the prevailing Nash equilibrium. If $0.39 > e$, then the price for security would be so inexpensive that it would be logical for all users (1–10) to be allocated to H_2. If $e > 3$, then security would be so expensive that no user would want to invest in security (and, as a result, all would be allocated to H_1). Within the allowable range given by (3.31), there are some interesting results. From $0.39 < e < 1.56$, there will be a mixed strategy in which User 4 will have a mixed Nash equilibrium while all other users will remain in their respective hypervisors. (U_1, U_2, and U_3 are allocated to H_1, while U_5, U_6, U_7, U_8, U_9, and U_{10} are allocated to H_2). The threshold of 1.56 is determined by the maximum value of e at which User 4 will potentially still pay for security (3.25). Past that point, investing in security will become too expensive, and thus $1.56 < e < 3$ will result in a pure Nash equilibrium in which U_1, U_2, U_3, and U_4 are allocated to H_1, while all other users are allocated to H_2.

To further show the critical role of User 4, in Figure 3.6 we have diagrammed the changing payoff structure for the attacker as the number of users on each hypervisor changes.

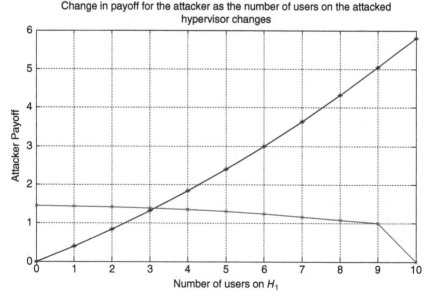

Change in payoff for the attacker as the number of users on the attacked hypervisor changes

Figure 3.6 Change in attacker's payoff with the number of users on the attacked hypervisor.

As can be seen, one line represents the attacker's payoff for attacking H_1, and another shows the attacker's payoff for attacking H_2. For example, the payoff for the attacker's targeting of H_2 if all users are located on H_1 is 0. If there is one user on H_2 (by default, it would be User 10), then the attacker's payoff is $q_I L_{10}$, which is 1.0. As the number of users on the hypervisor increases, the attacker payoff increases; the attacker payoff will decrease if the number of players decreases.

We can see that the payoffs for each strategy (attacking H_1 versus attacking H_2) intersect between the abscissa values of 3 and 4 (i.e., there are three or four players on H_1, and the remaining players are on H_2). However, users take on integer values, so this intersection reflects User 4's status as the pivoting user. At this intersection, the attacker becomes aware of which hypervisor User 4 has been allocated to, since User 4's hypervisor allocation stems from pursuit of a higher payoff. As stated previously, the strategy that U_4 chooses will depend on the prevailing value of e. Since $e = 0.4$, we will have a mixed Nash equilibrium.

In Figure 3.7, we show the reduced externality from the mixed-strategy placement of User 4 and the optimum placement of the other users.

The first and second bars represent the total externalities if all users were placed on H_1 and H_2, respectively, which we calculated by adding all the payoffs of the attacker that contained a factor of π. So the externality of all users on H_2 was calculated as the sum of $q_I \pi L_9 + \cdots + q_I \pi L_1$. The externality of all users on H_1 was calculated similarly. This aggregation of users would be an externality level characteristic of a (fairly standard) allocation method that prioritizes power consumption such that all users are clustered on as few hypervisors as possible.

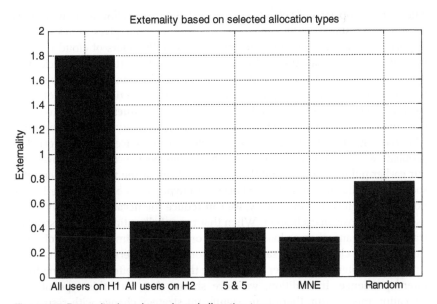

Figure 3.7 Externality based on selected allocation types.

The third bar represents the externality when five VMs are located on H_1 and five VMs are located on H_2; that scenario is supposed to represent another common allocation method, load balancing. Given our initial conditions, the attacker would attack H_1 with probability 1, and thus the externality is calculated as $q_N \pi L_1 + q_N \pi L_2 + q_N \pi L_3 + q_N \pi L_4$. With five users on H_1, the largest user (U_5) will be the direct target, while all the other players will be calculated in the externality figure, since they are all indirect targets.

The fourth bar shows the externality imposed onto other users in the instance of the mixed Nash equilibrium (MNE). The externality from a pure Nash equilibrium ($1.56 < e < 3$) will result in approximately the same value. As can be seen, the negative externality based on a Nash equilibrium allocation is the lowest among the five selected allocation methods, with a 20% externality reduction relative to the second-lowest value (third bar).

The fifth bar shows the externality imposed onto other users when random placement, another common VM allocation method, is used. We obtained this value by randomly allocating the VMs between H_1 and H_2 10 times and averaging the results. This allocation is the second worst for externalities, superior only to placement of all users in the less secure hypervisor, H_1. Our proposed allocation method based on mixed-strategy Nash equilibria provides a 56% reduction in externality relative to the random allocation method.

3.9 The Future

The unique properties of a cloud computing structure can allow for new avenues of attack. As stated earlier, the structure of the cloud gives attackers new opportunities to infiltrate and compromise users. Providers of cloud services may attempt to mitigate their vulnerabilities and fend off attempted attacks, but it is an ongoing battle. Many cloud providers may be significantly less secure than they believe, and many more clients are unsuspecting of the potential weaknesses that are inherent in the cloud. One of the issues we presented here was that of externality, whereby one user's lack of security may affect another who has significantly more to lose. While the existence of the externality problem within the context of game theory has been previously shown [6], our earlier publication [19] and this chapter represent the first attainment of a solution. By allowing users to be allocated to hypervisors based on whether or not they have invested in security, we can reduce the negative externality that users can impose on each other. When that type of allocation is used instead of traditional means, such as load balancing or energy optimization, we observe that users will cluster to the same hypervisor as other users based on the most similar loss potentials in order to minimize the negative effects stemming from interdependence. In addition, we have shown that with this type of VM allocation mechanism, there is no significant strategy change with respect to the value of π, giving further proof that the negative externalities are mitigated in this model. The lack of a strategy change holds even at extreme values such as $\pi \approx 0$ or $\pi \approx 1$. Those findings are supported by the data illustrated in Figure 3.7, which show that our allocation method resulted in the lowest amount of externality by a fair margin.

Thus, the Nash equilibrium strategy is more sensitive to initial conditions, such as the potential loss the users face or the cost of investment, than to the externalities. It is apparent that the value of e plays a crucial role in determining the Nash equilibrium, as shown in the numerical results in Section 3.8.5. For cloud providers, that information should be very useful in enabling them to set the value of e and predetermine the Nash equilibrium best suited for the maximum level of security and minimum externality. For end users, knowledge of the values of π and e can be crucial in helping them decide whether cloud computing is a useful tool that will help them attain cost reductions, or a dubious approach that is fraught with subtle, inherent dangers that have prevented it from proving its worth.

References

1 Cisco Systems, Inc. (2013) Cisco Global Cloud Index: forecast and methodology, 2012–2017. Available at https://www.cisco.com/web/mobile/tomorrow-starts-here/files/pdfs/Cisco_CGI_Forecast.pdf.

2 Horrigan, J.B., Use of cloud computing applications and services, Pew Internet & American Life Project, Pew Research Center, September 2008. Available at http://www.pewinternet.org/files/old-media//Files/Reports/2008/PIP_Cloud.Memo.pdf.pdf.

3 Vaughan-Nichols, S.J., Hypervisors: the cloud's potential security Achilles heel, ZDNet, March 29, 2014. Available at http://www.zdnet.com/article/hypervisors-the-clouds-potential-security-achilles-heel/.

4 Thales e-Security, Inc, Thales finds organizations more confident transferring sensitive data to the cloud despite data protection concerns, June 25, 2013. Available at https://www.thales-esecurity.com/company/press/news/2013/june/encryption-in-the-cloud.

5 Myerson, R.B. (1997) *Game Theory: Analysis of Conflict*, Harvard University Press, Cambridge, MA, p. 1.

6 Kamhoua, C.A., Kwiat, L., Kwiat, K.A., Park, J.S., Zhao, M., and Rodriguez, M. (2014) Game theoretic modeling of security and interdependency in a public cloud, in Proceedings of the 2014 IEEE 7th International Conference on Cloud Computing (CLOUD), Anchorage, Alaska, pp. 514–521.

7 Moseley, T.M., The Nation's Guardians: America's 21st century Air Force, CSAF White Paper, December 29, 2007. Available at http://www.au.af.mil/au/awc/awcgate/af/csaf_white_ppr_29dec07.pdf.

8 Carter, E.F. (1964) *Railways in Wartime*, Frederick Muller Limited, London, UK.

9 Kwiat, K. (2001) Can reliability and security be joined reliably and securely? in Proceedings of the 20th IEEE Symposium on Reliable Distributed Systems (SRDS), pp. 72–73.

10 Xu, J. and Fortes, J.A.B. (2010) Multi-objective virtual machine placement in virtualized data center environments, in Proceedings of the Green Computing and Communications (GreenCom) and 2010 IEEE/ACM International Conference on Cyber, Physical and Social Computing (CPSCom), Hangzhou, China, pp. 179–188.

11 Wei, G., Vasilakos, A.V., Zheng, Y. and Xiong, N. (2010) A game-theoretic method of fair resource allocation for cloud computing services. *The Journal of Supercomputing*, **54** (2), 252–269.

12 Jalaparti, V., Nguyen, G.D., Gupta, I., and Caesar, M., Cloud resource allocation games, University of Illinois at Urbana-Champaign, 2010. Available at http://hdl.handle.net/2142/17427.

13 Beloglazov, A. and Buyya, R. (2012) Optimal online deterministic algorithms and adaptive heuristics for energy and performance efficient dynamic consolidation of virtual machines in cloud data centers. *Concurrency and Computation: Practice & Experience*, **24** (13), 1397–1420.

14 Rao, N.S.V., Poole, S.W., He, F., Zhuang, J., Ma, C.Y.T., and Yau, D.K.Y. (2012) Cloud computing infrastructure robustness: a game theory approach, in Proceedings of 2012 International Conference on Computing, Networking and Communications (ICNC), Maui, Hawaii, pp. 34–38.

15 Künsemöller, J. and Karl, H. (2012) A game-theoretical approach to the benefits of cloud computing, in *Economics of Grids, Clouds, Systems, and Services: GECON 2011* (eds. K. Vanmechelen, J. Altmann, and O.F. Rana), Lecture Notes in Computer Science, vol. 7150, Springer, Heidelberg, Germany, pp. 148–160.

16 Han, Y., Alpcan, T., Chan, J., and Leckie, C. (2013) Security games for virtual machine allocation in cloud computing, in *Decision and Game Theory for Security: GameSec 2013* (eds. S.K. Das, C. Nita-Rotaru, and M. Kantarcioglu), Lecture Notes in Computer Science, vol. 8252, Springer International Publishing, Switzerland, pp. 99–118.

17 Ristenpart, T., Tromer, E., Shacham, H., and Savage, S. (2009) Hey, you, get off of my cloud: exploring information leakage in third-party compute clouds, in Proceedings of the 16th ACM Conference on Computer and Communications Security (CCS '09), Chicago, Illinois, pp. 199–212.

18 Hotelling, H. (1929) Stability in competition. *The Economic Journal*, **39** (153), 41–57.

19 Kwiat, L., Kamhoua, C.A., Kwiat, K.A., Tang, J., and Martin, A. (2015) Security-aware virtual machine allocation in the cloud: a game theoretic approach, in Proceedings of the 2015 IEEE 8th International Conference on Cloud Computing (CLOUD), New York, NY, pp. 556–563.

4

Detection and Security: Achieving Resiliency by Dynamic and Passive System Monitoring and Smart Access Control

Zbigniew Kalbarczyk,[1]
In collaboration with Rakesh Bobba,[2] Domenico Cotroneo,[3]
Fei Deng,[4] Zachary Estrada,[5] Jingwei Huang,[6] Jun Ho Huh,[7]
Ravishankar K. Iyer,[1] David M. Nicol,[8] Cuong Pham,[4] Antonio
Pecchia,[3] Aashish Sharma,[9] Gary Wang,[10] and Lok Yan[11]

[1]University of Illinois at Urbana-Champaign, Department of Electrical and Computer Engineering and Coordinated Science Laboratory, Urbana, IL, USA
[2]Oregon State University, School of Electrical Engineering and Computer Science, Corvallis, OR, USA
[3]Università degli Studi di Napoli Federico II, Dipartimento di Ingegneria Elettrica e delle Tecnologie dell'Informazione, Naples, Italy
[4]University of Illinois at Urbana-Champaign, Department of Electrical and Computer Engineering, Urbana, IL, USA
[5]Rose-Hulman Institute of Technology, Department of Electrical and Computer Engineering, Terre Haute, IN, USA; and University of Illinois at Urbana-Champaign, Department of Electrical and Computer Engineering, Urbana, IL, USA
[6]Old Dominion University, Department of Engineering Management and Systems Engineering, Norfolk, VA, USA; and University of Illinois at Urbana-Champaign, Information Trust Institute, Urbana, IL, USA
[7]Samsung Electronics, Samsung Research, Seoul, South Korea
[8]University of Illinois at Urbana-Champaign, Department of Electrical and Computer Engineering and Information Trust Institute, Urbana, IL, USA
[9]Lawrence Berkeley National Lab, Berkeley, CA, USA
[10]University of Illinois at Urbana-Champaign, Department of Computer Science, Urbana, IL, USA
[11]Air Force Research Laboratory, Rome, NY, USA

In this chapter, we discuss methods to address some of the challenges in achieving resilient cloud computing. The issues and potential solutions are brought about by examples of (i) active and passive monitoring as a way to provide situational awareness about a system and users' state and behavior; (ii) automated reasoning about system/application state based on observations from monitoring tools; (iii) coordination of monitoring and system activities to provide a robust response to accidental failures and malicious attacks; and (iv) use of smart access control methods to reduce the attack surface and limit the likelihood of an unauthorized access to the system. Case studies covering different application domains, for example, cloud computing, large computing infrastructure for scientific applications, and industrial control systems, are used to show both the practicality of the proposed approaches and their capabilities, for example, in terms of detection coverage and performance cost.

Assured Cloud Computing, First Edition. Edited by Roy H. Campbell, Charles A. Kamhoua, and Kevin A. Kwiat.

4.1 Introduction

Building resilient (i.e., reliable and secure) computing systems is hard due to growing system and application complexity and scale, but maintaining reliability and security is even harder. A *resilient system* is expected to maintain an acceptable level of service in the presence of internal and external disturbances. Design for resiliency is a multidisciplinary task that brings together experts in security, fault tolerance, and human factors, among others. Achieving resiliency requires mechanisms for efficient monitoring, detection, and recovery from failures due to malicious attacks and accidental faults with minimum negative impact on the delivered service.

Why is design for resiliency challenging?

- *Design and assessment:* Systems become untrustworthy because of a combination of human failures, hardware faults, software bugs, network problems, and inadequate balance between the cyber and the physical systems, for example, the network and control infrastructures.
- *Delivery of critical services:* Many systems, including cloud systems, large computing infrastructure, and cyber-physical systems (e.g., for energy delivery, transportation, communications, and heath care) are expected to provide uninterruptable services.
- *Interdependencies among systems:* Resiliency of one system may be conditioned on the availability of another system, for example, resiliency of a transportation system may depend heavily on the robust operation of an energy delivery infrastructure, or on a human-in-the-decision-loop, such that human intelligence plays a role in system remediation, service restoration, and recovery.

In this chapter, we discuss methods to address some of those challenges.

We will discuss examples of (1) active and passive monitoring as a way to provide situational awareness about a system and users' state and behavior, (2) automated reasoning about system/application state based on observations from monitoring tools, (3) coordination of monitoring and system activities to provide a robust response to accidental failures and malicious attacks, and (4) use of smart access control methods to reduce the attack surface and limit the likelihood of an unauthorized access to the system.

The methods, tools, and techniques we discuss are illustrated by case studies covering different application domains, for example, cloud computing, large computing infrastructure for scientific applications, and industrial control systems. The goal is to show both the practicality of the proposed approaches and their capabilities, for example, in terms of detection coverage and performance cost.

Specific examples encompass dynamic virtual machine monitoring using hypervisor probes, pitfalls of passive virtual machine monitoring, model-based identification of compromised users in large computing infrastructure, and system protection using a combination of attribute-based policies and role-based access control.

This chapter will highlight both cyber and cyber-physical examples that combine research expertise in security and system reliability with human factors, verification, and distributed systems, providing a truly integrated view of the relevant technologies.

4.2 Vision: Using Cloud Technology in Missions

Prolific failures have kept reliability a leading concern for customers considering the cloud [1]. Monitoring is especially important for security, since many attacks go undetected for long periods of time. For example, Trustwave surveyed 574 locations that were victims of cyberattacks [2]. Of those victims, 81% did not detect the attacks themselves; either a customer reported data misuse or a third-party audit uncovered a compromised system. When attacks were detected, the mean detection time was 86 days.

Cloud computing environments are often built on top of virtual machines (VMs) running on top of a hypervisor. A virtual machine is a complete computing system that runs on top of another system. The hypervisor is a privileged software component that manages the VMs. Typically, one can run multiple VMs on top of a single hypervisor, which is often how cloud providers distribute customers across multiple physical servers. As the low-level manager of VMs, the hypervisor has privileged access to those VMs, and this access is often supported by hardware-enforced isolation. The strong isolation between the hypervisor and VMs provides an opportunity for robust security monitoring. Because cloud environments are often built using hypervisor technology, VM monitoring can be used to protect cloud systems.

There has been significant research on virtual machine monitoring [3–8]. Existing VM monitoring systems require setup and configuration as part of the guest OS (operating system) boot process or modification of guest OS internals. In either case, the effect on the guest is the same: At the bare minimum, a VM reboot is necessary to adapt the monitoring system; in the worst case, the guest OS needs to be modified and recompiled. Operationally, these requirements are undesirable for a number of reasons, for example, increased downtime.

In addition to the lack of runtime reconfigurability, VM monitoring is at a disadvantage compared to traditional in-OS monitoring in terms of the information available to the monitors. VM monitoring operates at the hypervisor level, and therefore has access only to low-level hardware information, such as registers and memory, with limited semantic information on what the low-level information represents (e.g., what function is being called when the *0xabcd* instruction is executed). In the literature, the hypervisor's lack of semantic information about the guest OS is referred to as the *semantic gap*.

Given the current technology trends, one may say that virtual machines will be everywhere. Whether in enterprise computing or as the key building block in a cloud, most environments employ VMs to some extent. Consequently, ensuring resilient operation of the cloud in the presence of both accidental failures and malicious attacks is of primary importance.

4.3 State of the Art

The research community has produced a variety of techniques for virtual machine monitoring. Of particular note are the Lares [5] and SIM [7] approaches. Lares uses a memory-protected trampoline inserted by a driver in the guest VM. That trampoline issues a hypercall to notify a separate security VM that an event of interest has occurred. This approach requires modification to the guest OS (albeit in a trusted manner), so runtime adding and removing of hooks is not possible. Furthermore, a guest OS driver and trampoline are needed for every OS and version of OS supported by the monitoring infrastructure. The *Secure In-VM Monitoring* (SIM) approach uses a clever configuration of hardware-assisted virtualization (HAV) that prevents VM Exits when switching to a protected page inside the VM that performs monitoring. Since SIM does not incur VM Exits, it achieves low overhead. However, this method involves adding special entry and exit gates to the guest OS and hooks are placed in specific kernel locations. In addition to platforms built on top of open-source technology, similar dynamic monitoring solutions also exist in proprietary systems. For example, there are vprobes for VMware ESXi [9].

However, the existing approaches do not support dynamic (i.e., at runtime) addition/removal of monitoring probes. The reason is that hooks are statically inserted into the guest OS. In those systems, in order to support the bare minimum flexibility of application-specific monitoring, one would either have to maintain a set of guest OS kernels or use a scheme that modifies a running kernel from within the guest.

Furthermore, there is lack of techniques that are rooted in hardware invariants [8,10], and hence are enforced by hardware-generated events (e.g., VM Exit). In this context, the invariant is that a properly functioning virtual machine will generate VM Exits on privileged operations (an assumption that is essential for a "trap-and-emulate" VMM). To protect probes, we can use Intel's Extended Page Tables (EPT) or AMD's Nested Page Tables (NPT) and write-protect the pages that contain active probes. This write protection satisfies the security requirement where probes cannot be evaded by actors inside the VM and incurs a performance impact only when pages containing probes are written to (a rare event for code in memory). The existing framework does place the hypervisor at the root of trust, but well-known techniques exist for signing hypervisor code [11–13].

Previous researchers have utilized int3 for VMs in Xenprobes [14], which provides a guest OS kernel debugging interface for Xen VMs. Additionally, xenprobes can use an Out-of-line Execution Area (OEA) to execute the replaced instruction (instead of always executing in place with a single step, such as the hprobe prototype does). The OEA provides a performance boost, but it results in a more complex code base and carries the need to create and maintain a separate memory region for this area. The OEA requires an OS driver to allocate and configure the OEA at guest OS boot, and the number of OEAs are statically allocated at boot, placing a hard upper bound on the number of supported probes (which is acceptable for debugging, but not for dependability monitoring).

Ksplice [15], a rebootless kernel patching mechanism, can be used to provide a basis for VM monitoring. The Linux kernel is also scheduled to incorporate a rebootless patching feature [16]. Ksplice allows for live kernel patching by replacing a function call with a jump to a new patched version of that function. The planned Linux feature will use ftrace10 to switch to a new version of the function after some safety checks. While these techniques can be useful for patches that have been properly tested and worked through a QA cycle, many operators would be uneasy with an untested patch on a live OS. When considering newly reported vulnerabilities, probe-based approaches allow one to quickly deploy an out-of-band monitor to detect the vulnerability without modifying the control flow of a running kernel. This temporary monitoring could even be used to provide a stopgap measure while a rebootless patch is in QA testing: One could use the monitor immediately after a vulnerability is announced and until the patch is vetted and safe to use. A technique such as this would drastically reduce the vulnerable window and alleviate pressure to perform risky maintenance outside of critical windows. It should be noted that while our example focused on a kernel vulnerability, this emergency detector technique can be extended to a user space program.

The next-generation probe-based (or hook-based) active VM monitoring techniques should

- perform application-level hook-based monitoring from the hypervisor, paving the way for cloud-based monitoring as a service (where a cloud provider could give the user a set of probe-based detectors to choose from),
- require no modification of the guest OS, and
- allow adding/removing of monitors/detectors at runtime.

These features are indispensable in any system intended for production use.

4.4 Dynamic VM Monitoring Using Hypervisor Probes

Here we discuss active and passive monitoring as a way of providing situational awareness about the system and users' state and behavior in the context of large computing infrastructure, such as that in cloud computing. (We refer the interested reader to our earlier publication [17] for more details.) Virtual machines provide strong isolation that can be used to enhance reliability and security monitoring [3,8,10,18]. Previous VM monitoring systems require setup and configuration as part of the boot process, or modification of guest OS internals. In either case, the effect on the guest is the same: A VM reboot is necessary to adapt the system. This is undesirable for a number of reasons, such as increased downtime (discussed further in the next section). By using a dynamic monitoring system that requires no guest OS modifications or reboots, we can allow users to respond to new threats and failure modes quickly and effectively.

Monitoring systems can generally be split into two classes: those that perform passive monitoring, and those that perform active monitoring [19]. *Passive monitoring systems* are polling-based systems that periodically inspect the system's state. These systems are vulnerable to transient attacks that occur between monitoring checks [8]. Furthermore, constant polling of a system can be a source of unnecessary performance overhead. *Active monitoring systems* overcome these weaknesses since they are triggered only when events of interest occur. However, it is essential to ensure that an active monitoring system's event generation mechanism cannot be circumvented.

One class of active monitoring systems is that of hook-based systems, in which the monitor places hooks inside the target application or OS [5]. A *hook* is a mechanism used to generate an event when the target executes a particular instruction. When the target's execution reaches the hook, control is transferred to the monitoring system, which can record the event and/or inspect the system's state. Once the monitor has finished processing the event, it returns control to the target system, and execution continues until the next event. Hook-based techniques are robust against failures and attacks inside the target when the monitoring system is properly isolated from the target system.

We find dynamic hook-based systems attractive for dependability monitoring, as they can be easily adapted: Once the hook delivery mechanism is functional, implementation of a new monitor involves merely adding a hook location and deciding how to process the event. In this context, *dynamic* refers to the ability to add and remove hooks without disrupting the control flow of the target. This is particularly important in real-world use, where monitoring needs to be configured for multiple applications and operational environments. In addition to supporting a variety of environments, monitoring must also be responsive to changes in those environments.

We created the *hprobe* framework, a dynamic hook-based VM reliability and security monitoring solution. The key contributions of the hprobe framework are that it is loosely coupled with the target VM, can inspect both the OS and user applications, and supports runtime insertion and removal of hooks. Those qualities mean that hprobe is a VM monitoring solution that is suitable for use on an actual production system.

4.4.1 Design

Hook-Based Monitoring

An illustration of a hook-based monitoring system adapted from the formal model presented in Lares [5] is shown in Figure 4.1. In hook-based monitoring, a

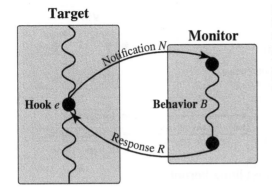

Figure 4.1 Hook-based monitoring. A hook is triggered by event *e*, and control is transferred to the monitor through notification *N*. The monitor processes *e* with a behavior *B* and returns control to the target with a response *R*.

monitor takes control of the target after the target reaches a hook. In the case of hypervisor-based VM monitoring, the target is a virtual machine, and the monitor can run in the hypervisor [4], in a separate security VM [5] or in the same VM [7]. Regardless of the separation mechanism used, one must ensure that the monitor is resilient to tampering from within the target VM and that the monitor has access to all relevant state of that VM (e.g., hardware, memory). Furthermore, a VM monitoring system should be able to trigger upon the execution of any instruction, be it in the guest OS or in an application.

If a monitoring system can capture all relevant events, it follows that the monitoring system should be dynamic. This is important in the fast-changing landscape of IT security and reliability. As new vulnerabilities and bugs are discovered, one will inevitably need to account for them. The value of a static monitoring system decreases drastically over time unless periodic software updates are issued. However, in many VM monitoring solutions [3,5,7,8], the downtime caused by such software updates is unacceptable, particularly when the schedule is unpredictable (e.g., updates for security vulnerabilities). Dynamic monitors can also provide performance improvement relative to statically configured monitoring; one can monitor just an event of interest, or a general class of events (e.g., a single system call versus all system calls). Furthermore, it is possible to construct dynamic detectors that change during execution (e.g., a hook can be used to add or remove other hooks). Static monitoring systems also present a subtle design flaw: A configuration change in the monitoring system can affect the control flow of the target system (e.g., by requiring a restart).

In line with dynamism and loose coupling with the target system, the detector must also be simple in its implementation. If a system is overly complex and difficult to extend, the value of that system is drastically reduced, as using it requires much effort. In fact, such a system will simply not be used. DNSSEC1 and SELinux2 can serve as instructive examples; they provide valuable security features (e.g., authentication and access control), they were both released around the year 2000, and to this day are still disabled in many environments. Furthermore, a simpler implementation should yield a smaller attack surface [20].

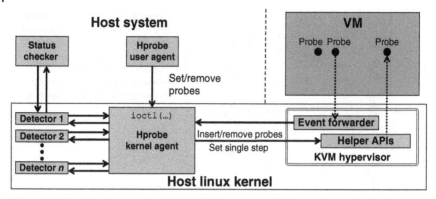

Figure 4.2 Hprobes integrated with the KVM hypervisor. The Event Forwarder has been added to KVM and communicates with a separate kernel agent through Helper APIs. Detectors can be implemented as kernel modules either in the host OS or in user space. In the latter case ioctl functions are used to communicate with the kernel agent.

4.4.2 Prototype Implementation

Integration with KVM

The hprobe prototype was inspired by the Linux kernel profiling feature kprobes [21], which has been used for real-time system analysis [22]. The operating principle behind our prototype is to use VM Exits to trap the VM's execution and transfer control to monitoring functionality in the hypervisor. Our implementation leverages HAV, and the prototype framework is built on the KVM hypervisor [23]. The prototype's architecture is shown in Figure 4.2. The modifications to KVM itself make up the Event Forwarder, which is a set of callbacks inserted into KVM's VM Exit handlers. The Event Forwarder uses Helper APIs to communicate with a separate hprobe kernel agent. The hprobe kernel agent is a loadable kernel module that is the workhorse of the framework. The kernel agent provides an interface to detectors for inserting and removing probes. This interface is accessible by kernel modules through a kernel API in the host OS (which is also the hypervisor, since KVM itself is a kernel module) or by user programs via an ioctl interface.

The execution of an hprobe-based detector is illustrated in Figure 4.3. A probe is added by rewriting the instruction in memory at the target address with int3, saving the original instruction, and adding the target address to a doubly linked list of active probes. This process happens at runtime and requires no application or guest OS restart. The int3 instruction generates an exception when executed. With HAV properly configured, this exception generates a VM Exit event, at which point the hypervisor intervenes (step 1). The hypervisor uses the Event Forwarder to pass the exception to the hprobe kernel agent, which traverses the list of active probes and verifies that the int3 was generated by an hprobe. If it was, the hprobe kernel agent reports the event and optionally calls an hprobe handler function that can be associated with the probe. If the exception

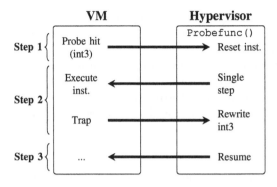

Figure 4.3 A probe hit in the hprobe prototype. Right-facing arrows are VM Exits, and left-facing arrows are VM Entries. When int3 is executed, the hypervisor takes control. The hypervisor optionally executes a probe handler (probefunc()) and places the CPU into single-step mode. It then executes the original instruction and does a VM Entry to resume the VM. After the guest executes the original instruction, it traps back into the hypervisor, and the hypervisor will write the int3 before allowing the VM to continue as usual.

does not belong to an hprobe (e.g., it was generated by running gdb or kprobes inside the VM), the int3 is passed back to KVM to be handled as usual. Each hprobe handler performs a user-defined monitoring function and runs in the host OS. When the handler returns, the hypervisor replaces the int3 instruction with the original opcode and puts the CPU in single-step mode. Once the original instruction executes, a single-step exception is generated, causing another VM Exit event [21] (step 2). At this point, the hprobe kernel agent rewrites the int3 and performs a VM Entry, and the VM resumes its execution (step 3). This single-step and instruction-rewrite process ensures that the probe is always caught. If one wishes to protect the probes from being overwritten by the guest, the page containing the probe can be write-protected. Although the prototype was implemented using KVM, the concept will extend to any hypervisor that can trap on similar exceptions. Note that instead of int3, we could use any other instruction that generates VM Exits (e.g., hypercalls or illegal instructions). We chose int3 because it is well-supported and has a single-byte opcode.

Limitations
This prototype is useful for a large class of monitoring use cases; however, it does have a few limitations:

- Hprobes trigger only on instruction execution. If one is interested in monitoring data access events (e.g., triggering every time a particular address is read from or written to), hprobes will not provide a clean way to do so. One would need to place a probe at every instruction that modifies the data (potentially every instruction that modifies any data, if addresses are affected by user input). More cleanly, one could use an hprobe at the beginning and end of a critical section to turn on and off page protection for data relevant to that critical section, capturing the events in a manner similar to that of livewire [3] but with the flexibility of hprobes.

- Hprobes leverage VM Exits, resulting in nonoptimal performance. However, the simpler, more robust implementation that hprobes offer, with trust rooted in HAV, is worth this trade-off.
- Probes cannot be fully hidden from the VM. Even with clever use of Extended Page Tables (EPT) tricks to hide the existence of a probe when reading from its location, a timing side-channel would still exist, since an attacker could observe that the probed instruction takes longer than expected to complete.

4.4.3 Example Detectors

In this section, we present sample reliability and security detectors built upon the hprobe prototype framework. These detectors are unique to the hprobe framework and cannot be implemented on any other current VM monitoring system.

4.4.3.1 Emergency Exploit Detector

Most system operators fear zero-day vulnerabilities, as there is little that can be done about them until the vendor or maintainer of the software releases a fix. Furthermore, even after a vulnerability is made public, a patch takes time to be developed and must be put through a QA cycle. The challenge is even greater in environments with high availability concerns and stringent change control requirements; even if a patch is available, many times it is not possible to restart the system or service until a regular maintenance window. This leaves operators with a difficult decision: risk damage from restarting a system with a new patch, or risk damage from running an unpatched system.

Consider the CVE-2008-0600 vulnerability, which resulted in a local root exploit through the vmsplice() system call [24,25]. This example represents a highly dangerous buffer overflow, since a successful exploit allows one to arbitrarily execute code in ring 0 using a program that is publicly available on the Internet. Since this exploit involves the base kernel code (i.e., not a loadable module), patching it would require installation of a new kernel followed by a system reboot. As discussed earlier, in many operational cases, a system reboot or OS patch can be conducted only during a predetermined maintenance window. Furthermore, many organizations would be hesitant to run a fresh kernel image on production systems without having gone through a proper testing cycle first.

The vmsplice() system call is used to perform a zero-copy map of user memory into a pipe. At a high level, the CVE-2008-0600 vmsplice() constructs specially crafted compound page structures in user space. A compound page is a structure that allows one to treat a set of pages as a single data structure. Every compound page structure has a pointer to a destructor function that handles the cleanup of those underlying pages. The exploit works by using an integer overflow to corrupt the kernel stack such that it references the compound page structures crafted in user space. Before calling vmsplice(), the exploit closes the pipe, so that when the system call runs, it deallocates the pages, resulting in a call to the compound pages' destructor function. The destructor is set to privilege escalation shellcode that allows an attacker to hijack the system.

The emergency detector works by checking the arguments of a system call for a potential integer overflow. This differs in functionality from the upstream patch, which checks whether the memory region (specified by the struct iovec argument) is accessible to the user program. One could write a probe handler that performs a similar function by checking whether the entire region referred to by the struct iovec pointer + iov_len is in the appropriate range (e.g., by walking the page tables that belong to that process). However, a temporary measure to protect against an attack should be as lightweight and simple as possible to avoid unpredictable side effects. One major benefit of using an hprobe handler is that developing such a detector does not require a deep understanding of the vulnerability; the developer of the emergency detector only needs to understand that there is an integer overflow in an argument. This is far simpler than developing and maintaining a patch for a core kernel function (a system call), especially when reasoning about the risk of running a home-patched kernel (a process that would void most enterprise support agreements).

Our solution uses a monitoring system that resides outside of the VM and relies on a hardware-enforced int3 event. A would-be attacker cannot circumvent this event without having first compromised the hypervisor or modified the guest's kernel code. He or she could do so with a code injection attack that causes a different sys_vmsplice() system call handler to be invoked. However, it is unlikely that an attacker who already has the privileges necessary for code injection into the kernel would have anything to gain by exploiting a local privilege escalation vulnerability. While the proposed emergency detector cannot defeat an attacker who has previously obtained root access, its ease of rapid deployment sufficiently mitigates this risk.

Since no reboot is required and the detector can be used in a "read-only" monitoring mode (only reporting the attack, and otherwise not taking an action), the risk of using this detector on a running production system is minimal. To test the CVE-2008-0600 detector, we used a CENTOS5 VM and the publicly available exploit. As an unprivileged user, we ran an exploit script on the unpatched OS and were able to obtain root access. With the monitor in place, all attempts to obtain root access using the exploit code were detected.

4.4.3.2 Application Heartbeat Detector

One of the most basic reliability techniques used to monitor computing system liveness is a heartbeat detector. In that class of detector, a periodic signal is sent to an external monitor to indicate that the system is functioning properly. A heartbeat serves as an illustrative example of how an hprobe-based reliability detector can be implemented. Using hprobes, we can construct a monitor that directly measures the application's execution. That is, since probes are triggered by application execution itself, they can be viewed as a mechanism for direct

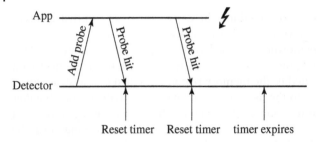

Figure 4.4 Application heartbeat detector. A probe is inserted in a critical periodic section of the application (e.g., the main loop). During normal execution, a timer is continuously reset. In the presence of a failure (such as an I/O hang), the timer expires and failure is declared.

validation that the application is functioning correctly. Many applications execute a repetitive code block that is periodically reentered (e.g., a Monte Carlo simulation that runs with a main loop, or an HTTP server that constantly listens for new connections). If one profiles the application, it is possible to identify a period (defined in units of time or using a counter, like the number of instructions) at which this code block is reentered. During correct operation of the application, one can expect that the code block will be executed at the profiled interval.

The hprobe-based application heartbeat detector is built on the principle described in the previous paragraph and illustrated in Figure 4.4. The test detector (i.e., one of the detectors on the left side of Figure 4.2) is a kernel module that is installed in the host OS. An hprobe is inserted at the start of the code block that is expected to be periodically reentered. When the hprobe is inserted, a delayed workqueue5 is scheduled for the timeout corresponding to the reentry period for the code block. When the timeout expires, the workqueue function is executed and declares failure. (If the user desires a more aggressive watchdog-style detector, it is possible to have the hprobe handler perform an action such as restart of the application or VM.) During correct operation (i.e., when the hprobe is hit), the workqueue is canceled and a new workqueue is scheduled for the same interval, starting a new timeout period. This continues until the application finishes or the user no longer desires to monitor it and removes the hprobe. If having an hprobe hit on every iteration of the main loop is too costly, one can ensure that the probe is active for an acceptable time interval; it can be added/removed until a desirable performance is achieved. (The detection latency would still be low, as a tight loop would have a small timeout value.)

We use the open-source Path Integral Quantum Monte Carlo (pi-qmc) simulator [26] as a test application. This application represents a long-running scientific program that can take many hours or days to complete. As is typical with scientific computing applications, pi-qmc has a large main loop. Since Monte Carlo simulation involves repeated sampling and therefore repeated execution of the same functions, we need to run the main loop only a handful of times to determine the time per iteration. After determining the expected duration of each iteration, we set the heartbeat timeout to twice the expected value, set the detector to a statement at the end of the main loop, injected hangs (e.g., SIGSTOP), and crashed the application (e.g., SIGKILL). All crashes (including any VM crashes that occurred after the timer was executed in the hypervisor) were detected.

4.4.4 Performance

All of our microbenchmarks and detector performance evaluations were conducted on a Dell PowerEdge R720 server with dual-socket Intel Xeon E5-2660 "Sandy Bridge" 2.20 GHz CPUs (3.0 GHz turbo boost). To obtain runtime measurements, we added an extra hypercall to KVM that starts and stops a timer inside the host OS. This allows us to obtain measurements independent of VM clock jitter. To ensure consistency among measurements, the test VMs were rebooted between two subsequent measurements.

4.4.4.1 Microbenchmarks

We performed microbenchmarks that estimated the latency of a single hprobe, which is the time from execution of int3 by the VM until the VM is resumed (steps 1–3 in Figure 4.3). We ran these microbenchmarks without a probe handler function to determine the lower bound of hprobe-based detector overhead. Since the round-trip latency of an individual VM Exit on Sandy Bridge CPUs has been estimated to take roughly 290 ns [27] and our hypercall measurement scheme induces additional VM Exits, it would be difficult to accurately measure the individual probe latency. Instead, we obtained a mean round-trip latency by repeatedly executing a probed function a large number of times (one million) and dividing by the total time taken for those executions. That approach helped remove jitter due to timer inaccuracies as well as the actual latency of the hypercall measurement system itself. For the test probe function, we added a no-op kernel module to the guest OS that creates a dummy no-op device with an ioctl that calls a noop_func() kernel function that performs no useful work (return 0). First, we inserted an hprobe at the noop_func()'s location. Our microbenchmarking application started by issuing a hypercall to start the timer and then an ioctl against the no-op device. When the no-op module in the guest OS received the ioctl, it called noop_func() one million times. Afterward, another hypercall was issued from the benchmarking application to read the timer value.

For the microbenchmarking experiment, we used a 32-bit Ubuntu 14.04 guest and measured 1000 samples. The mean latency (across samples) was found to be 2.6 μs. In addition to the Sandy Bridge CPU, we have also included data for an older generation 2.66 GHz Xeon E5430 "Harpertown" processor (running the

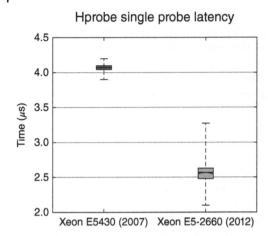

Figure 4.5 Single probe latency. (CPUs' release years are in parentheses.) The E5-2660's larger range can be attributed to "Turbo Boost," whereby the clock scales from 2.2 to 3.0 GHz. The shaded area is the quartile range (25–75 percentile); whiskers are the minimum/maximum; the center is the mean; and the notches in the middle represent the 95% confidence interval of the mean.

same kernel, KVM version, and VM image), which had a mean latency of 4.1 µs. The distribution of latencies for these experiments is shown in Figure 4.5. The remainder of the benchmarks presented used the Sandy Bridge E5-2660. The hprobe prototype requires multiple VM Exits per probe hit. However, in many practical cases, the flexibility of dynamic monitoring and the reduced maintenance costs resulting from a simple implementation outweigh that cost. The flexibility can increase performance in many practical cases by allowing one to add and remove probes throughout the VM's lifetime. Furthermore, CPU manufacturers are constantly working to reduce the impact of VM Exits; for example, Intel's VT-x saw an 80% reduction in VM Exit latency over its first 6 years [27].

4.4.4.2 Detector Performance
In addition to microbenchmarking individual probes, we measured the overhead of the example hprobe-based detectors presented in the previous section. All measurements in this section were obtained using the hypercall-based timer.

1) *Emergency Exploit Detector:* Our integer overflow detector that protects against the CVE-2008-0600 vmsplice() vulnerability is extremely lightweight. Unless vmsplice() is used, the overhead of the detector is zero since the probe will not be executed. The vmsplice() system call is rare (at least in the open-source repositories that we searched), so zero overhead is overwhelmingly the common case. One application that does use vmsplice() is Checkpoint/ Restart in User space (CRIU). CRIU uses vmsplice() to capture the state of open file descriptors referring to pipes. We used the Folding@Home molecular dynamics simulator [28] and the pi-qmc Monte Carlo simulator from earlier as test programs. We ran these applications in a 64-bit Ubuntu 14.04 VM. At each sample, we allowed the application to warm up (load input data and start the main simulation) and then checkpointed it. The timing

Table 4.1 CVE-2008-0600 Detector w/CRIU.

Application	Runtime (s)	95% CI (s)	Overhead (%)
F@H Normal	0.221	0.00922	0
F@H w/Detector	0.228	0.0122	3.30
F@H w/Naïve Detector	0.253	0.00851	14.4
pi-qmc Normal	0.137	0.00635	0
pi-qmc w/Detector	0.140	0.00736	1.73
pi-qmc w/Naïve Detector	0.125	0.00513	11.1

hypercalls were inserted into CRIU to measure how long it takes to dump the application. This was repeated 100 times for each case with and without the detector, and the results are tabulated in Table 4.1. In the table, we can see that there is a slight difference between the mean checkpoint times (roughly 3.3% for F@H and 1.7% for pi-qmc) and that the variance in the experiment with the detector active is higher for the Folding@Home case. Sys_vmsplice() was called 28 times when Folding@Home was being checkpointed, and 11 times for pi-qmc. We can attribute this difference to the negative cache effects of the context switch when probes are being activated.

2) *Application Heartbeat Detector*: We used the pi-qmc simulator to measure the performance overhead of the application watchdog detector. The pi-qmc simulator allows configuration of its internal sampling, and we utilized this feature to vary the length of the main loop. In order to determine how the detector impacts performance, we measured the total runtime of each iteration of the main loop when the probe was inserted and ran the program for 15 min. The results of our experiments are shown in Figure 4.6.

From Figure 4.6, we can see that the detector did not affect performance in a statistically significant way. The reason is that pi-qmc, like many scientific computing applications, does a large amount of work in each iteration of its main loop. However, by setting the threshold of the detector to a conservative value (like twice the mean runtime), one can achieve fault detection in a far more acceptable timeframe than with other methods, like manual inspection. Furthermore, this detector goes beyond checking whether the process is still running; it can detect any fault that causes a main loop iteration to halt (e.g., a disk I/O hang, a network outage when using MPI, or a software bug that does not lead to a crash).

4.4.5 Summary

The hprobe framework is characterized by its simplicity, dynamism, and ability to perform application-level monitoring. Our prototype for this framework uses

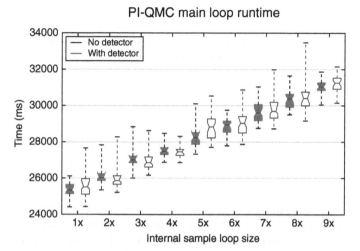

Figure 4.6 Benchmarking of the application watchdog detector. The horizontal axis indicates the scaling of an internal loop in the target pi-qmc program. The vertical axis shows a distribution of the completion times for the iterations of the main loop. The boxplot characteristics are the same as in Figure 4.5.

hardware-assisted virtualization and satisfies protection requirements presented in the literature. We find that compared to past work, the simplicity with which the detectors can be implemented and inserted/removed at runtime allows us to develop monitoring solutions quickly. Based on our experience, this framework is appropriate for use in real-world environments. Through use of our sample detectors, we have found that the framework is suitable for detecting bugs and random faults, and for use as a stopgap measure against vulnerabilities.

4.5 Hypervisor Introspection: A Technique for Evading Passive Virtual Machine Monitoring

While dynamic monitoring (wherein monitoring checks are triggered by certain events, for example, access of specific hardware registers or memory regions) has its advantages, it comes with additional costs in terms of implementation complexity and performance overhead. In consequence, one may ask, why not use passive monitoring (wherein a monitoring check is invoked on a predefined interval, for example, to determine what processes are running every second), which could be less costly?

To address the trade-offs between active and passive monitoring, in this section, we discuss some drawbacks of passive monitoring. In particular, we show that it is possible for a guest VM to recognize the presence of a passive VMI (virtual machine introspection) system and its monitoring frequency through a timing side-channel. We call our technique *Hypervisor Introspection* (HI). We also present an insider attack scenario that leverages HI to evade a passive VMI monitoring system. Finally, we discuss current state-of-the-art defenses against side-channel attacks in cloud environments and their shortcomings against HI.

We refer the interested reader to our earlier publication [29] for more details on the material covered in this section.

4.5.1 Hypervisor Introspection

We developed a VMI monitor against which to test Hypervisor Introspection, and we used the side-channel attack to evade the passive VMI. The test system was a Dell PowerEdge 1950 server with 16 GB of memory and an Intel Xeon E5430 processor running at 2.66 GHz. The server was running Ubuntu 12.04 with kernel version 3.13. The hypervisor used was QEMU/KVM version 1.2.0, and the guest VMs were running Ubuntu 12.04 with kernel version 3.11.

4.5.1.1 VMI Monitor

In order to test the effectiveness of HI, we implemented a VMI monitor. To do so, we used LibVMI [18]. LibVMI is a software library that helps with the development of VMI monitors. It focuses on abstracting the process of accessing a guest VM's volatile memory. The volatile memory of a VM contains information about the guest VM's OS, such as kernel data structures, which can be examined to determine runtime details such as running processes or active kernel modules.

Because the polling rate of the VMI monitor is directly related to the performance overhead of the monitor, the polling rate must be chosen carefully so as not to introduce an unacceptable decrease in VM performance. We configured the VMI monitor to poll the guest VM every second. This rate introduced only a 5% overhead in VM performance, based on benchmark results from UnixBench.

4.5.1.2 VM Suspend Side-Channel

Because HI revolves around making measurements from a side-channel and inferring hypervisor activity, we first had to identify the actual side-channel to be exploited. We noted that whenever the hypervisor wants to perform a monitoring check on a guest VM, the VM has to be paused so that a consistent view of the hardware state can be obtained. If an observer can detect these VM suspends, then that observer might be able to learn about the monitoring checks performed by the hypervisor. We call this approach the *VM suspend side-channel* (see Figure 4.7). We came up with two potential methods for measuring the VM suspends: network-based timing measurements and host-based, in-VM timing measurements. Since the network is a noisy medium because of dropped packets and routing inconsistencies, we focused on performing in-VM measurements to detect VM suspends.

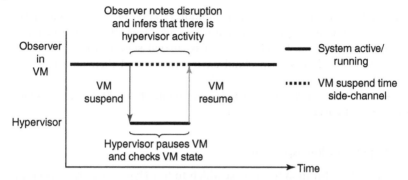

Figure 4.7 Illustration of the VM suspend side-channel in which an observer notes disruptions in VM activity to determine when the hypervisor is performing some action on the VM.

4.5.1.3 Limitations of Hypervisor Introspection

HI has two limitations: the low accuracy of the monitoring intervals it measures, and the difficulty of determining the threshold value for identifying VM suspends. Our testing of HI showed that it is capable of determining monitoring intervals as short as 0.1 s, as we determined by testing HI against increasingly frequent monitoring checks; the maximum frequency resolved by HI was one check every 0.1 s (10 Hz).

Because the threshold value for HI is found through empirical testing, development of HI is not straightforward for any given system. Our testing of HI indicated that threshold values ranging from 5 to 32 ms yielded the same accuracy when we were performing HI and detecting VM suspends. As future work, the threshold value may be correlated with various system specifications, such as kernel version, CPU model/frequency, and system load. After threshold values on various systems are found via empirical testing, a formal relationship between the various system specifications and the threshold value may be derived so that a threshold can be obtained without testing.

4.5.2 Evading VMI with Hypervisor Introspection

This section discusses applications of Hypervisor Introspection to hide malicious activity from a passive VMI system.

4.5.2.1 Insider Attack Model and Assumptions

We present an insider threat attack model in which the insider already has administrator (i.e., root) access to VMs running in a company's public IaaS cloud. The insider knows that he will be leaving the company soon but would like to maintain a presence on the VMs to which he has access. The insider does not have access to the underlying hypervisor hosting the VMs but knows that the

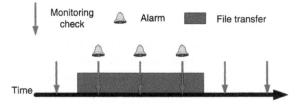

Figure 4.8 Illustration of an unauthorized file transfer's detection by the VMI system.

company is utilizing some form of passive VMI monitoring. We also assume that the company's VMI monitor is similar to the monitor we implemented, which regularly checks for unauthorized processes. As insiders have full control over their VMs, the company relies on VMI for monitoring, so changes to the VM, such as kernel modifications or creations of new files, are not detected. In this attack model, the insider can utilize HI to detect invocations of the monitor and to hide malicious activities from the VMI monitor.

4.5.2.2 Large File Transfer

Attackers commonly want to exfiltrate data out of the network after compromising a system. The attacker may leverage various tools and protocols to accomplish this, such as secure copy (SCP), file transfer protocol (FTP), the attacker's own custom utility, or any number of other file transfer methods. A passive VMI system may detect attempts at data exfiltration by maintaining a restrictive whitelist of processes that are allowed to run with a network socket open. A restricted VM would have a whitelist with zero or few entries.

For transfer of sufficiently large files via a nonwhitelisted process, the transfer time would be longer than the monitoring interval, and the VMI monitor would detect the file transfer. This scenario is illustrated in Figure 4.8. We tested the effectiveness of our VMI monitor in this scenario by transferring a 250 MB file over TCP from the monitored VM to a coresident VM, and our VMI monitor detected the Python process running with an open socket and signaled an alarm.

It is possible to transfer a large file over a network without triggering any VMI alarms by splitting the file into chunks and transferring the chunks individually between monitoring checks. This process is illustrated in Figure 4.9. Thus, HI can be used to time the transfer of each chunk to evade the VMI system.

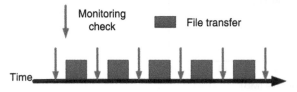

Figure 4.9 Illustration of how a large file can be transferred in chunks by a nonwhitelisted process without triggering the passive VMI monitor.

We implemented this attack by writing a Python script that transfers a chunk of a file each time it is executed. The offset of the current chunk is written to an "offset log" file after each execution, so that the next execution knows which chunk to transfer. We extended our HI kernel module with the user mode helper kernel function that is invoked by the Python script after a suspected VM suspends. With our modified HI kernel module and Python script, we were able to transfer the same 250 MB file without being detected by our VMI monitor.

On our test system, naïve transfer of the entire 250 MB file at once over a TCP connection between two coresident VMs took on average (over 100 tests) 4.905 s, with a standard deviation of 0.414 s. Using our technique with 8 MB chunks, the entire transfer took 32 calls to the Python script, and thus 32 monitoring intervals. Because each monitoring interval is 1 s long, the entire transfer took around 32 s, which is about eight times the transfer time of the naïve method.

We note that the transfer speed of the large file when this method is used depends on the polling frequency of the VMI monitor. Lower polling frequencies lead to longer transfer times because each chunk transfer is triggered by a monitoring check, so chunks would be transferred less frequently. However, after learning the polling frequency through HI, the attacker can increase the chunk size to counteract the delay. Increasing the chunk size is risky, as it may end up triggering the VMI system, so the attacker would likely verify a larger chunk size on his own system before performing the attack. Figure 4.10 shows the trade-offs involved in increasing the chunk size for the file transfer. Increased chunk size leads to a higher detection rate, but the whole file is transferred in fewer monitoring checks.

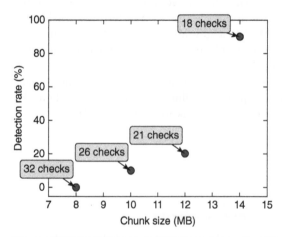

Figure 4.10 Chunk size versus detection rate in using HI to hide a large file transfer. Note the trade-off between number of monitoring checks (i.e., transfer time) and detection rate.

4.5.3 Defenses against Hypervisor Introspection

In this section, we discuss current state-of-the-art defenses against side-channel attacks in virtual environments and their shortcomings in defending against HI. We also discuss a potential defense against HI that aims to address the shortcomings of the current state-of-the-art defenses.

4.5.3.1 Introducing Noise to VM Clocks

Because HI relies on fine-grained timing measurements to determine occurrences of VM suspends, it follows that reducing the accuracy or granularity of VM clocks could prevent HI from working. Previous research has looked at reducing the granularity or accuracy of time sources to prevent cross-VM side-channel attacks. Although that work aimed to address cross-VM side-channel attacks, much of it is somewhat applicable to the hindering of HI.

Vattikonda *et al.* explored the possibility of fuzzing timers to reduce the granularity of measurements needed for side-channel attacks [24]. They modified the Xen hypervisor to perturb the x86 RDTSC instruction by rounding it off by 4096 cycles. Because the RDTSC instruction is commonly used to obtain high-resolution times tamps, side-channel attacks may be prevented. An additional benefit of fuzzing the RDTSC instruction is that timing system calls, such as gettimeofday and clock gettime, are fuzzed as well. Although HI relies on one of these system calls for time stamping, the perturbations caused only a 2 μs change in the true RDTSC value. HI needs measurements on the order of milliseconds, so the fuzzing does not perturb the RDTSC value enough to hinder HI.

Li *et al.* developed a system called StopWatch, in which the VM clock was replaced with a virtual clock that constantly skews [25]. Because the virtual clock depends only on the number of instructions executed, it hid VM suspends from in-VM timing measurements. However, applications with real-time requirements would not be able to use a virtual clock. In addition, StopWatch has a worst-case performance of 2.8× for workloads that require heavy network usage, which might not be acceptable for high-performance workloads.

4.5.3.2 Scheduler-Based Defenses

Recently, scheduling policies have been explored as another means to prevent cross-VM, cache-based, side-channel attacks. Varadarajan *et al.* proposed using the scheduler to force each process to run for a minimum runtime before another process is allowed to run [30]. If all processes are forced to run for a certain amount of time without being preempted, an attacker would obtain less information from observing a process's cache usage. Adjusting the scheduling policy could prevent part of our HI technique from working if the minimum runtime were greater than the VM suspend threshold. If that were the case, then process scheduling could not be used as one of the events observed for the in-VM timing measurements. However, many other events that occur during normal OS operation could still be observed, such as network operations or memory allocation and deallocation. An attacker could also artificially spawn processes that purposely utilize specific OS operations to increase the frequency of the events and improve the granularity of the measurements needed for HI. Thus, changing the scheduling policy might hinder HI, but it would not altogether block it. Further, enforcing a minimum runtime could degrade performance because CPU-intensive workloads would have to compete with less intensive workloads.

4.5.3.3 Randomized Monitoring Interval

Because HI looks for regular VM suspends to determine when monitoring checks occur, it may seem that simply randomizing the monitoring interval would prevent HI. This is not the case because randomized intervals have a lower bound on the duration between monitoring checks. A patient attacker could use HI to establish the lower bound on the monitoring interval and craft his attacks around that lower bound. Thus, the randomized monitoring interval forces the attacker to be inefficient, but it cannot prevent the attacker from evading the VMI system.

For example, consider a passive VMI system that polls the guest VM on a randomized interval that lasts anywhere from 1 to 10 s, inclusive. Assuming that the monitoring interval is a discrete random variable (i.e., the monitoring interval is an integer), one would expect to observe 10 monitoring checks before the smallest possible interval is seen.

Even if the monitoring interval is a uniform random variable (i.e., the monitoring interval can be a noninteger value), the attacker can transform the problem into the discrete-value case by taking the floor of the observed monitoring interval. In the case of an interval that lasts anywhere from 1 to 10 s, flooring the value leaves 10 possible values. Thus, the expected number of observed monitoring checks before the minimum is obtained is still 10.

The range of the random monitoring intervals is directly related to how difficult it is for an attacker to establish the minimum possible monitoring interval. Because security improves with a lower monitoring interval (and higher monitor frequency), the range of monitoring interval values would remain small in practice to improve security, but that would also make it easier for an attacker to determine the lower bound of the randomized monitoring interval through HI.

We reconfigured our VMI monitor to use a randomized monitoring interval between 0 and 2 s that would change after each monitoring check. We chose this interval because it kept the performance overhead around 5% (based on the same UnixBench benchmarks from earlier), and we expected the lower monitoring intervals (under 1 s) to detect the attack described in the earlier section. We performed the attack 30 times and found that the reconfigured monitor was able to detect the large file transfer 70% of the time.

We were surprised that some large file transfer attacks succeeded against the randomized monitoring defense. Some of the large file transfers went undetected because the randomized intervals that triggered the chunk transfers in those cases were long enough to thwart detection. However, the majority of the large file transfers were detected with randomized monitoring. Based on these tests, we argue that a randomized monitoring interval is sufficient for preventing a large file transfer attack but not a backdoor shell attack. Further, randomized monitoring does not prevent HI from learning that there is a passive VMI system in place, and HI can be used to learn the distribution sampled by the randomized monitor.

4.5.4 Summary

We discussed Hypervisor Introspection as a technique to determine the presence of and evade a passive VMI system through a timing side-channel. Through HI, we demonstrated that hypervisor activity is not perfectly isolated from the guest VM. In addition, we showed an example insider threat attack model that utilizes HI to hide malicious activity from a realistic, passive VMI system. We also showed that passive VMI monitoring has some inherent weaknesses that can be avoided by using active VMI monitoring.

4.6 Identifying Compromised Users in Shared Computing Infrastructures

System monitoring (both dynamic and passive) requires smart and effective techniques that can accurately, and in a timely fashion, detect potential malicious activities in the system. One way to achieve efficiency is to build the detectors based on past data on security incidents. In this section, we discuss a data-driven approach for accurate (with a low false-positive rate) identification of compromised users in a large computing infrastructure.

In the case of shared computing infrastructure, users get remote access by providing their credentials (e.g., username and password) through a public network and using well-established authentication protocols, for example, SSH. However, user credentials can be stolen via phishing or social engineering techniques and made available to attackers via the cybersecurity black market. By using stolen credentials, an attacker can masquerade as a legitimate user and penetrate a system in effect as an insider.

An access to a system performed with stolen credentials is hard to detect and may lead to serious consequences, such as the attackers obtaining root-level privileges on the machines of the system or breach of privacy, for example, e-mail access. Therefore, timely detection of ongoing suspicious activities is crucial for secure system operations. Thus, computing infrastructures are currently equipped with multiple monitoring tools (e.g., intrusion detection systems (IDS) and file integrity monitors) that allow system administrators to detect suspicious activities. However, the need to ensure high coverage in detecting attacks calls for accurate and highly sensitive monitoring, which in turn leads to a large number of false positives. Furthermore, the heterogeneity and large volume of the collected security data make it hard for the security team to conduct timely and meaningful forensic analysis. In this section, we consider credential-stealing incidents, that is, incidents initiated by means of stolen credentials. We refer the interested reader to our earlier publication [31] for more details on the material covered here.

4.6.1 Target System and Security Data

We performed a study of data on credential-stealing incidents collected at the National Center for Supercomputing Applications (NCSA) during the 2008–2010 timeframe. Credential-stealing incidents can occur in virtually any network that is accessible from the Internet (e.g., social networking sites, e-mail systems, or corporate networks that allow VPN access) or from an intranet or a business network managed by an IT department within a corporation. For that reason, the key findings of this study can be used to drive the design of better defensive mechanisms in organizations other than NCSA.

The NCSA computing infrastructure consists of about 5000 machines (including high-performance clusters, small research clusters, and production systems such as mail and file servers) accessed by worldwide users. Figure 4.11 provides a high-level overview of the target system. Users log into the system remotely by forwarding their credentials through the public network via the SSH protocol. Credentials might have been stolen. Thus, an attacker can masquerade as a legitimate user and penetrate the system. The security monitoring tools deployed in the NCSA network infrastructure are responsible for (1) alerting about malicious activities in the system, for example, a login from a blacklisted IP address or the occurrence of suspicious file downloads; and (2) collecting

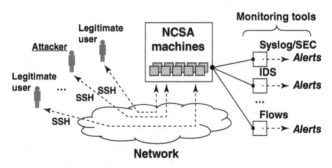

Figure 4.11 Overview of the target system.

(in security logs) data relevant to detected incidents. An in-depth analysis of the security logs can then be used to identify compromised users.

4.6.1.1 Data and Alerts

The log produced by the target machines, which is collected with the syslog protocol, is used to identify the users that access the infrastructure.

In the following, we describe the alerts produced by the monitoring tools when incidents that involve stolen credentials occur. As detailed below, we have assigned to the alerts IDs that are used throughout the remainder of this section.

First, we will discuss alerts that can be triggered when a login violates a user's profile. The detection of potential violations is done by checking login and profile data against rules set in the Simple Event Correlator (SEC):

- *Unknown address* (A1): A login comes from a previously unknown IP address, that is, the user has never before logged in from that IP, according to his or her profile.
- *Multiple login* (A2): The same external IP address is used by multiple users to log into the system.
- *Command anomaly* (A3): A suspicious command is executed by the user.
- *Unknown authentication* (A10): According to the profile data, the user has never before logged into the system using that authentication mechanism.
- *Anomalous host* (A11): The login is reported by a node within the infrastructure that has never before been used by the user.
- *Last login > 90 days* (A12): The last login performed by the user occurred more than 90 days before the current one.

In most cases, the occurrence of a profile alert does not by itself provide definitive proof that a user has been compromised. For example, A1 is raised each time a user (legitimate or not) logs in for the first time from a remote site that is not stored in the profile. In order to increase the chances of correctly detecting compromised users, the analysis of profile alerts is combined with the data provided by the security tools, for example, IDS and NetFlows (Figure 4.11), which are available in the NCSA network infrastructure. The security alerts used in the detection process are as follows:

- *HotClusterConn* (A4): A node of the computing infrastructure performs a download, although it is expected never to do so. An additional alert (A13) is introduced to indicate that the downloaded file exhibits a *sensitive* extension, for example, `.c`, `.sh`, `.bin`, or `.tgz`;
- *HTTP* (A5) and *FTP* (A9) *sensitive URI*: These alerts are triggered upon detection of well-known exploits, rootkits, and malware.
- *Watchlist* (A7): The user logs in from a blacklisted IP address (the list of suspicious addresses is held by and distributed among security professionals).
- *Suspicious download* (A14): A node of the computing infrastructure downloads a file with a sensitive extension.

Finally, two further alerts, A6 and A8, were designed by combining profile and security data. Alert A6 is generated whenever the remote IP address used to perform a login is involved in subsequent anomalous activities, such as a suspicious file download. Similarly, the alert A8 is generated if a user responsible for a multiple login is potentially related to other alerts in the security logs.

It should be noted that an event could trigger more than one alert. For example, the download of a file with a sensitive extension, if performed by a node that is not supposed to download any files, can trigger alerts A4, A13, and A14. While the occurrence of a profile alert leads to an initial level of *suspiciousness* about a user, a set of subsequent notifications, such as command anomalies or suspicious downloads, might actually be the symptoms of ongoing system misuse. Correlation of multiple data sources is valuable to improving detection capabilities and ruling out potential false alarms.

4.6.1.2 Automating the Analysis of Alerts

The timely investigation of alerts is crucial in identifying compromised users and initiating proper recovery actions. However, the analysis can become time-consuming because of the need to correlate alerts coming from multiple sources. In order to automate the alert analysis, we developed a software tool to (1) parse the content of heterogeneous security logs and (2) produce a representation of the security data more suitable for facilitating the Bayesian network approach.

Given the data logs (both syslogs and logs produced by the security tools), the tool returns a *user/alerts table*, which provides (1) the list of users that logged into the system during the time when the logs were being collected and (2) a set of 14-bit vectors (one for each user), with each bit of the vectors representing one of the alerts introduced in the previous section. Given a user in the system, a bit in the vector assumes value 1 if at least one alert of that type (observed in the security log) has *potentially* been triggered by that user. In order to illustrate the concept, Table 4.2 shows a hypothetical user/alerts table. For example, a binary vector of [10010000001000] is associated with *user_1*, which indicates that during the observation period, *user_1* was potentially responsible for triggering

Table 4.2 Example of user/alerts table.

Users	Alerts													
	A1	A2	A3	A4	A5	A6	A7	A8	A9	A10	A11	A12	A13	A14
user_1	1	0	0	1	0	0	0	0	0	0	1	0	0	0
user_2	0	0	0	0	0	0	0	0	0	0	0	0	0	0
⋮														
user_N	0	0	0	0	0	0	1	0	0	0	0	1	0	0

three alerts: *unknown address* (A1), *HotClusterConn* (A4), and *anomalous host* (A11).

4.6.2 Overview of the Data

Our study analyzed data related to 20 security incidents at NCSA that were initiated using stolen credentials and that occurred during 2008–2010. The NCSA security team comprehensively investigated each incident. The key findings of the investigations are summarized in Ref. [32]. The ground truth, that is, information about the actually compromised users, reports describing the system misuse, and proposed countermeasures, is available for each incident considered in the study. That detailed knowledge makes it possible to validate the effectiveness of the proposed approach. Of the 20 incidents, 3 were detected by means of a third-party notification (i.e., someone outside NCSA reported that anomalous activities had occurred) and 1 incident was a false positive (i.e., the conclusion that the user had been compromised was erroneous).

In order to characterize the ability of each alert A_i ($1 \leq i \leq 14$) to detect compromised users, we estimated the average number of users per day that are flagged as potentially responsible for A_i. We did so by analyzing the logs collected on the day the incident occurred. Figure 4.12 shows the average number of users per day flagged as potentially responsible for each alert considered in our analysis. Note that there is significant variability in the occurrence frequencies of different alerts. For example, the *watchlist* (A7) alert was observed for only two users during the entire observation period (which spanned the 16 incidents analyzed in this study). In both cases, the user was actually compromised. In the context of our analysis, *watchlist* is considered a reliable alert, since it has a small number of occurrences (0.125 users/day) and no false positives. On the other end of the spectrum is the *HotClusterConn* (A4)

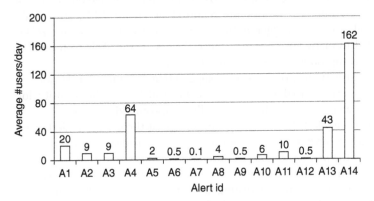

Figure 4.12 Average number of users (per day) flagged as potentially responsible for each alert A_i.

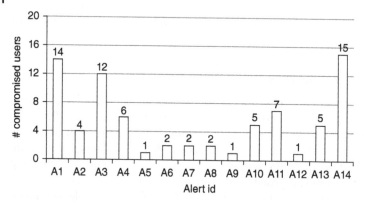

Figure 4.13 Number of compromised users that triggered an alert A_j.

alert, which has a high occurrence frequency (64 users/day) and a relatively high false-positive rate. (Most of the users flagged as responsible for this alert have not actually been compromised.)

The detection capability is another important feature of the alerts. The detection capability is determined by extracting (from the user/alerts tables generated for the 16 analyzed incidents) the 14-bit vectors for the compromised users. Recall that the actually compromised users (i.e., the ground truth) are known for each incident. There are 20 compromised users in the incident data. Figure 4.13 shows how many compromised users (y-axis) are responsible for the specific alert types (x-axis). Comparison of these data with those presented in Figure 4.12 indicates that the alerts with the largest numbers of potentially responsible users are likely to be observed when the user is actually compromised. For example, around 20 users per day triggered an *unknown address* (A1) alert (see Figure 4.12); however, while most of these alerts turned out to be false positives, in 14 out of 20 cases (as reported in Figure 4.13), an actually compromised user had triggered the alert. Similarly, the *HotClusterConn* (A4) alert led to many false positives; nevertheless, it was likely to have been triggered by compromised users (6 out 20 cases).

The inconclusive nature of alerts could suggest that it might be hard to identify compromised users based on the alert data available in the security log. However, our analysis shows that an actually compromised user will be related to more than one alert. In our study, an average of three unique alerts were related to each compromised user for each analyzed incident, such as the joint occurrence of the *unknown address, command anomaly*, and *HotClusterConn* alerts. In consequence, it is feasible to correlate multiple alerts by means of statistical techniques in order to distinguish between cases in which unreliable alerts can be discarded and cases in which an alert provides stronger evidence that a user has been compromised.

4.6.3 Approach

We processed security logs to produce (1) the list of users that logged into the system at any time throughout the time interval during which the logs were being collected and (2) the bit vector reporting alerts potentially related to a given user. Using that information, our objective was to estimate the probability that a user was compromised.

4.6.3.1 The Model: Bayesian Network

We have proposed a data-driven Bayesian network approach [33,34] to facilitate identification of compromised users in the target infrastructure. A Bayesian network is a direct acyclic graph in which each node of the network represents a variable of interest in the reference domain. The network makes it possible to estimate the probability of one or more hypothesis variable(s) given the evidence provided by a set of information variables. In the context of this work, the hypothesis variable is "the user is compromised," while the information variables are the alerts related to the user.

We modeled the problem using a naïve Bayesian network, that is, a network in which a single hypothesis node is connected to each information variable. It was assumed that no connections exist among the information variables. The structure of the network is shown in Figure 4.14. A naïve network estimates the probability of the hypothesis variable by assuming the independence of the information variables. In other words, given a system user, the presence of an alert in a bit vector does not affect the probability of observing the other alerts. A set of vectors (the ones composing the training set described later in the chapter) was used to validate the assumption of independence of information variables. For each combination of the alerts (A_i, A_j) (for $i, j \in \{1, 2, 3, \ldots, 14\}$), we counted how many vectors in the training set contained A_i, A_j, or both A_i and A_j. Then, the chi-squared test was applied to verify the null hypothesis H_0, that is,

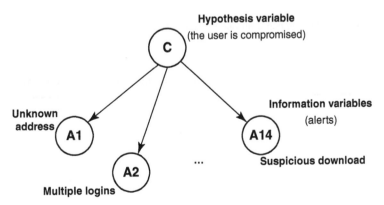

Figure 4.14 Adopted Bayesian network.

alerts A_i and A_j are independent of each other. Of the 91 possible combinations of alert pairs, in 28 cases (around 30% of the combinations) the chi-squared test (with a 95% significance level) indicated that H_0 had to be rejected. Regardless, in the majority of cases, the assumption of alerts' independence held.

It has to be noted that assuming the independence of a pair of dependent alerts leads to overestimation of the statistical evidence provided by the joint occurrence of the alerts. Thus, because the dependencies among the information variables are neglected, it might happen that the *a posteriori* probability of the hypothesis variable "the user is compromised" is greater than the real value. Consequently, the assumption of alerts' independence makes the analysis more conservative; however, more importantly, it does not compromise the detection capability of the adopted Bayesian network.

4.6.3.2 Training of the Bayesian Network

We used the logs collected for a subset of the available incidents, that is, 5 out of 16, as a training set for the network. Let T denote the adopted training set. T consists of 717 users (and corresponding bit vectors) and includes 6 compromised users. T contains the minimum number of training incidents, ensuring that almost all the alerts are covered by the actually compromised users, as shown in Table 4.3 (where *comp_i* denotes a compromised user). The strategy taken in this study has two main advantages: (1) It makes it possible to analyze the performance of the Bayesian network with conservative assumptions, that is, only a few training incidents are considered. (2) It does not bias the results, because all the adopted alerts are represented by the training set. Although other criteria might have been adopted, this selection of the training set was reasonable for our preliminary analysis of the performance of the proposed approach.

The training stage makes it possible to tune the network parameters, that is, (1) the *a priori probability* of the hypothesis variable, and (2) the *conditional probability table* (CPT) for each information variable A_i. The *a priori* probability

Table 4.3 Alerts related to the compromised users for the incidents in the training set.

Users	Alerts													
	A1	A2	A3	A4	A5	A6	A7	A8	A9	A10	A11	A12	A13	A14
comp_1			1									1	1	
comp_2	1		1							1	1			
comp_3		1		1			1					1	1	
comp_4	1				1									
comp_5				1		1							1	
comp_6				1										1

Table 4.4 Structure of the CPT for each alert A_i.

Alert (A_i)	Compromised (C)	
	True	False
True	$P(Ai|C)$	$P(Ai|\neg C)$
False	$P(\neg Ai|C)$	$P(\neg Ai|\neg C)$

of the hypothesis node C (Figure 4.14) is estimated as $P(C) = 6/717 = 0.008$. Calculation of CPTs requires additional effort. Four parameters must be estimated for each alert, as shown in Table 4.4. For example, $P(Ai|C)$ denotes the probability that an alert of type A_i is related to a particular user, given that the user is compromised. Similarly, $P(Ai|\neg C)$ represents the probability that an alert of type A_i is related to a user, given that the user is *not* compromised.

The probability values of the CPTs are estimated as follows. The overall number of users in the training set T is divided into two disjoint subsets: *good* and *compromised*. Let G and C denote the two subsets, respectively. Note that $|T| = |G| + |C|$. For each alert Ai, let Li be the set of users (good or compromised) in the training set T that exhibits that type of alert, for example, all the users in T responsible for a *command anomaly* alert. $P(Ai|C)$ is the ratio $|Li \cap C|/|C|$, that is, the cardinality of the intersection of Li and C divided by the cardinality of C. Similarly, $P(Ai|\neg C) = |Li \cap G|/|G|$, that is, the cardinality of the intersection of Li and G divided by the cardinality of G. $P(\neg A_i|C)$ and $P(\neg A_i|\neg C)$ are the complement to 1 of $P(Ai|C)$ and $P(Ai|\neg C)$, respectively.

Table 4.5 summarizes the obtained results. It can be noted that $P(Ai|C)$ assumes a relatively high value, which depends on the number of compromised users included in the training set T. Furthermore, $P(Ai|\neg C)$ is a measure of the quality of the alerts, in terms of the number of false positives they are likely to raise. For example, there is a high chance that an uncompromised user will be responsible for an *unknown address* (A1) or *HotClusterConn* (A4) alert. Similarly, the chances of observing a *watchlist* (A7) alert if the user is not compromised are extremely low.

Table 4.5 Values of $P(A_i|C)$ and $P(A_i|\neg C)$ computed for each alert.

	A1	A2	A3	A4	A5	A6	A7
$P(Ai|C)$	0.333	0.166	0.166	0.500	0.166	0.166	0.166
$P(Ai|\neg C)$	0.042	0.022	0.012	0.303	0.021	0.001	0.001
	A8	A9	A10	A11	A12	A13	A14
$P(Ai|C)$	0.166	0.001	0.166	0.166	0.001	0.500	0.833
$P(Ai|\neg C)$	0.019	0.001	0.011	0.012	0.001	0.240	0.527

By means of the proposed Bayesian network, given a user and the related vector of alerts, it is possible to perform the query "What is the probability $P(C)$ that the user will be compromised, given that the user is responsible for 0 or more alerts?" In the following, we analyze how $P(C)$ varies across the incidents and investigate the possibility of using the network as the *decision tool*.

4.6.4 Analysis of the Incidents

We used the incident data that were not used to train the network to assess the effectiveness of the proposed approach. A list of the incidents is given in the first column of Table 4.6. The analysis consisted in computing the probability $P(C)$ that each user logged into the NCSA machines during the day the given incident occurred. Table 4.6 (the last column) gives the $P(C)$ values computed for the actually compromised users. (Recall that the ground truth is known from the data.) The results of this analysis allow selection of a *classification threshold*, that is, a value for $P(C)$ that discriminates between compromised and uncompromised users.

4.6.4.1 Sample Incident
Example: Incident #5 (see Table 4.6) occurred on July 24, 2009. On the day the incident occurred, 476 users logged into the system. We estimated $P(C)$ for each bit vector (i.e., user) in the user/alerts table computed for this incident. Figure 4.15 gives a histogram of the numbers of users with respect to the

Table 4.6 List of the incidents and $P(C)$ values observed for the compromised users.

	Incident		
ID	Date	# Compromised users	$P(C)$ [a]
1	May 3, 2010	1	16.8%
2	Sep. 8, 2009	1	99.3%
3	Aug. 19, 2009	2	13.5%; 13.5%
4	Aug.13, 2009	1	4.3%
5	Jul. 24, 2009	1	28.5%
6	May 16, 2009	1	0.7%
7	Apr. 22, 2009	1	1.2%
8	Nov. 3, 2008	1	28.5%
9	Sep.7, 2008	3	99.7%; 99.9%; 76.8%
10	Jul. 12, 2008	1	44.6%
11	Jun. 19, 2008	1	18.1%

[a] Assumed by the compromised users.

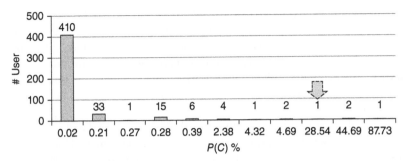

Figure 4.15 Example 1: Histogram of the number of users for the observed *P(C)* values (incident #5, occurred on July 24, 2009).

observed $P(C)$ values. For the majority of users (410 out of 476), the computed $P(C)$ is 0.02%. A closer look into the data reveals that none of these users were responsible for alerts observed in the security logs, and hence we can conclude that those users were *not compromised*.

In all the other cases, that is, 66 out of 476, at least one alert (in the security log) can be associated with each user. These 66 users are considered *potentially compromised*. For example, during the day the incident occurred, 33 users triggered a *multiple login* alert. However, 24 of those alerts were false positives caused by a training class that was using the GPU cluster at NCSA, for which all the users were logged in from the same IP address. Therefore, the presence of the *multiple login* alert alone does not provide strong enough evidence that a user is compromised, and the Bayesian network returns a small value of $P(C)$ (i.e., 0.21%). The compromised user in incident #5 exhibits a value of $P(C)$ around 28.54% (dotted arrow in Figure 4.15). An *unknown address* alert, a *multiple logins* alert, and a *command anomaly* alert are associated with the user.

We also observed cases in which a large value of $P(C)$ was computed for an uncompromised user. For example, a user that jointly triggered *unknown address*, *multiple login*, *unknown authentication*, and *anomalous host* alerts resulted in a probability value of 87.73%. Nevertheless, it has to be noted that the number of false indications produced by the Bayesian network is small. In incident #5, only for three uncompromised users was the $P(C)$ value greater than the one computed for the actually compromised user (to the right of the dotted arrow in Figure 4.15). For this incident, the proposed approach would have brought the number of manual investigations of potentially compromised users from 66 down to 4.

4.6.4.2 Discussion
The described analysis was conducted for each incident in the validation set. The analysis results reported in Table 4.6 reveal that $P(C)$, for different incidents, varies across a large range of values.

One can see that in all but two cases (incidents #6 and #7), the $P(C)$ values for the compromised users were relatively large. For incident #6, a single alert, *HotClusterConn* (with a sensitive extension), was associated with the compromised user. However, this alert, if observed alone, is quite unreliable. As shown in Figure 4.12 (A13), around 43 users per day are potentially responsible for this type of alert. Our Bayesian network returns a very small value for $P(C)$, 0.7%.

As discussed, most of the *suspicious* users, that is, the ones related to at least one alert, are not actually compromised: $P(C)$ is generally small. This finding suggests that the Bayesian network can be used to remove the noise (false positives) induced by the alerts. In other words, *it is feasible to define a classification threshold, that is, (C), that will allow suppression of a significant fraction of false positives while still identifying all compromised users.*

4.6.5 Supporting Decisions with the Bayesian Network Approach

In this section, we discuss how our Bayesian network can be used to discriminate between compromised and uncompromised users by means of a classification threshold: if the alerts related to a user result in a value of $P(C)$ greater than the classification threshold, we assume that the user is compromised.

4.6.5.1 Analysis of the Incidents

According to the results provided in Table 4.6, the *minimum* classification threshold that allows detection of all the compromised users is 0.7% (the value of $P(C)$ observed for the compromised user in incident #6). We used this value to quantify the effectiveness of the Bayesian network approach.

Table 4.7 summarizes the obtained results. Column 1 lists the IDs of incidents in the validation set. (The IDs are the same as in Table 4.6.) For each incident, column 2 reports the number of users that logged into the system during the day the incident occurred, and column 3 provides the number of *suspicious* users, that is, the ones related to at least one alert.

Column 4 gives the number of *actionable* users, that is, the subset of the suspicious users whose probability of having been compromised is ≥0.7%. For each incident, the ratio between the numbers of actionable and suspicious users quantifies the effectiveness of the proposed approach in reducing the number of false indications/positives. That ratio is reported in column 5. For example, in the worst case (represented by incident #3, in which 134 out of 312 suspicious users are actionable) the tool removed around 58% ((1−134/312)*100%) of false indications. As shown in the last row of Table 4.7, the average ratio (estimated across all the incidents) is around 0.20 when the classification threshold is set to 0.7%. In other words, for the analyzed data set, *the Bayesian network approach automatically removed around 80% of false positives (i.e., claims that an uncompromised user was compromised).*

Table 4.7 Analysis of the incidents with the Bayesian network approach.

Incident			TH. [≥0.7]		TH. [≥1.2]		TH [≥4.3]	
ID	#users	#susp	#actn.	ratio	#actn.	ratio	#actn.	ratio
1	305	252	41	0.16	32	0.12	19	0.07
2	477	122	29	0.23	27	0.22	14	0.11
3	353	312	134	0.42	39	0.12	33	0.10
4	309	203	34	0.16	28	0.13	13	0.06
5	476	66	11	0.16	11	0.16	7	0.10
6	491	7	1	0.14	0	0	0	0
7	446	201	41	0.20	37	0.18	10	0.05
8	447	251	62	0.24	50	0.20	43	0.17
9	497	422	137	0.32	85	0.20	58	0.13
10	193	118	9	0.07	3	0.02	3	0.02
11	380	280	3	0.01	3	0.01	2	0.01
avg.	398	221	50	0.20	29	0.12	19	0.07

The effectiveness of the proposed strategy is bounded by the need to select a relatively low threshold. In fact, 0.7% is a conservative value, which makes it possible to avoid false negatives (wherein actually compromised users are missed). We analyzed how the number of actionable users varies when the value of the classification threshold increases. Results are reported in columns 6 and 7, and in columns 8 and 9, of Table 4.7, for the threshold values 1.2 and 4.3%, respectively. According to Table 4.6, 1.2 and 4.3 are the next two smallest $P(C)$ values observed in the analysis. The compromised user of incident #6 went undetected when the threshold was 1.2%. However, the network removed around 88% of false positives. Similarly, when the classification threshold was set to 4.3%, the compromised users of incidents #6 and #7 were undetected. In that case, the network removed around 93% of false positives.

Analysis results also suggest that the Bayesian network approach can help system administrators by directing them toward the users that are likely to be compromised. After the security logs have been collected, the tool can be used to obtain a list of users exhibiting a particularly large value of $P(C)$, for example, all the users whose probability of being compromised is ≥4.3%. This procedure reduces the work burden of the administrators. As indicated in the last row of Table 4.7, on average, around 19 users out of 398, that is, only 4% of all users that log into NCSA during a normal day of operation, surpass the 4.3% classification threshold. If an actually compromised user is not detected (i.e., the classification

Table 4.8 Analysis of the borderline cases with the Bayesian network approach.

Event			TH. [≥0.7]			
Type (date)	#users	#susp	#actn.	*ratio*	P(C)max	P(C)^a)
ex. notif. #1 Apr. 21, 2009	386	176	26	0.15	4.6%	0.02%
ex. notif. #2 Mar. 18, 2009	269	179	63	0.35	96.3%	0.02%
ex. notif. #3 Feb.9, 2009	289	28	3	0.11	1.7%	0.4%
False positive Nov. 3, 2008	447	251	62	0.24	92.9%	92.9%
norm. day #1 Jun. 30, 2010	323	227	25	0.11	87.7%	—
norm. day #2 Jul. 25, 2010	154	88	28	0.32	41.7%	—
New incident Oct. 29, 2010	358	159	32	0.20	65%	9.4%

a) Assumed by the compromised users (if applicable).

threshold is set to a large value), the analyst can decrease the threshold in order to augment the set of possible suspicious users to be investigated.

4.6.5.2 Analysis of the Borderline Cases

We also assess the effectiveness of the network for *borderline* cases, such as incidents reported by third parties, or for *normal* days, that is, days on which no incidents occurred. In conducting the analysis, we used 0.7% as the classification threshold. The obtained results are summarized in Table 4.8. The meanings of columns 2, 3, 4, and 5 are the same as in Table 4.7. Furthermore, for each case (when applicable), columns 6 and 7 report the maximum observed $P(C)$ and the $P(C)$ for the compromised user, respectively. The main findings of the analysis are discussed in the following.

External Notifications

Three of the analyzed incidents were missed/undetected by the NCSA monitoring tools and were discovered only because of notifications from external sources, that is, third parties. Our tool was used to analyze the logs collected during the days when the undetected incidents occurred, and for each incident, the bit vectors in the user/alerts table were queried against the network. All compromised users were undetected with the proposed approach. The values of $P(C)$ for the compromised users are 0.02% (external notifications #1 and #2) and

0.4% (external notification #3), which are below the assumed classification threshold of 0.7%. In the first two cases, no alerts in the log seem to have been generated by the compromised users. In the third case, only the *command anomaly* alert was observed, and this alert, if observed alone, does not provide strong evidence that a user is really compromised. Since the Bayesian network approach relies on the low-level monitoring infrastructure, when *no alert is triggered*, it is not feasible to identify a compromised user.

False Positive

On November 3, 2008, the NCSA security team was alerted about a login performed by a user that triggered an *unknown address*, a *command anomaly*, an *unknown authentication*, and an *anomalous host* alert. The user/alerts table obtained from the logs collected during that day confirms the joint occurrence of the alerts. The computed value of $P(C)$ is 92.9%, and hence it is reasonable to assume that the user was compromised. However, system administrators contacted the owner of the account, who confirmed his/her activity and verified that the login was legitimate.

These cases reinforce our earlier finding that alert correlation improves our ability to identify compromised users; however, the deficiencies of the low-level monitoring infrastructure (missing events or false notifications) can produce misleading analysis results.

Normal Days

We queried the network with the user/alerts tables obtained from the logs collected during two normal days of operation, June 30, 2010 and July 25, 2010 (see Table 4.8). The numbers of *actionable* users, that is, the users whose probability of being compromised was $\geq 0.7\%$, were small (25 and 28, respectively). It can be noted that some users exhibited a high $P(C)$. For example, during the normal day #1 (June 30, 2010), the $P(C)$ was 87.7% for one user that jointly exhibited an *unknown address*, a *multiple login*, an *unknown authentication*, and an *anomalous host* alert. Again, the proposed strategy reduces the number of false indications due to untrusted alerts, but if the user is potentially responsible for multiple alerts, the $P(C)$ will be high even if the user is not actually compromised.

New Incident

We analyzed data collected during an incident that occurred on October 29, 2010. The incident is not included in either the training or the validation data set of the network. During the day when the incident occurred, 358 users logged into NCSA machines. Among them, 159 users raised alerts. The Bayesian network approach allowed us to reduce the initial set of 159 suspicious users to 32 actionable users. The compromised user was correctly included in the actionable set and detected. (Three alerts, that is, *unknown address, command anomaly*, and *HotClusterConn*, were raised by the compromised user, and the $P(C)$ was 9.4%).

4.6.6 Conclusion

We discussed a Bayesian network approach to support the detection of compromised users in shared computing infrastructures. The approach has been validated by means of real incident data collected during 3 years at NCSA. The results demonstrate that the Bayesian network approach is a valuable strategy for driving the investigative efforts of security personnel. Furthermore, it is able to significantly reduce the number of false positives (by 80%, with respect to the analyzed data). We also observed that the deficiencies of the underlying monitoring tools could affect the effectiveness of the proposed network.

4.7 Integrating Attribute-Based Policies into Role-Based Access Control

While dynamic (active) and passive monitoring techniques provide runtime awareness of system activities, access control plays an essential role in preventing potentially malicious actors from entering a system at its "entrance gate."

Toward that end, in this section we discuss a framework that uses attribute-based access control (ABAC) policies to create a more traditional role-based access control (RBAC). RBAC has been widely used but has weaknesses: It is labor-intensive and time-consuming to build a model instance, and a pure RBAC system lacks flexibility to efficiently adapt to changing users, objects, and security policies. In particular, it is impractical to manually make (and maintain) user-to-role assignments and role-to-permission assignments in an industrial context characterized by a large number of users and/or security objects.

Here we discuss a new approach that integrates into RBAC attribute-based policies that were designed to support RBAC model building for large-scale applications. We modeled RBAC in two layers. One layer, called the *aboveground level*, is a traditional RBAC model extended with environment constraints. It retains the simplicity of RBAC and allows routine operations and policy review. In the second layer, called the *underground layer*, we focus on how to construct attribute-based policies to automatically create the primary RBAC model on the aboveground level. This second layer adopts the advantages of ABAC, eases the difficulty of RBAC model building (particularly for large-scale applications), and provides flexibility to adapt to dynamic applications. Thus, the proposed approach combines the advantages of RBAC and ABAC.

Prior work (e.g., Ref. [35,36]) focused on rule-based automatic user-role assignment, addressing the difficulty of user-role assignment when there is a large, dynamic population of users. We extended this by defining attribute-based policies for both user-role assignment and role-permission assignment. This extension addresses the challenge posed by the large numbers of security objects found in industrial control systems (ICS). Much of the prior research considered

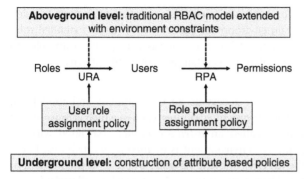

Figure 4.16 A two-layered framework integrating attribute-based policies into RBAC.

only the attributes of users; we considered the attributes of users, roles, objects, and the environment.

We refer the interested reader to our earlier publications [29,37] for more details on the material covered in this section.

4.7.1 Framework Description

In this section, we present the two-layered framework that integrates attribute-based policies into RBAC. Figure 4.16 outlines the architecture of the framework.

4.7.2 Aboveground Level: Tables

We use first-order logic to make formal descriptions and follow the convention that all unbound variables are universally quantified in the largest scope. The aboveground level is a simple and standard RBAC model but extended with constraints on attributes of the environment. We use the notion of an *environment* to represent the context of a user's access, such as the time of access, the access device, the system's operational mode, and so forth. The model is formally described as a tuple,

$$\mathcal{M} = (U, R, P, O, OP, EP, URA^e, RPA^e),$$

in which U is a set of *users*, where a user could be either a human being or an autonomous software agent. R is a set of *roles*, where a role reflects a job function and is associated with a set of permissions. O is a set of *objects*, which are the resources protected by access control. OP is a set of *operators*, which represent a specific type of operations. P is a set of *permissions* that are defined by legitimate operations, each of which comprises an operator and an object. EP is a set of predefined *environment state patterns*. We use environment state to model the context of a user's access. Each environment state pattern (called *environment pattern* hereafter) defines a set of environment states. URA^e is the extended user-role assignment relation, which in essence is a mapping from users to roles, and is associated with certain environment patterns. RPA^e is the extended role-permission assignment relation, which in essence is a mapping from roles to permissions, and is also associated with certain environment patterns.

4.7.2.1 Environment

We represent the context of an access as an environment and model the environment as a vector of environment attributes, each of which is represented by an environment variable (called an *attribute name*) associated with an attribute value in a domain. An environment is defined by n attributes. Let $v_i \in D_i$, $i = 1, \ldots, n$, be the ith environment variable, where D_i is the domain of that environment variable. Then, a vector $(v_1, \ldots, v_n)$, in which all variables are instantiated, is called an *environment state* (denoted by s). The set of all possible environment states is denoted by E. The choice of environment attributes (and hence the environment state) is domain-dependent. Environment attributes, particularly the dynamic attributes, are gathered by an access control engine at runtime.

Example (environment state): Assume that the environment is defined by three attributes: mode, access location, and access time. Then, mode = "normal," access location = "station 1," and access time = "8:00 a.m. Monday" together is an environment state.

An environment pattern, denoted by e, is treated as a unique pattern in domain EP, but is semantically defined by a first-order logical expression of assertions involving environment attributes. An environment pattern defines a set of environment states in which every environment state satisfies the environment pattern, that is,

$$\{(v_1, \ldots, v_n) | e(v_1, \ldots, v_n)\}.$$

Hereafter, we will sometimes directly use e to denote the set of environment states defined by the environment pattern.

4.7.2.2 User-Role Assignments

A particular user-role assignment associates a user, a role, and an environment pattern:

$$URA^e \subseteq U \times R \times EP, \tag{4.1}$$

where EP is the set of all environment patterns that have been defined for the system of interest.

The semantics of a user-role assignment, $(u, r, e) \in URA^e$, is defined as

$$\text{match}(s, e) \rightarrow \text{has role}(u, r), \tag{4.2}$$

which states that if the real environment state s matches the given environment pattern e, then user u is assigned to role r. We assume that the RBAC engine can understand the semantics of each environment pattern as defined.

Basic RBAC models define user-role assignments simply as mappings from users to roles,

$$URA \subseteq U \times R. \tag{4.3}$$

We have extended this notion with a dependency on the environment, as a means of integrating certain extended RBAC features of context and constraints. The environment pattern associated with a user-role assignment is the environment-dependent condition that is sufficient for the assignment. This feature can be regarded as constrained user-role assignment. If there are no constraints on user-role assignments, the associated environment patterns are simply empty, so the model becomes the common one. The relation between URA^e and URA is given as

$$(u, r) \in URA \leftrightarrow (\exists e, (u, r, e) \in URA^e). \tag{4.4}$$

Following this formalism, an example user-role assignment with an environment extension may be expressed as follows:

User id: com:ab:zn1:amy; **Role id:** manager.zone1; **Environment pattern:** Device = "Station_1.2" & Time = "Weekday"
User id: com:ab:zn1:ben; **Role id:** operator.zone1; **Environment pattern:** Mode = "emergency"

4.7.2.3 Role-Permission Assignments

A role-permission assignment associates a role, a permission, and an environment pattern. Thus, the set of all such assignments is a subset,

$$RPA^e \subseteq R \times P \times EP. \tag{4.5}$$

The semantics of a role-permission assignment, $(r, p, e) \in RPA^e$, is defined as

$$\text{match}(s, e) \rightarrow \text{has permission}(r, p), \tag{4.6}$$

which states that if the real environment state s matches the pattern e, then permission p is assigned to role r.

Similar to the approach for user-role assignments, we also extended the common role-permission assignment with environment patterns. The relation between RPA^e and RPA is given as

$$(r, p) \in RPA \leftrightarrow (\exists e, (r, p, e) \in RPA^e). \tag{4.7}$$

As with user-role assignments, the role-permission assignment may also be organized in tabular fashion.

4.7.3 Underground Level: Policies

The underground level of the RBAC model focuses on the security policies used to construct the aboveground level of the RBAC model. We have attempted to explicitly represent the implicit knowledge used to construct an RBAC model, and to integrate the extensions to standard RBAC models in an attribute-based paradigm.

In the following, we treat all users, roles, permissions, operators, and security objects as "objects" (in the sense of object-oriented design), each of which has certain attributes. The notation *obj.attr*, or equivalently *attr (obj)*, denotes the attribute *attr* of object *obj*.

The attributes needed in RBAC are typically domain-dependent and need to be customized for each specific target system. Some examples of attributes are as follows. The attributes of users may include "ID," "department," "security clearance," "knowledge domain," "academic degree," or "professional certificate." A role may have attributes such as "name"; "type," reflecting job function types such as "manager," "engineer," or "operator"; "security level"; "professional requirements"; and "direct superior roles" and "direct subordinate roles" (if role hierarchy is modeled). Objects may have attributes such "ID," "type," "security level," and "state." Operators may have attributes such as "name" and "type." The environment attributes may include "access time," "access location," "access application," "system mode," and "target value," among others.

4.7.3.1 Role-Permission Assignment Policy

The role-permission assignment policy is a set of rules. Each rule has the following structure:

```
rule id {
target {
role pattern; permission pattern {
operator pattern;
object pattern;
};
environment pattern;
} condition;
decision.
}
```

where all of the patterns and the condition are FOL (first-order logic) expressions; an environment pattern defines a set of environment states; a role pattern defines a set of roles by specifying their common attributes; a permission pattern consisting of an operator pattern and one object pattern defines a set of permissions by specifying their common features with respect to the attributes of the operator and the object in a permission; an operator pattern defines a set of operators (or operation types); each object pattern defines a set of objects; the target defines the range of (role, permission, environment pattern) triples to which this rule applies; and the condition is a logical expression defining a relation among the attributes of the roles, the permissions (operators and objects), and the environment. Rules of this form state that when the condition is true, a role covered by the role pattern can be assigned with a permission covered by the permission pattern in the specified environment pattern.

4.7.3.2 User-Role Assignment Policy

User-role assignment is highly dependent on business rules and constraints. In our view, the task of assigning users to roles can be approached like the role-permission assignment problem, in terms of policies that enforce those rules and constraints. Such policies would be formulated in terms of user and role attributes, and would be crafted to enforce things like separation of duty. However, unlike role-permission assignment, user assignment may have to balance competing or conflicting policy rules against each other. Correspondingly, a complete policy-oriented formulation will need to specify how to combine rules and arrive at a final assignment decision. In what we present below, an attribute-based user-role assignment policy is used only to identify potential assignments. A rule-combining algorithm is used for making the final assignments.

The user-role assignment rule is similar to the role-permission assignment rule and consists of

```
rule id {
target {
user pattern;
role pattern; environment pattern;
}
condition;
decision.
}
```

where a user/role/environment pattern defines a set of users/roles/environment states by specifying the common attributes; all of the patterns and the condition are expressions in first-order logic; and (unlike the case in the role-permission assignment policy) the decision of a rule marks a (user, role, environment pattern) triple as a potential assignment.

4.7.4 Case Study: Large-Scale ICS

This section explains how the proposed framework might be applied to construction of an RBAC model for an industrial control system (ICS).

Problem: The target application domain of ICS has the following features. There are a very large number of security objects (on the order of millions), among which there are complex relations; on the other hand, many objects and operations applied on them are similar, and there are patterns to follow. Security objects are organized in hierarchical structures. Each role or security object may have a security level. Users dynamically change over time as business changes; security objects also change over time because of device replacement or maintenance. Access to control processes and devices is through some human–machine interfaces and software applications. Each protected point of access to a control system, called a *point* or *control block*, contains information about the status of a control process or device and is used to set target control values or control parameters; all those points are important parts of the security objects to be protected by the target access control system. The runtime operation environment (with dynamic attributes), for example, access location and/or access time, is a sensitive and important factor in access control. Different zones have similar structures, that is, roles, operations, and objects are similar in different zones. Zones and devices may have operation "modes." Finally, control stations play an important role in access control.

A specific challenge we face is that of constructing an RBAC framework that is compatible with the access control mechanisms used in modern ICS, which are a mixture of different mechanisms that include station-based, group-based, attribute-based, and (simplified) lattice-based mechanisms. The underlying reason for the compatibility requirement is the practical need for an incremental transition path. Another major challenge is that of how to define roles and assign fine-grained permissions to roles for a large-scale application effectively, efficiently, and in an automated fashion, as much as possible.

4.7.4.1 RBAC Model-Building Process

In the following, we briefly explain how to apply the proposed framework in building an RBAC model instance for an ICS. More detailed discussion of RBAC for ICS can be found in Ref. [32].

a) *Identify security objects and object hierarchy*: In a plant, security engineers identify devices and points to be protected by RBAC as a set of security objects and organize those objects in hierarchical structures (usually an acyclic graph). In ICS, it is common to find that the control devices and other assets involved in monitoring and controlling the physical process and in overall management of a plant are organized in a hierarchical and modular manner as per ISA-88 and ISA-95 standards. This hierarchical structure that groups objects is called an *object hierarchy* in this chapter.

b) *Identify operation types*: The types of the operations applied to each security object are identified, for example, a specific type of control parameter may be of type "read" and "reset." At a minimum, an operation type could be defined by just an operator; in general, an operation type can be defined by a permission pattern.

c) *Identify role templates*: ICS tend to have very well-defined job functions, with well-defined access needs, to monitor and control the physical processes. For example, "Operator" and "Engineer" are very well-understood job functions in the ICS community, even though there might be some variations in their functions across different types of ICS. In the world of electric power, "Operator" is actually a licensed position. Such well-defined job functions with well-defined access needs are good candidates for use in defining "roles." However, in practice, users who perform such well-defined job functions are limited in scope to either particular subprocesses or particular geographical areas, in accordance with the organization of their plants. In other words, the job functions identified in ICS, such as "Operator" and "Engineer," perform certain types of operations; the objects to which those operations can apply are dependent on the attributes of the users and those objects. From this point of view, we defined in our previous work [25,32] a role template and proto-permission that leverage some common characteristics of ICS to simplify the creation and management of roles. Here we show how our framework can be used to represent these concepts and realize a manageable RBAC solution for ICS. Unlike a permission that comprises a pair of an operator and an object, a proto-permission consists of an operator and the object type of the operand. Thus, a proto-permission represents just an operation type, rather than a specific operation on a specific object. Proto-permission is a specific form of permission pattern. Unlike a "role" in RBAC, a *role template* is associated with a set of proto-permissions. A role template tells what types of operations can be performed by a role created from the role template. Let RT denote the set of all role templates, and let PP denote the set of proto-permissions. Every role template has an attribute (or, equivalently, function) of a proto-permission set, denoted by *pps*; every proto-permission has an attribute (or function) of an operator, and an attribute (or function) of an object type, denoted by *objType*.

We use the role template to formally represent each well-defined job function in an ICS, and use proto-permissions to represent each allowed operation type associated with a role template. Assume that the plant has a number of basic types of job functions, such as "Operator," "Engineer," and "Manager." They are identified as role templates. Each of them is associated with a set of operation types, for example, "Engineer" has operation type "reset parameter" on objects of type "XYZ"; "Manager" has operation type "view schedule" on "System"; and so forth. All identified operation types should be covered by role templates.

d) *Identify roles and their privilege ranges*: Based on business workflow analysis, a number of roles can be identified for each role template. Each role has an assigned working (access) environment pattern, such as access through station B in zone 1 in daytime. Furthermore, as discussed earlier, in practice each role has access only to a certain range of objects; this range of accessible objects is called the *privilege range* of the role. Formally,

$$pr : R \rightarrow 2^O. \tag{4.8}$$

A privilege range is used to define the boundary of objects for which a role is responsible, and is used as constraints in role-permission assignment. A role has access to an object only if the object is within the role's privilege range. A privilege range can be defined over an object hierarchy. An object hierarchy (denoted by *OH*) is simply a subset of the power set of all objects considered, that is, $OH \subseteq 2^O$. A node in the object hierarchy is called an *object group*, which is a subset of the objects, that is, $og \in 2^O$. In an object hierarchy, if one object group is a child of another in *OH*, it means that the former is a subset of the latter.

e) *Analyze security policies and identify attributes*: The security engineers analyze all security policies and requirements applied to the plant and list all of the attributes of users, roles, objects, and the access environment. For example, privilege range is a critical attribute of a role. A particular attribute that needs to be considered is *security level*. It is common to assign a security object a "security level," requiring a subject of access to have a corresponding security level.

f) *Develop attribute-based policies*: Security engineers construct the underground level of an RBAC model by developing attribute-based policies for role-permission assignment and user-role assignment.

Let us consider a simple yet general case: (1) A role can perform only the types of operations specified by the proto-permissions of the role's template. (2) There is a privilege range constraint, such that a role can access an object only if that object is within the privilege range of that role. (3) There is a security-level constraint, in that to access an object, the role's security level needs to be greater than or equivalent to the one of the object. This can be represented with the attribute-based policy for role-permission assignment:

```
rule{
target:{} condition:{
memberOf(o, r.pr);
r.securityLevel >= o.securityLevel; memberOf(pp, r.template.pps)
 and op = pp.op and o.type = pp.objType;} decision: add
(r, p(op,o), \phi) in RPAe.
}
```

g) *Create RBAC assignment tables*: Use the specified attribute-based policies to create role-permission assignment and user-role assignment in the aboveground level.

This task can be illustrated by considering how the above example rule would be used to make a role-permission assignment. For each pair of role and permission, if the pair or the role's assigned working environment did not match the target of the rule, then this pair would be skipped; otherwise, the condition part of the rule would continue to be evaluated. If the condition is found to be true, then the permission would be assigned to the role in that assigned environment, in the role-permission assignment table of the above-ground level. Consider a role "Engineer Chem Zone1 Daytime" and a permission "reset parameter T" on object "point 1.2.7" (i.e., it represents the seventh point in zone 1 sector 2), in the environment pattern stated in the example rule. This (role, permission) pair is within the target. Assume that object "point 1.2.7" is within the privilege range of the role and the access station. If the role's security level dominates the object's, and the professional domains match, then the permission is assigned to the role in the specified environment pattern. Consider another role, "Engineer Chem Zone2 Daytime," which has privilege in zone 2 that does not cover "point 1.2.7"; thus, permission would not be granted to it for this role.

h) *Repeat the process*: If the above logical verification and review failed, it would be necessary to go back to an earlier step to revise the RBAC model, and then verify and review it again.

4.7.4.2 Discussion of Case Study

Migration to RBAC from a legacy system could be a great challenge for ICS. The proposed framework could support both building of an RBAC model for ICS and the migration. We highlight some major features as follows.

Expressibility

The proposed framework is general enough to cover the required features of the targeted access control systems in ICS [30]. User groups are modeled by role template, and the types of operations conducted by a user group are modeled by proto-permissions. Station-based access constraints, application access constraints, temporal constraints, "mode" constraints, "parameter range" constraints, and others are modeled as environment constraints.

Support for Role Engineering

Role engineering is widely recognized as difficult. It is an even greater challenge for ICS due to the large-scale and dynamic features of the system. The proposed framework enables automatic user-role assignment and role-permission assignment, through attribute-based polices. This approach could be a great help in overcoming the problems of manual user-role assignment and role-permission assignment for large-scale applications in ICS.

Simplicity

The proposed framework can integrate the existing mechanisms and concepts in ICS uniformly in the form of attribute-based policies, thus avoiding the complexity that would be caused by ad hoc representation and management. Attribute-based policies can express security policies and requirements in a straightforward manner, making them easier to construct and maintain. The simplicity of the aboveground level eases RBAC model review.

Flexibility
The attribute-based policies have the flexibility to adapt to dynamically changing users, objects, security policies and requirements, and even business processes.

Verifiability
The logic representation of the attribute-based policies provides a basis for formal verification of an RBAC model.

4.7.5 Concluding Remarks

We discussed an approach we developed to combine ABAC and RBAC, bringing together the advantages of both models. We developed our model in two levels: aboveground and underground. The aboveground level is a simple and standard RBAC model extended with environment constraints, which retains the simplicity of RBAC and supports straightforward security administration and review. In the underground level, we explicitly represent the knowledge for RBAC model building as attribute-based policies, which are used to automatically create the simple RBAC model in the aboveground level. The attribute-based policies bring the advantages of ABAC: They are easy to build and easy to change for a dynamic application. We showed how the proposed approach can be applied to RBAC system design for large-scale ICS applications.

4.8 The Future

Cloud computing allows users to obtain scalable computing resources, but with a rapidly changing landscape of attack and failure modes, the effort to protect these complex systems is increasing. As we discussed in this chapter, cloud computing environments are built with VMs running on top of a hypervisor, and VM monitoring plays an essential role in achieving resiliency. However, existing VM monitoring systems are frequently insufficient for cloud environments, as those monitoring systems require extensive user involvement when handling multiple operating system (OS) versions. Cloud VMs can be heterogeneous, and therefore the guest OS parameters needed for monitoring can vary across different VMs and must be obtained in some way. Past work involves running code inside the VM, which may be unacceptable for a cloud environment.

We envisage that this problem will be solved by recognizing that there are common OS design patterns that can be used to infer monitoring parameters from the guest OS. We can extract information about the cloud user's guest OS with the user's existing VM image and knowledge of OS design patterns as the only inputs to analysis. As a proof of concept, we have been developing VM monitors by applying this technique. Specifically, we implemented sample monitors that include a return-to-user attack detector and a process-based keylogger detector.

Another important aspect of delivering robust and efficient monitoring and protection against accidental failures and malicious attacks is our ability to validate (using formal and experimental methods) the detection capabilities of the proposed mechanisms and strategies. Toward that end we require development of validation frameworks that integrate the use of tools such as model checkers (for formal analysis and symbolic execution of software) and fault/attack injectors (for experimental assessment).

Further exploration of all these ideas is needed to make *Reliability and Security as a Service* an actual offering from cloud providers.

References

1 Hernandez, P., Skype, AWS outages rekindle cloud reliability concerns, *eWeek*, Sep. 22, 2015. Available at http://www.eweek.com/cloud/skype-aws-outages-rekindle-cloud-reliability-concerns.html.

2 2015 Trustwave Global Security Report, Trustwave Holdings, Inc., 2015. Available at https://www.trustwave.com/Resources/Library/Documents/2015-Trustwave-Global-Security-Report/.

3 Garfinkel, T. and Rosenblum, M. (2003) A virtual machine introspection based architecture for intrusion detection, in Proceedings of the 10th Network and Distributed System Security Symposium, pp. 191–206. Available at http://www.isoc.org/isoc/conferences/ndss/03/proceedings/.

4 Payne, B.D., Carbone, M.D.P.de A., and Lee, W. (2007) Secure and flexible monitoring of virtual machines, in Proceedings of the 23rd Annual Computer Security Applications Conference, pp. 385–397.

5 Payne, B.D. *et al.* (2008) Lares: an architecture for secure active monitoring using virtualization, in Proceedings of the 2008 IEEE Symposium on Security and Privacy, pp. 233–247.

6 Jones, S.T., Arpaci-Dusseau, A.C., and Arpaci-Dusseau, R.H. (2008) VMM-based hidden process detection and identification using Lycosid, in Proceedings of the 4th ACM SIGPLAN/SIGOPS International Conference on Virtual Execution Environments, pp. 91–100.

7 Sharif, M.I. *et al.* (2009) Secure in-VM monitoring using hardware virtualization, in Proceedings of the 16th ACM Conference on Computer and Communications Security, pp. 477–487.

8 Pham, C., Estrada, Z., Cao, P., Kalbarczyk, Z., and Iyer, R.K. (2014) Reliability and security monitoring of virtual machines using hardware architectural invariants, Proceedings of the 44th Annual IEEE/IFIP International Conference on Dependable Systems and Networks, pp. 13–24.

9 Carbone, M. *et al.* (2014) VProbes: deep observability into the ESXi hypervisor. *VMware Technical Journal*, **3** (1). Available at https://labs.vmware .com/vmtj/vprobes-deep-observability-into-the-esxi-hypervisor.

10 Jones, S.T., Arpaci-Dusseau, A.C., and Arpaci-Dusseau, R.H. (2006) Antfarm: tracking processes in a virtual machine environment, in Proceedings of the USENIX Annual Technical Conference, pp. 1–14. Available at https://www .usenix.org/legacy/events/usenix06/tech/full_papers/jones/jones.pdf.

11 Azab, A.M., Ning, P., and Zhang, X. (2011) SICE: a hardware-level strongly isolated computing environment for x86 multi-core platforms, in Proceedings of the 18th ACM Conference on Computer and Communications Security, pp. 375–388.

12 Liu, Z. *et al.* (2013) CPU transparent protection of OS kernel and hypervisor integrity with programmable DRAM. *ACM SIGARCH Computer Architecture News*, **41** (3), 392–403.

13 Zhou, Z. *et al.* (2012) Building verifiable trusted path on commodity x86 computers, in Proceedings of the 2012 IEEE Symposium on Security and Privacy (SP), pp. 616–630.

14 Quynh, N.A. and Suzaki, K. (2007) Xenprobes, a lightweight user-space probing framework for Xen Virtual Machine, in Proceedings of the 2007 USENIX Annual Technical Conference, Available at https://www.usenix.org/ legacy/events/usenix07/tech/full_papers/quynh/quynh.pdf.

15 Arnold, J. and Kaashoek, M.F. (2009) Ksplice: automatic rebootless kernel updates, in Proceedings of the 4th ACM European Conference on Computer Systems, pp. 187–198.

16 Vaughan-Nichols, S.J., No reboot patching comes to Linux 4.0, ZDNet, March 3, 2015. Available at http://www.zdnet.com/article/no-reboot-patching-comes-to-linux-4-0/.

17 Estrada, Z.J., Pham, C., Deng, F., Yan, L., Kalbarczyk, Z., and Iyer, R.K., Dynamic VM dependability monitoring using hypervisor probes, in Proceedings of the 2015 11th European Dependable Computing Conference, pp. 61–72.

18 Payne, B.D. (2012) Simplifying Virtual Machine Introspection Using LibVMI, Sandia Report SAND2012-7818, Sandia National Laboratories. Available at http://prod.sandia.gov/techlib/access-control.cgi/2012/127818.pdf.

19 Bishop, M. (1989) A model of security monitoring, in Proceedings of the 5th Annual Computer Security Applications Conference, pp. 46–52.

20 Manadhata, P.K. and Wing, J.M. (2011) An attack surface metric. *IEEE Transactions on Software Engineering*, **37** (3), 371–386.

21 Krishnakumar, R. (2005) Kernel korner – kprobes: a kernel debugger. *Linux Journal*, **2005** (133), 1–11.

22 Feng, W. *et al.* (2007) High-fidelity monitoring in virtual computing environments, in Proceedings of the ACM International Conference on Virtual Computing Initiative, Research Triangle Park, NC.

23 Kivity, A. *et al.* (2007) kvm: the Linux virtual machine monitor, in Proceedings of the Linux Symposium, vol. 1, pp. 225–230.

24 Vattikonda, B.C., Das, S., and Shacham, H. (2011) Eliminating fine grained timers in Xen, in Proceedings of the 3rd ACM Workshop on Cloud Computing Security, pp. 41–46.

25 Li, P., Gao, D., and Reiter, M.K. (2013) Mitigating access-driven timing channels in clouds using StopWatch, in Proceedings of the 43rd Annual IEEE/ IFIP International Conference on Dependable Systems and Networks, pp. 1–12.

26 Gilbert, M.J. and Shumway, J. (2009) Probing quantum coherent states in bilayer graphene. *Journal of Computational Electronics*, **8** (2), 51–59.

27 Agesen, O. *et al.* Software techniques for avoiding hardware virtualization exits, in Proceedings of the 2012 USENIX Annual Technical Conference, pp. 373–385. Available at https://www.usenix.org/system/files/conference/atc12/ atc12-final158.pdf.

28 Larson, S.M. *et al.* (2002) Folding@ home and genome@ home: using distributed computing to tackle previously intractable problems in computational biology, in *Computational Genomics* (ed. R. Grant), Horizon Press.

29 Wang, G., Estrada, Z.J., Pham, C., Kalbarczyk, Z., and Iyer, R.K. (2015) Hypervisor Introspection: a technique for evading passive virtual machine monitoring, in Proceedings of the 9th USENIX Workshop on Offensive Technologies, Available at https://www.usenix.org/node/191959.

30 Varadarajan, V., Ristenpart, T., and Swift, M. (2014) Scheduler-based defenses against cross-VM side-channels, in Proceedings of the 23rd USENIX Security Symposium, pp. 687–702. Available at https://www.usenix.org/system/files/ conference/usenixsecurity14/sec14-paper-varadarajan.pdf.

31 Pecchia, A., Sharma, A., Kalbarczyk, Z., Cotroneo, D., and Iyer, R.K. (2011) Identifying compromised users in shared computing infrastructures: a data-driven Bayesian network approach, in Proceedings of the IEEE 30th International Symposium on Reliable Distributed Systems, pp. 127–136.

32 Sharma, A., Kalbarczyk, Z., Barlow, J., and Iyer, R. (2011) Analysis of security data from a large computing organization, in Proceedings of the IEEE/IFIP 41st International Conference on Dependable Systems and Networks, pp. 506–517.

33 Jensen, F.V. (2001) *Bayesian Networks and Decision Graphs*, Springer, New York, NY.

34 Pearl, J. (1988) *Probabilistic Reasoning in Intelligent Systems: Networks of Plausible Inference*, Morgan Kaufmann, San Francisco, CA.

35 Al-Kahtani, M.A. and Sandhu, R. (2002) A model for attribute-based user-role assignment, in Proceedings of the Annual Computer Security Applications Conference. Available at https://www.acsac.org/2002/papers/95.pdf.

36 Kern, A. and Walhorn, C. (2005) Rule support for role-based access control, in Proceedings of the 10th ACM Symposium on Access Control Models and Technologies, pp. 130–138.

37 Huang, J., Nicol, D.M., Bobba, R., and Huh, J.H. (2012) A framework integrating attribute-based policies into role-based access control, in Proceedings of the 17th ACM Symposium on Access Control Models and Technologies, pp. 187–196.

5

Scalability, Workloads, and Performance: Replication, Popularity, Modeling, and Geo-Distributed File Stores

Roy H. Campbell,[1] Shadi A. Noghabi,[1] and Cristina L. Abad[2]

[1]*Department of Computer Science, University of Illinois at Urbana-Champaign, Urbana, IL, USA*
[2]*Escuela Superior Politecnica del Litoral, ESPOL, Guayaquil, Ecuador*

This chapter explores the problems of scalability of cloud computing systems. Scalability allows a cloud application to change in size, volume, or geographical distribution while meeting the needs of the cloud customer. A practical approach to scaling cloud applications is to improve the availability of the application by replicating the resources and files used; this includes creating multiple copies of the application across many nodes in the cloud. Replication improves availability through use of redundant resources, services, networks, file systems, and nodes, but also creates problems with respect to clients' ability to observe consistency as they are served from the multiple copies. Variability in data sizes, volumes, and the homogeneity and performance of the cloud components (disks, memory, networks, and processors) can impact scalability. Evaluating scalability is difficult, especially when there is a large degree of variability. That leads to the need to estimate how applications will scale on clouds based on probabilistic estimates of job load and performance. Scaling can have many different dimensions and properties. The emergence of low-latency worldwide services and the desire to have higher fault tolerance and reliability have led to the design of geo-distributed storage with replicas in multiple locations. At the end of this chapter, we consider scalability in terms of the issues involved with cloud services that are geo-distributed and also study, as a case example, scalable geo-distributed storage.

5.1 Introduction

Cloud computing system scalability has many dimensions, including size, volume, velocity, and geographical distribution, which must be handled while continuing to meet the needs of the cloud customer. Here, we will address scalability and the related design issues in a number of steps, covering, in our opinion, the most important current issues. First, we address the size and volume scalability problem by examining how many replicas to allocate for each file in a cloud file system and where to place them, using probabilistic sampling and a

Assured Cloud Computing, First Edition. Edited by Roy H. Campbell, Charles A. Kamhoua, and Kevin A. Kwiat.

competitive aging algorithm independently at each node. We discuss a statistical metadata workload model that captures the relevant characteristics of a workload (the attributes of its metadata, i.e., directories, file sizes, and number of files) and is suitable for synthetic workload generation. Then, we examine traces of file access in real workloads, characterizing popularity, temporal locality, and arrival patterns of the workloads. In particular, we show how traces of workloads from storage, feature animation, and streaming media can be used to derive synthetic workloads that may be used to help design cloud computing file systems.

Next, we introduce and analyze a set of complementary mechanisms that enhance workload management decisions for processing MapReduce jobs with deadlines. The three mechanisms we consider are the following: (i) a policy for job ordering in the processing queue; (ii) a mechanism for allocating a tailored number of map and reduce slots to each job with a completion time requirement; and (iii) a mechanism for allocating and deallocating (if necessary) spare resources in the system among the active jobs.

Finally, we examine a solution to building a geo-distributed cloud storage service giving, as an example, an implementation on which we collaborated: LinkedIn's Ambry. This solution offers an entirely decentralized replication protocol, eliminating any leader election overheads, bottlenecks, and single points of failure. In this geo-distributed system case, data replication and the consistency of data are simplified by making the data in the storage immutable (written once and never modified). This makes the behavior of the system easier to analyze with respect to the issues of scalability. Other chapters in this volume examine the problems of consistency and correctness of design in storage that is not immutable.

5.2 Vision: Using Cloud Technology in Missions

The success of cloud-based applications and services has increased the confidence and willingness of federal government organizations to move mission-critical applications to the cloud. A mission-critical application is typically one that is essential to the successful operation of that government organization and can involve timeliness, availability, reliability, and security. Unfortunately, the situation is unpredictable and ever-evolving as more cloud solutions are adopted and more cloud applications are developed. *Scalability* of applications and services on the cloud is a key concern and refers to the ability of the system to accommodate larger loads just by adding resources either vertically by making hardware perform better (scale-up) or horizontally by adding additional nodes (scale-out). Further, as clouds become more ubiquitous and global, geo-distributed storage and processing become part of the scaling issue. This chapter's vision is to consider the tools and methodologies that can underlie the test and design of cloud computing applications to make it possible to build them more reliably, and potentially to guide solutions that scale within the limitations of the resources of the cloud. We discuss how these tools can guide the development of geo-distributed storage and processing for clouds. Such tools and methodologies would enhance the ability to create mission-oriented cloud applications and services and help accelerate the adoption of cloud solutions by the government.

The concept of *assured cloud computing* encompasses our ability to provide computation and storage for mission-oriented applications and services. This involves the design, implementation, and evaluation of dependable cloud architectures that can provide assurances with respect to security, reliability, and timeliness of computations or services. Example applications include dependable big data applications; data analytics; high-velocity, high-volume stream processing; real-time computation; control of huge cyber-physical systems such as power systems; and critical computations for rescue and recovery.

Scalability concerns accommodation of larger loads through addition of resources. *Elasticity* is the ability to best fit the resources needed to cope with loads dynamically as the loads change. Typically, elasticity relates to scaling out appropriately and efficiently. When the load increases, the system should scale by adding more resources, and when demand wanes, the system should shrink back and remove unneeded resources. Elasticity is mostly important in cloud environments in which, on the one hand, customers who pay per use don't want to pay for resources they do not currently need, and on the other hand, it is necessary to meet demands when they rise.

When a platform or architecture *scales*, the hardware costs increase linearly with demand. For example, if one server can handle 50 users, 2 servers can handle 100 users, and 10 servers can handle 500 users. If, every time a thousand users were added to a system, the system needed to double the number of servers, then it can be said that the design does *not* scale, and the service organization would quickly run out of money as the user count grew.

Elasticity is how well your architecture can adapt to a changing workload in real time. For example, if one user logs on to a website every hour, this could be handled by one server. However, if 50,000 users all log on at the same time, can the architecture quickly (and possibly automatically) provision new Web servers on the fly to handle this load? If so, it can be said that the design is *elastic*.

Vertical scaling can essentially resize your server with no change to your code. It is the ability to increase the capacity of existing hardware or software by adding resources. Vertical scaling is limited by the fact that you can get only as big as the size of the server. *Horizontal scaling* affords the ability to scale wider to deal with traffic. It is the ability to connect multiple hardware or software entities, such as servers, so that they work as a single logical unit. This kind of scaling cannot be implemented at a moment's notice.

The availability, reliability, and dependability of applications and services are one aspect of providing mission-oriented computation and storage. The variability of demand for those services is another aspect, and it depends on the success of the system in meeting concerns of scalability and elasticity. For example, in an emergency scenario, a set of rescue services might need to be elastically scaled from a few occurrences a day to hundreds or thousands. Once the emergency is over, can the resources used to provide the services be quickly released to reduce costs? Similarly, scalability may determine whether an existing e-mail system can be expanded to coordinate the activities of a new six-month task force organized to cope with a rescue mission. *Scalable, elastic, mission-oriented cloud-based systems require understanding of what the workload of the system will be under the new circumstances and whether the architecture of the system allows that workload to operate within its quality of service or service-level agreements.* The difficulty here is that example workloads might not be available until the applications are deployed and are in use. The difficulty is compounded when the application has a huge number of users and its timeliness and availability are very sensitive to the mission it is accomplishing. Understanding the nature of the workloads and the ability to create synthetic workloads and volume and stress testing benchmarks that can be used in testing and performance measurement become crucial elements for system planners, designers, and implementers. Workloads and synthetic workloads can identify bottlenecks, resource constraints, high latencies, and what-if demand scenarios while allowing comparisons. An example of how synthetic and real workloads help in real system deployment is given in Ref. [1], which describes how synthetic workloads can be scaled and used to predict the performance of different cluster sizes and hardware architectures for MapReduce loads. However, in some applications that involve personal or private data, example workloads may not even be available, making synthetic workloads essential. Representing those workloads by synthetic traces that do not have private or personal information may then be the only way to evaluate designs and test performance.

Our goal in this chapter, then, is to describe technology that makes it possible to construct scalable systems that can meet the missions to which they are assigned.

5.3 State of the Art

Many of the experimental methodologies for evaluating large-scale systems are surveyed in Ref. [2]. As the systems evolve, these methodologies also change. The evaluations often use example workloads, traces, and synthetic traces. Much effort has gone into the collection of workloads and synthetic traces for cloud computing, mainly because researchers studying the performance of new cloud solutions have not had easy access to appropriate data sets [3–5]. Many of the larger cloud companies, including Google, Microsoft, Twitter, and Yahoo!, have published actual and trace-based workloads that do not have private or personal information in an effort to help research. In our research, Twitter and Yahoo! permitted University of Illinois students who interned with those companies to build trace-based workloads that do not have private or personal information from more sensitive internal workload data sets [4,6]. The research described in this chapter has built on the existing body of work and examines many scalability issues as well as the production of traces.

5.4 Data Replication in a Cloud File System

Cloud computing systems allow the economical colocation of large clusters of computing systems, fault-tolerant data storage [7,8], frameworks for data-intensive applications [9,10], and huge data sets. *Data locality*, or placing of data as close as possible to computation, is a common practice in the cloud to support high-performance, data-intensive computation economically [11,12]. Current cluster computing systems use uniform data replication to ensure data availability and fault tolerance in the event of failures [13–18], to improve data locality by placing a job at the same node as its data [9], and to achieve load balancing by distributing work across the replicas. Data locality is beneficial as the amount of data being processed in data centers keeps growing at a tremendous pace, exceeding increases in the available bandwidth provided by the network hardware in the data centers [19].

Our research introduced DARE [7], a distributed data replication and place-ment algorithm that adapts to changes in workload. We assume that the scheduler used is one that is oblivious to the data replication policy, such as the first-in, first-out (FIFO) scheduler or the Fair scheduler in Hadoop systems, so our algorithm will be compatible with existing schedulers. We implemented and evaluated our algorithm using the Hadoop framework [10], Apache's open-source implementation of MapReduce [9]. In the tested implementation, when local data are not available, a node retrieves data from a remote node in order to process the assigned task and discards the data once the task is completed. The algorithm takes advantage of existing remote data retrievals and selects a subset of the data to be inserted into the file system, hence creating a replica without consuming extra network and computation resources.

Each node runs the algorithm independently to create replicas of data that are likely to be heavily accessed during a short period of time. Observations from a 4000-node production Yahoo! cluster log indicate that the popularity of files follows a heavy-tailed distribution [20]. This makes it possible to predict file popularity from the number of accesses that have already occurred; for a heavy-tailed distribution of popularity, the more a file has been accessed, the more future accesses it is likely to receive.

From the point of view of an individual data node, the algorithm comes down to quickly identifying the most popular set of data and creating replicas for this set. Popularity means that a piece of data receives not only a large *number* of accesses but also a high *intensity* of accesses. We observe that this is the same as the problem of heavy hitter detection in network monitoring: In order to detect flows occupying the largest bandwidth, we need to identify flows that are both *fast* and *large*. In addition, the popularity of data is relative: We want to create replicas for files that are *more* popular than others. Hence, algorithms based on a hard threshold of number of accesses do not work well.

We designed a probabilistic dynamic replication algorithm with the following features:

1) Each node sample assigned tasks and uses the ElephantTrap [21] structure to replicate popular files in a distributed manner. Experiments on dedicated Hadoop clusters and virtualized EC2 clusters showed more than sevenfold improvement of data locality for the FIFO scheduler and 70% improvement for the Fair scheduler. Our algorithm, when used with the Fair scheduler – which increases locality by introducing a small delay when a job that is scheduled to run cannot execute a local task, allowing other jobs to launch tasks instead – can lead to locality levels close to 100% for some workloads.

2) Data with correlated accesses are distributed over different nodes as new replicas are created and old replicas expire. This helps data locality and reduces job turnaround time by 16% in dedicated clusters and 19% in virtualized public clouds. Job slowdown is reduced by 20% and 25%, respectively.

3) By taking advantage of existing remote data retrievals, the algorithm incurs no extra network usage. Thrashing is minimized through use of sampling and a competitive aging algorithm, which produces data locality comparable to that of a greedy least recently used (LRU) algorithm but with only 50% of the disk writes of the latter.

The contribution of the research is twofold. First, our analysis examined existing production systems to obtain effective bandwidth and data popularity distributions, and to uncover characteristics of access patterns. Second, we proposed the distributed dynamic data replication algorithm, which significantly improves data locality and task completion times.

5.4.1 MapReduce Clusters

MapReduce clusters [9,10] offer a distributed computing platform suitable for data-intensive applications. MapReduce was originally proposed by Google, and its most widely deployed implementation, Hadoop, is used by many companies, including Facebook, Yahoo!, and Twitter.

MapReduce uses a divide-and-conquer approach in which input data are divided into fixed-size units processed independently and in parallel by *map* tasks. The *map* tasks are executed in a distributed manner across the nodes in the cluster. After the *map* tasks are executed, their output is shuffled, sorted, and then processed in parallel by one or more *reduce* tasks.

To provide cost-effective, fault-tolerant, fast movement of data into and out of the compute nodes, the compute nodes use a distributed file system (GFS [22] for Google's MapReduce and HDFS [16] for Hadoop).

MapReduce clusters use a master–slave design for the compute and storage systems. For the sake of simplicity, we will use the HDFS terminology to refer to the components of the distributed file system, where *name node* refers to the master node and *data node* refers to the slave. The master file system node handles the metadata operations, while the slaves handle the reads/writes initiated by clients. Files are divided into fixed-sized blocks, each stored at a different HDFS data node. Files are read-only, but appends may be performed in some implementations.

MapReduce clusters use a configurable number of replicas per file (three by default). While this replica policy makes sense for availability, it is ineffective for locality and load balancing when access patterns of data are not uniform.

As it turns out, the data access patterns (or popularity distributions) of files in MapReduce clusters are not uniform; it is common for some files to be much more popular than others (e.g., job configuration files during initial job stages), while some may be significantly unpopular (e.g., old log files are rarely processed). For job files, popularity can be predicted (e.g., launching a job creates a hotspot), so a solution adopted in currently implemented systems is to have the framework automatically increase the replication factor for these files [9,16]. For other cases, the current approach is to manually increase or decrease the number of replicas for a file by using organization heuristics based on data access patterns. For example, Facebook dereplicates aged data, which can have a lower number of replicas (as low as one copy) than other data [3]. The manual approach described above is not scalable and can be error-prone.

Within this chapter, the term *file* denotes the smallest granularity of data that can be accessed by a MapReduce job. A file is composed of N fixed-size data *blocks* (of 64–256 MB).

File Popularity

The data access distribution, or file popularity, of data in a MapReduce cluster can be nonuniform, as our analysis of a Yahoo! cluster [20] shows [7]. Figure 5.1, which illustrates data generated during that analysis, shows a heavy-tailed distribution in which some files are significantly more popular than others. In this case, the reason is that the cluster is used mainly to perform different types of analysis on a common (time-varying) data set. Similar results were obtained by analyzing the 64-node CCT Hadoop production cluster, and have previously been observed by the developers of Microsoft's Bing [11]. This suggests that a uniform increase in the number of replicas is not an adequate way of improving locality and achieving load balancing.

In our observations, we found that around 80% of the accesses of a file occur during its first day of life. In our analysis, typically, 50% of a file's accesses occur

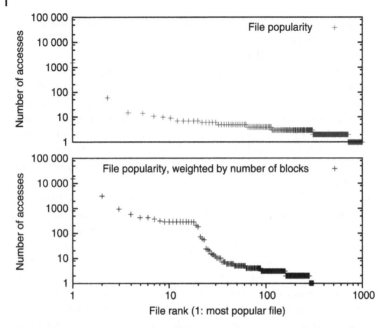

Figure 5.1 Number of accesses per file. Files are ranked according to their popularity (measured first by the number of accesses to the file and second by the number of accesses weighted by the number of 128 MB blocks in the file.)

within 10 h following its creation. A similar finding has been presented by Fan *et al.* [23]; it was obtained from the Yahoo! M45 research cluster, for which the authors found that 50% of the accesses of a block occurred within 1 min after its creation.

Given those observations, the research goal has been to create an adaptive replication scheme that seeks to increase data locality by replicating "popular" data while keeping a minimum number of replicas for unpopular data. In addition, the scheme should (i) dynamically adapt to changes in file access patterns, (ii) use a replication budget to limit the extra storage consumed by the replicas, and (iii) impose a low network overhead.

The solution described in DARE is a greedy reactive scheme that takes advantage of existing data retrievals to avoid incurring any extra network traffic [7], as follows: When a map task is launched, its data can be local or remote to the node (i.e., located in a different node). In the case of remote data, the original MapReduce framework fetches and processes the data, without keeping a local copy for future tasks. With DARE, when a map task processes remote data, the data are inserted into the HDFS at the node that fetched them.

Data are replicated at the granularity of a block. DARE uses a replication budget to limit the extra storage consumed by the dynamically replicated data.

The budget is configurable, but a value between 10% and 20% proved reasonable. To avoid completely filling the storage space assigned to dynamically created replicas, an eviction mechanism can employ an LRU or least frequently used (LFU) strategy to free up storage. Whether a file should be cached locally is probabilistically determined using a scheme similar to the ElephantTrap [21]. If cached, the number of replicas for the data is automatically increased by one, without incurring explicit network traffic. Because file caching is determined probabilistically, the algorithm is more stable and avoids the possibility of thrashing when the cache is full and eviction is required.

In summary, data access patterns in MapReduce clusters are heavy-tailed, with some files being considerably more popular than others. For nonuniform data access patterns, current replication mechanisms that replicate files a fixed number of times are inadequate, can create suboptimal task locality, and hinder the performance of MapReduce clusters. DARE [7] is an adaptive data replication mechanism that can improve data locality by more than seven times for a FIFO scheduler and 70% for the Fair scheduler, without incurring extra networking overhead. Turnaround time and slowdown are improved by 19% and 25%, respectively. The scheme is scheduler-agnostic and can be used in parallel with other schemes, such as Zaharia *et al.*'s delay scheduling [12], that aim to improve locality.

Several efforts have dealt with the specific case of dynamic replication in MapReduce clusters, including CDRM [24] and Scarlett [11]. In Ref. [24], Wei *et al.* presented CDRM, a "cost-effective dynamic replication management scheme" for cloud storage clusters. CDRM is a replica placement scheme for Hadoop that aims to improve file availability by centrally determining the ideal number of replicas for a file, along with an adequate placement strategy based on the blocking probability. The effects of increasing locality are not studied. In this chapter, we consider the case of maximizing the (weighted) overall availability of files given a replication budget and a set of file class weights. We propose an autonomic replication number computation algorithm that assigns more replicas for the files belonging to the highest-priority classes and fewer replicas for files in lower-priority classes, without exceeding the replication budget. Parallel to our work, Ananthanarayanan *et al.* [11] proposed Scarlett, an offline system that replicates blocks based on their observed probability in a previous epoch. Scarlett computes a replication factor for each file and creates budget-limited replicas distributed throughout the cluster with the goal of minimizing hotspots. Replicas are aged to make space for new replicas. While Scarlett uses a proactive replication scheme that periodically replicates files based on predicted popularity, we proposed a reactive approach that is able to adapt to popularity changes at smaller time scales and can help alleviate recurrent as well as nonrecurrent hotspots. Zaharia *et al.*'s delay scheduling [17] increases locality by delaying, for a small amount of time, a map task that – without the delay – would have run nonlocally. DARE is scheduler-agnostic and can work together with this and other scheduling techniques that try to increase locality. The delay scheduling technique is currently part of Hadoop's Fair scheduler, one of the two schedulers used in our evaluations.

5.4.1.1 File Popularity, Temporal Locality, and Arrival Patterns

The growth of data analytics for big data encourages the design of next-generation storage systems to handle peta- and exascale storage requirements. As demonstrated by DARE [7], a better understanding of the workloads for big data becomes critical for proper design and tuning. The workloads of enterprise storage systems [25], Web servers [26], and media server clusters [27] have been extensively studied in the past. There have been several studies of jobs and the workload created by jobs in big data clusters [28,29] but few storage-system-level studies [30]. A few recent studies have provided us with some limited insight on the access patterns in MapReduce scenarios [7,11,23]. However, these have been limited to features of interest to the researchers for their specific projects, such as block age at time of access [23] and file popularity [7,11]. Parallel to that work, other researchers did a large-scale characterization of MapReduce workloads, including some insights on data access patterns [31]. Their work concentrates on interactive query workloads and does not study the batch type of workload used in many production systems. Furthermore, the logs they processed are those of the Hadoop scheduler, and for this reason do not provide access to information such as the age of the files in the system, or the time when a file was deleted.

In the work we described in a prior study [3], we explored a frequently used application of big data storage clusters: those that are dedicated to supporting a mix of MapReduce jobs. Specifically, we studied the file access patterns of two multipetabyte Hadoop clusters at Yahoo! across several dimensions, with a focus on popularity, temporal locality, and arrival patterns. We analyzed two 6-month traces, which together contained more than 940 million creates and 12 billion file open events.

We identified unique properties of the workloads and made the following key observations:

- Workloads are dominated by high file churn (a high rate of creates/deletes), which leads to 80–90% of files' being accessed at most 10 times during a 6-month period.
- A small percentage of files are highly popular: less than 3% of the files account for 34–39% of the accesses (opens).
- Young files account for a high percentage of accesses, but a small percentage of bytes stored. For example, 79–85% of accesses target files that are at most one day old, yet add up to 1.87–2.21% of the bytes stored.
- The observed request interarrivals (opens, creates, and deletes) are bursty and exhibit self-similar behavior.

- The files are very short-lived: 90% of the file deletions target files that are 22.27 min to 1.25 h old.

Derived from those key observations and a knowledge of the domain and application-level workloads running on the clusters, we highlight the following insights and implications for storage system design and tuning:

- The peculiarities observed are mostly derived from the short lives of the files and the high file churn.
- File churn is a result of typical MapReduce workflows; a high-level job is decomposed into multiple MapReduce jobs, which are arranged in a directed acyclic graph (DAG). Each of these (sub)jobs writes its final output to the storage system, but the output that interests the user is the output of the last job in the graph. The output of the (sub)jobs is deleted soon after it is consumed.
- The high rates of change in file popularity whereby a small number of files are very popular, interarrivals are bursty, and files are short-lived prompt research on appropriate storage media and tiered storage approaches.
- Caching young files or placing them on a fast storage tier could lead to performance improvement at a low cost.
- "Inactive storage" (due to data retention policies and dead projects) constitutes a significant percentage of stored bytes and files; timely recovery of files and appropriate choice of replication mechanisms and media for passive data can lead to improved storage utilization.
- Our findings call for a model of file popularity that accommodates dynamic change.

Perhaps the work most similar to ours (in approach) is that of Cherkasova and Gupta [27], who characterized enterprise media server workloads. An analysis of the influence of new files and file life span was made, but they did not possess file creation and deletion time stamps, so they consider a file to be "new" the first time it is accessed and its lifetime to "end" the last time it is accessed. No analysis on the burstiness of requests was made.

Our work complements prior research by providing a better understanding of one type of big data workload: filling of gaps at the storage level. The workload characterization, key observations, and implications for storage system design are important contributions. More studies of big data storage workloads and their implications should be encouraged so that storage system designers can validate their designs, and deployed systems can be properly tuned [30].

For the case of the workloads we studied, the analysis demonstrated how traditional popularity metrics (e.g., the percentage of the file population that accounts for 90% of the frequency counts – in this case, accesses) can be misleading and make it harder to understand what those numbers imply about the popularity of the population (files). In our analysis, the problem arose from the high percentage of short-lived (and thus infrequently accessed) files. New or adapted models and metrics are needed to better express popularity under these conditions.

The high rate of change in file populations has some interesting implications for the design of the storage systems: Does it make sense to handle the short-lived files in the same way as longer-lived files? Tiered storage systems that combine different types of storage media for different types of files can be tailored to these workloads for improved performance. While the burstiness and autocorrelations in the request arrivals may be a result of typical MapReduce workloads in which multiple tasks are launched within some small time window (where all of the tasks are operating on different parts of the same large file or set of related files), a characterization of the autocorrelations is relevant independent of the MapReduce workload that produced them, for the following reasons.

- Such a characterization allows researchers to reproduce the workload in simulation or real tests without having to use an application workload generator (e.g., Apache GridMix [29] or SWIM [31] for MapReduce). This is useful because current MapReduce workload generators execute MapReduce jobs on a real cluster, meaning that researchers who lack access to a large cluster cannot perform large-scale studies that could otherwise be performed at the simulation level.
- Current MapReduce workload generators (and published models) have overlooked the data access patterns, so their use in evaluating a storage system would be limited.
- Some of the autocorrelations present may also be evident in other big data workloads, such as bag-of-tasks parallel jobs in high-performance computing (HPC). If that is the case, our characterization (and future models that could be proposed) could be useful for designers of storage systems targeted at the HPC community.

5.4.1.2 Synthetic Workloads for Big Data

Our research that used data analytics to investigate the behavior of big data systems [3,7] allowed us to propose new algorithms and data structures to improve their performance when Hadoop and HDFS are used. In general, it is difficult to obtain real traces of systems. Often, when data are available, the traces must be de-identified to be used for research. However, workload generation can often be used in simulations and real experiments to help reveal how a system reacts to variations in the load [32]. Such experiments can be used to validate new designs, find potential bottlenecks, evaluate performance, and do capacity planning based on observed or predicted workloads. Workload generators can replay real traces or do model-based synthetic workload generation. Real traces capture observed behavior and may even include nonstandard or undiscovered (but possibly important) properties of the load [33]. However, real trace-based approaches treat the workload as a "black box" [32]. Modifying a particular workload parameter or dimension is difficult, making such approaches inappropriate for sensitivity and what-if analysis. Sharing of traces can be hard because of their size and privacy concerns. Other problems include those of scaling to a different system size and describing and comparing traces in terms that can be understood by implementers [33].

Model-based synthetic workload generation can be used to facilitate testing while modifying a particular dimension of the workload, and can model expected future demands. For that reason, synthetic workload generators have been used extensively to evaluate the performance of storage systems [4,33], media streaming servers [34,35], and Web caching systems [32,36]. Synthetic workload generators can issue requests on a real system [32,33] or generate synthetic traces that can be used in simulations or replayed on actual systems [35,37].

Our research on this topic [38] focused on synthetic generation of object request streams, where the object can be of different types depending on context, like files [4], disk blocks [33], Web documents [36], and media sessions [35].

Two important characteristics of object request streams are popularity (access counts) and temporal reference locality (i.e., the phenomenon that a recently accessed object is likely to be accessed again in the near future) [35]. While highly popular objects are likely to be accessed again soon, temporal locality can also arise when the interarrival times are highly skewed, even if the object is unpopular [39].

For the purpose of synthetic workload generation, it is desirable to simultaneously reproduce the access counts and the request interarrivals of each individual object, as both of these dimensions can affect system performance. However, single-distribution approaches – which summarize the behavior of different types of objects with a single distribution per dimension – cannot accurately reproduce both at the same time. In particular, the common practice of collapsing the per-object interarrival distributions into a single system-wide distribution (instead of individual per-object distributions) obscures the identity of the object being accessed, thus homogenizing the otherwise distinct per-object behavior [37].

As big data applications lead to emerging workloads and these workloads keep growing in scale, the need for workload generators that can scale up the workload and/or facilitate its modification based on predicted behavior is increasingly urgent.

Motivated by previous observations about big data file request streams [3,23,40], we set the following goals for our model and synthetic generation process [41]:

- *Support for dynamic object populations:* Most previous models consider static object populations. Several workloads, including storage systems that support MapReduce jobs [3] and media server sessions [35], have dynamic populations with high object churn.

- *Fast generation:* Traces in the big data domain can be large (e.g., 1.6 GB for a 1-day trace with millions of objects). A single machine should be able to generate a synthetic trace modeled after the original one without suffering from memory or performance constraints.
- *Type awareness:* Request streams are composed of accesses to different objects (types), each of which may have distinct access patterns. We want to reproduce these access patterns.
- *Workload agnosticism:* The big data community is creating new workloads (e.g., key-value stores [42], and batch and interactive MapReduce jobs [3]). Our model should not make workload-dependent assumptions that may render it unsuitable for emerging workloads.

Our research [41] initially considered a stationary segment [33] of the workload. It used a model based on a set of delayed renewal processes (one per object in the stream) in which the system-wide popularity distribution asymptotically emerges through explicit reproduction of the per-object request arrivals and active span (time during which an object is accessed). However, this model is unscalable, as it is heavy on resources (needs to keep track of millions of objects).

Instead, we built a lightweight version of the model that uses unsupervised statistical clustering to identify groups of objects with similar behavior and significantly reduce the model space by modeling "types of objects" instead of individual objects. As a result, the clustered model is suitable for synthetic generation.

Our synthetic trace generator uses this lightweight model, and we evaluated it across several dimensions. Using a big data storage (HDFS [16]) workload from Yahoo!, we validated our approach by demonstrating its ability to approximate the original request interarrivals and popularity distributions. (The supremum distance between the real and synthetic cumulative distribution functions – CDFs – was under 2%.) Workloads from other domains included traces for ANIM and MULTIMEDIA. ANIM is a 24-h NFS trace from a feature animation company that supports rendering of objects, obtained in 2007 [43]. MULTI-MEDIA is a 1-month trace generated using the Medisyn streaming media service workload generator from HP Labs [35]. Both were also modeled successfully (with only a 1.3–2.6% distance between the real and synthetic CDFs). Through a case study in Web caching and a case study in the big data domain (on load in a replicated distributed storage system), we next showed how our synthetic traces can be used in place of the real traces (with results within 5.5% of the expected or real results), outperforming previous models.

Our model can accommodate the appearance and disappearance of objects at any time during the request stream (making it appropriate for workloads with high object churn) and is suitable for synthetic workload generation. Experiments have shown that we can generate a 1-day trace with more than 60 million object requests in under 3 min. Furthermore, our assumptions are minimal, since the renewal process theory does not require that the model be fit to a particular interarrival distribution, or to a particular popularity distribution.

In addition, the use of unsupervised statistical clustering leads to autonomic "type-awareness" that does not depend on expert domain knowledge or introduce human biases. The statistical clustering finds objects with similar behavior, enabling type-aware trace generation, scaling, and "what-if" analysis. (For example, in a storage system, what if the short-lived files were to increase in proportion to the other types of files?)

Concretely, our technical contributions [41] are (i) we provide a model based on a set of delayed renewal processes in which the system-wide popularity distribution asymptotically emerges through explicit reproduction of the per-object request interarrivals and active span; (ii) we use clustering to build a lightweight clustered variant of the model, suitable for synthetic workload generation; and (iii) we show that clustering enables workload-agnostic type-awareness, which can be exploited during scaling, what-if, and sensitivity analysis.

5.4.2 Related Work

Synthetic workload generators are a potentially powerful approach, and several synthetic workload generators have been proposed for Web request streams [32,36]. The temporal locality of requests is modeled using a stack distance model of references that assumes that each file is introduced at the start of the trace. Although this approach is suitable for static file populations, it is inadequate for populations with high file churn [35]. Our approach is a little more flexible in that it considers files with delayed introduction.

ProWGen [36] was developed to enable investigation of the sensitivity of Web proxy cache replacement policies to three workload characteristics: the slope of the Zipf-like document popularity distribution, the degree of temporal locality in the document request stream, and the correlation (if any) between document size and popularity. Instead of attempting to accurately reproduce real workloads, ProWGen's goal is to allow the generation of workloads that differ in one chosen characteristic at a time, thus enabling sensitivity analysis of the differing characteristics. Further, through domain knowledge of Web request streams, the authors note that a commonly observed workload is that of "one-timers," or files accessed only once in the request stream. One-timers are singled out as a special type of file whose numbers can be increased or decreased as an adjustment in relation to other types of files. In contrast, we were able to approximate the percentage of one-timers in the HDFS workload without explicitly modeling them. (Those approximations were 9.79% of the real workload and 10.69% of our synthetic trace, when the number of types of files was chosen to be 400.)

GISMO [34] and MediSyn [35] model and reproduce media server sessions, including their arrival patterns and per-session characteristics. For session arrivals, both generators have the primary goals of (i) reproducing the file popularity and (ii) distributing the accesses throughout the day based on observed diurnal or seasonal patterns (e.g., percentage of accesses to a file that occur during a specific time slot). In addition, MediSyn [35] uses a file introduction process to model accesses to new files, and explicitly considers two types of files that differ in their access patterns: regular files and news-like files. Our work allows the synthetic workload generation of objects with different types of behavior without prior domain knowledge.

In earlier work, we developed Mimesis [4], a synthetic workload generator for namespace metadata traces. While Mimesis is able to generate traces that mimic the original workload with respect to the statistical parameters included with it (arrivals, file creations and deletions, and age at time of access), reproducing the file popularity was left for future work.

Chen *et al.* [40] proposed the use of multidimensional statistical correlation (k-means) to obtain storage system access patterns and design insights at the user, application, file, and directory levels. However, the clustering was used for synthetic workload generation.

Hong *et al.* [38] used clustering to identify representative trace segments to be used for synthetic trace reconstruction, thus achieving trace compression ratios of 75–90%. However, the process of fitting trace segments, instead of individual files based on their behavior, neither facilitates deeper understanding of the behavior of the objects in the workload nor enables what-if or sensitivity analysis.

Ware *et al.* [37] proposed the use of two-level arrival processes to model bursty accesses in file system workloads. In their implementation, objects are files, and accesses are any system calls issued on a file (e.g., read, write, lookup, or create). Their model uses three independent per-file distributions: interarrivals to bursts of accesses, intraburst interarrival times, and distribution of burst lengths. A two-level synthetic generation process (in which burst arrivals are the first level, and intraburst accesses to an object are the second level) is used to reproduce bursts of accesses to a single file. However, the authors do not distinguish between the access to the first burst and the accesses to subsequent bursts and, as a consequence, are unable to model file churn. In addition, the authors use one-dimensional hierarchical clustering to identify bursts of accesses in a trace of per-file accesses. The trace generation process is similar to ours: one arrival process per file. However, the size of the systems they modeled (the largest ~567 files out of a total of 8000) did not require a mechanism to reduce the model size. We are considering systems two orders of magnitude larger, so a mechanism to reduce the model size is necessary. The approach of modeling intraburst arrivals independent of interburst arrivals can be combined with our delayed first arrival plus clustering of similar objects approach to capture per-file burstiness.

5.4.3 Contribution from Our Approach to Generating Big Data Request Streams Using Clustered Renewal Processes

The model we presented in Ref. [4] supports an analysis and synthetic generation of object request streams. The model is based on a set of delayed renewal processes, where each process represents one object in the original request stream. Each process in the model has its own request interarrival distribution, which, combined with the time of the first access to the object plus the period during which requests to the object are issued, can be used to approximate the number of arrivals or renewals observed in the original trace. Key contributions of this work in Ref. [33] include the following:

A lightweight version of the model that uses unsupervised statistical clustering to significantly reduce the number of interarrival distributions needed for representing the events, thus making the model suitable for synthetic trace generation.

The model is able to produce synthetic traces that approximate the original interarrival, popularity, and span distributions within 2% of the original CDFs. Through two case studies, we showed that the synthetic traces generated by our model can be used in place of the original workload and produce results that approximate the expected (real) results.

Our model is suitable for request streams with a large number of objects and a dynamic object population. Furthermore, the statistical clustering enables autonomic type-aware trace generation, which facilitates sensitivity and "what-if" analysis.

5.4.3.1 Scalable Geo-Distributed Storage

The emergence of low-latency worldwide services and the desire to have high fault tolerance and reliability necessitate geo-distributed storage with replicas in multiple locations. Social networks connecting friends from all around the world (such as LinkedIn and Facebook) or file sharing services (such as YouTube) are examples of such services. In these systems, hundreds of millions of users continually upload and view billions of diverse objects, from photos and videos to documents and slides. In YouTube alone, hundreds of hours of videos (approximately hundreds of GBs) are uploaded, and hundreds of thousands of hours of videos are viewed per minute by a billion people from all around the globe [44]. These objects must be stored and served with low latency and high throughput in a geo-distributed system while operating at large scale. Scalability in this environment is particularly important, both because of the growth in request rates and since data rarely get deleted (e.g., photos in your Facebook album).

In collaboration with LinkedIn, we developed "Ambry," a scalable geo-distributed object store [45]. For over 2.5 years, Ambry has been the main storage mechanism for all of LinkedIn's media objects across all four of its data centers, serving more than 450 million users. It is a production-quality system for storing large sets of immutable data (called *blobs*). Ambry is designed in a decentralized way and leverages techniques such as logical blob grouping, asynchronous replication, rebalancing mechanisms, zero-cost failure detection, and OS caching. Our experimental results show that Ambry reaches high throughput (reaching up to 88% of the network bandwidth) and low latency (serving 1 MB objects in less than 50 ms), works efficiently across multiple geo-distributed data centers, and improves the imbalance among disks by a factor of 8X–10X while moving minimal data.

In a geo-distributed environment, locality becomes a key. Inter-data-center links, going across the continent, are orders of magnitude slower and more expensive than intra-data-center links. Thus, moving data around becomes a significant and limiting factor, and data centers should be as independent as possible. To alleviate this issue, Ambry uses a mechanism called *asynchronous writes* to leverage data locality. In asynchronous writes, a put request is performed synchronously only among replicas of the local data center; that is, the data in the replicas are stored either in or as close by the data center receiving the request as possible. The request is counted as successfully finished at this point. Later on, the data are asynchronously replicated to other data centers by means of a lightweight background replication algorithm.

Ambry randomly groups its large immutable data or blobs into virtual units called *partitions*. Each partition is physically placed on machines according to a separate and independent algorithm. Thus, logical and physical placements are decoupled, making data movement transparent, simplifying rebalancing, and avoiding rehashing of data during cluster expansion [45].

A partition is implemented as an append-only log in a pre-allocated large file. Partitions are fixed-size during the lifetime of the system. The partition size is chosen so that the overhead of managing a partition, that is, the additional data structures maintained per partition, is negligible, and the time for failure recovery and rebuild is practical. Typically, 100 GB partitions are used. Rebuilding may be done in parallel from multiple replicas, which allows the 100 GB partitions to be rebuilt in a few minutes. In order to provide high availability and fault tolerance, each partition is replicated on multiple data store nodes through a greedy approach based on available disk space.

Ambry uses an entirely decentralized replication protocol, eliminating any leader election overheads, bottlenecks, and single points of failure. In this procedure, each replica individually acts as a master and syncs up with other replicas in an all-to-all fashion. Synchronization occurs using an asynchronous two-phase replication protocol. This protocol is *pull-based*, whereby each replica independently requests missing data from other replicas. It operates as follows:

- *First phase:* This phase finds any data that have been missing since the last synchronization point. Unique IDs are requested for all of the data written since the latest syncing offset. The data missing locally are then filtered out.
- *Second phase:* This phase replicates the missing data. A request for only the missing data is sent. Then, any missing data are transferred and appended to the replica.

In order to operate on a large scale and with geo-distribution, a system must be scalable. One main design principle in Ambry is to remove any master or manager. Ambry is a completely decentralized system with an active–active design, that is, data can be read or written from any of the replicas. However, load imbalance is inevitable during expansion (scale-out); new machines are empty while old machines have years-old and unpopular data. Ambry uses a non-intrusive rebalancing strategy based on popularity, access patterns, and size. This strategy uses spare network bandwidth to move data around in the background.

5.4.4 Related Work

The design of Ambry is inspired by log-structured file systems (LFS) [46,47]. These systems are optimized for write throughput in that they sequentially write in log-like data structures and rely on the OS cache for reads. Although these single-machine file systems suffer from fragmentation issues and cleaning overhead, the core ideas are very relevant, especially since blobs are immutable. The main differences between the ideas of a log-structured file system and Ambry's approach are the skewed data access pattern in Ambry's workload and a few additional optimizations used by Ambry, such as segmented indexing and Bloom filters.

In large file systems, metadata and small files need efficient management to reduce disk seeks [48] by using combinations of log-structured file systems (for metadata and small data), fast file systems (for large data) [49], and stores for the initial segment of data in the index block [50]. Our system resolves this issue by using in-memory segmented indexing plus Bloom filters and batching techniques.

Distributed File Systems
NFS [51] and AFS [52], GFS [22], HDFS [16], and Ceph [53] manage large amounts of data, data sharing, and reliable handling of failures. However, all these systems suffer from the high metadata overhead and additional capabilities (e.g., nested directories and permissions) unnecessary for a simple blob store. In many systems (e.g., HDFS, GFS, and NFS), a separate single metadata server increases metadata overhead, is a single point of failure, and limits scalability. Metadata [53] can be distributed or cached [54]. Although these systems reduce

the overhead for accessing metadata, each small object still has a large amount of metadata (usually stored on disk), decreasing the effective throughput of the system.

Distributed Data Stores

Key-value stores [8,55–57] handle a large number of requests per second in a distributed manner, but currently are unable to handle massively large objects (tens of MBs to GBs) efficiently, and add overhead to provide consistency that is unnecessary in Ambry. Certain systems [55–57] hash data directly to machines, and that can create large data movement whenever nodes are added/deleted.

PNUTS [58] and Spanner [59] are scalable, geographically distributed systems, where PNUTS maintains load balance as well. However, both systems provide more features and stronger guarantees than needed in a simple immutable blob store.

Blob Stores

A concept similar to that of partitions in Ambry has been used in other systems. Haystack uses logical volumes [60]; Twitter's blob store uses virtual buckets [6]; and the Petal file system introduces virtual disks [61]. Ambry can reuse some of these optimizations, like the additional internal caching in Haystack. However, neither Haystack nor Twitter's blob store tackle the problem of load imbalance. Further, Haystack uses synchronous writes across all replicas, impacting efficiency in a geo-distributed setting.

Facebook has also designed f4 [62], a blob store that uses erasure coding to reduce the replication factor of old data (that have become cold). Despite the novel ideas in this system, which potentially can be included in Ambry, our main focus is on both new and old data. Oracle's Database [63] and Windows Azure Storage (WAS) [64] also store mutable blobs, and WAS is even optimized for a geo-distributed environment. However, they both provide additional functionalities, such as support for many data types other than blobs, strong consistency guarantees, and modification to blobs, that are not needed in our use case.

5.4.5 Summary of Ambry

To summarize, our contributions with Ambry were as follows: (i) we designed and developed an industry-scale object store optimized for a geo-distributed environment, (ii) we minimized cross-data-center traffic by using asynchronous writes that update local data-center storage and propagate updates to storage at more remote sites in the background, (iii) we developed a two-phase background replication mechanism, and (iv) we developed a load-balancing mechanism that returns the system to a balanced state after expansion.

5.5 Summary

Clouds are complex systems involving difficult scalability concerns. Availability, workload considerations, performance, locality, and geo-distribution are some of the topics discussed in this chapter. We examine many of the key issues related to those topics that concern the design of applications and storage for clouds, and we show how knowledge of the behavior of a cloud is important in matching its services to user and cloud provider requirements. The scalability concerns of a cloud impact the behavioral analysis of its operation because example workloads may be enormous, involve large periods of time and huge volumes of data, process high-velocity data, or contain data for which there are privacy concerns. Building of tools that can aid in modeling of cloud workloads and be employed in the design of appropriate clouds is of considerable interest to researchers and developers.

5.6 The Future

This chapter discussed how, for example, the design of systems of applications in clouds in the social networking area is becoming influenced by extreme work-loads that involve billions of users and petabytes of data. In the future, cloud processing will continue to push the computation toward the data, an attribute we have seen earlier in many applications such as MapReduce and graph processing. *Edge computing* is an optimization for building large, potentially geographically distributed clouds by pushing data processing out to the edge of the networks, toward the source of the data. In our research, we have analyzed applications such as Ambry that allow cloud computing to stretch across the world and enable geographically distributed computation. Follow-on work after Ambry's development has examined stream processing solutions for cloud computing in the form of LinkedIn's Samza [65]. In Samza, information gathered from clients may consist of large, complex media objects that need low-latency data analysis. Samza reduces latency by not writing these media objects to long-term storage until they have been stream-processed. The reliability concerns for Samza include how to recover any data that are only stored in volatile random access memory if there is a fault (e.g., a power failure) in stream processing. Samza uses fault tolerance in the form of both redundant processing and fast restart to avoid lost data and minimize delay.

Looking to the future, the performance of cloud-based large machine learning computations is likely to become a major issue in cloud computing. In the few months prior to the time of this writing, production implementations of TensorFlow [66] and other deep-learning systems have come online at Google, Amazon, and Microsoft [67–69]. The models from these systems are used for inferencing in both clouds and local devices such as cell phones. These systems, when coupled with Edge learning systems or smartphones, form complex distributed learning systems that require performance analysis and evaluation. Ubiquitous sensors, autonomous vehicles that exchange state information about traffic conditions, and a host of close-to-real-time and health applications continue to expand the boundaries of cloud computing.

The future is exciting and difficult to predict. Regardless of whether edge computing will create "Cloudlet" solutions, how stream processing will influence design, or what new technology is used in these future cloud extensions, the techniques discussed in this chapter – measurement, modeling, and optimization based on performance – will still govern the design of such systems.

References

1 Verma, A., Cherkasova, L., and Campbell, R.H. (2014) Profiling and evaluating hardware choices for MapReduce environments: an application-aware approach. *Performance Evaluation*, **79**, 328–344.

2 Gustedt, J., Jeannot, E., and Quinson, M. (2009) Experimental methodologies for large-scale systems: a survey. *Parallel Processing Letters*, **19** (3), 399–418.

3 Abad, C.L., Roberts, A.N., Lu, A.Y., and Campbell, R.H. (2012) A storage-centric analysis of MapReduce workloads: file popularity, temporal locality and arrival patterns, in Proceedings of the IEEE International Symposium on Workload Characterization (IISWC), pp. 100–109.

4 Abad, C.L., Luu, H., Roberts, N., Lee, K., Lu, Y., and Campbell, R.H. (2012) Metadata traces and workload models for evaluating big storage systems, in Proceedings of the IEEE 5th International Conference on Utility and Cloud Computing (UCC), pp. 125–132.

5 Reiss, C., Tumanov, A., Ganger, G.R., Katz, R.H., and Kozuch, M.A. (2012) Heterogeneity and dynamicity of clouds at scale: Google trace analysis, Proceedings of the 3rd ACM Symposium on Cloud Computing, Article No. 7.

6 Twitter. (2012) "Blobstore: Twitter's in-house photo storage system." Available at https://blog.twitter.com/engineering/en_us/a/2012/blobstore-twitter-s-in-house-photo-storage-system.html (accessed March 2016).

7 Abad, C.L., Lu, Y., and Campbell, R.H. (2011) DARE: adaptive data replication for efficient cluster scheduling, in Proceedings of the IEEE International Conference on Cluster Computing, pp. 159–168.

8 Chang, F., Dean, J., Ghemawat, S., Hsieh, W.C., Wallach, D.A., Burrows, M., Chandra, T., Fikes, A. and Gruber, R.E. (2008) Bigtable: a distributed storage system for structured data. *ACM Transactions on Computer Systems*, **26** (2), Article No. 4.

9 Dean, J. and Ghemawat, S. (2004) MapReduce: simplified data processing on large clusters, in Proceedings of the USENIX Symposium on Operating Systems Design and Implementation (OSDI), pp. 137–150.

10 Apache Hadoop (2011) Available at http://hadoop.apache.org/ (accessed June 2011).

11 Ananthanarayanan, G., Agarwal, S., Kandula, S., Greenberg, A., Stoica, I., Harlan, D., and Harris, E. (2011) Scarlett: coping with skewed popularity content in MapReduce clusters, in Proceedings of the 6th European Conference on Computer Systems (EuroSys), pp. 287–300.

12 Zaharia, M., Borthakur, D., Sen Sarma, J., Elmeleegy, K., Shenker, S., and Stoica, I. (2010) Delay scheduling: a simple technique for achieving locality and fairness in cluster scheduling, in Proceedings of the 5th European Conference on Computer Systems (EuroSys), pp. 265–278.

13 Satyanarayanan, M. (1990) A survey of distributed file systems. *Annual Review of Computer Science*, **4**, 73–104.

14 Wei, Q., Veeravalli, B., Gong, B., Zeng, L., and Feng, D. (2010) CDRM: a cost-effective dynamic replication management scheme for cloud storage cluster, in Proceedings of the IEEE International Conference on Cluster Computing (CLUSTER), pp. 188–196.

15 Xiong, J., Li, J., Tang, R., and Hu, Y. (2008) Improving data availability for a cluster file system through replication, in Proceedings of the IEEE International Symposium on Parallel and Distributed Processing (IPDPS).

16 Shvachko, K., Kuang, H., Radia, S., and Chansler, R. (2010) The Hadoop distributed file system, in Proceedings of the IEEE Symposium on Mass Storage Systems and Technologies (MSST), pp. 1–10.

17 Ford, D., Labelle, F., Popovici, F.I., Stokely, M., Truong, V.-A., Barroso, L., Grimes, C., and Quinlan, S. (2010) Availability in globally distributed storage systems, in Proceedings of the 9th USENIX Symposium on Operating Systems Design and Implementation (OSDI). Available at https://www.usenix.org/legacy/event/osdi10/tech/.

18 Terrace, J. and Freedman, M.J. (2009) Object storage on CRAQ: high-throughput chain replication for read-mostly workloads, in Proceedings of the USENIX Annual Technical Conference. Available at https://www.usenix.org/conference/usenix-09/object-storage-craq-high-throughput-chain-replication-read-mostly-workloads.

19 Hey, T., Tansley, S., and Tolle, K. (eds.) (2009) *The Fourth Paradigm: Data-Intensive Scientific Discovery*, Microsoft Research.

20 "Yahoo! Webscope dataset ydata-hdfs-audit-logs-v1 0," direct distribution, February 2011. Available at https://webscope.sandbox.yahoo.com/catalog.php?datatype=s.

21 Lu, Y., Prabhakar, B., and Bonomi, F. (2007) ElephantTrap: a low cost device for identifying large flows, in Proceedings of the 15th Annual IEEE Symposium on High-Performance Interconnects (HOTI), pp. 99–108.

22 Ghemawat, S., Gobioff, H., and Leung, S.-T. (2003) The Google file system, in Proceedings of the 19th ACM Symposium on Operating Systems Principles (SOSP'03), pp. 29–43.

23 Fan, B., Tantisiriroj, W., Xiao, L., and Gibson, G. (2009) DiskReduce: RAID for data-intensive scalable computing, in Proceedings of the 4th Annual Workshop on Petascale Data Storage (PDSW), pp. 6–10.

24 Wei, Q., Veeravalli, B., Gong, B., Zeng, L., and Feng, D. (2010) CDRM: a cost-effective dynamic replication management scheme for cloud storage cluster, in Proceedings of the IEEE International Conference on Cluster Computing, pp. 188–196.

25 Chen, Y., Srinivasan, K., Goodson, G., and Katz, R. (2011) Design implications for enterprise storage systems via multi-dimensional trace analysis, in Proceedings of the 23rd ACM Symposium on Operating Systems Principles (SOSP), pp. 43–56.

26 Breslau, L., Cao, P., Fan, L., Phillips, G., and Shenker, S. (1999) Web caching and Zipf-like distributions: evidence and implications, in Proceedings of INFOCOM '99: 18th Annual Joint Conference of the IEEE Computer and Communications Societies, pp. 126–134.

27 Cherkasova, L. and Gupta, M. (2004) Analysis of enterprise media server workloads: access patterns, locality, content evolution, and rates of change. *IEEE/ACM Transactions on Networking*, 12 (5), 781–794.

28 Li, H. and Wolters, L. (2007) Towards a better understanding of workload dynamics on data-intensive clusters and grids, in Proceedings of the IEEE International Parallel and Distributed Processing Symposium (IPDPS), pp. 1–10.

29 Chen, Y., Ganapathi, A., Griffith, R., and Katz, R. (2011) The case for evaluating MapReduce performance using workload suites, in Proceedings of the IEEE 19th Annual International Symposium on Modelling, Analysis, and Simulation of Computer and Telecommunication Systems (MASCOTS), pp. 390–399.

30 Pan, F., Yue, Y., Xiong, J., and Hao, D. (2014) I/O characterization of big data workloads in data centers, in *Big Data Benchmarks, Performance Optimization, and Emerging Hardware: 4th and 5th Workshops, BPOE 2014, Salt Lake City, USA, March 1, 2014 and Hangzhou, China, September 5, 2014, Revised Selected Papers* (eds J. Zhan, R. Han, and C. Weng), LNCS, vol. 8807, Springer, pp. 85–97.

31 Chen, Y., Alspaugh, S., and Katz, R. (2012) Interactive analytical processing in big data systems: a cross-industry study of MapReduce workloads. *Proceedings of the VLDB Endowment*, 5 (12), 1802–1813.

32 Barford, P. and Crovella, M. (1998) Generating representative Web workloads for network and server performance evaluation. *ACM SIGMETRICS Performance Evaluation Review*, 26 (1), 151–160.

33 Tarasov, V., Kumar, K., Ma, J., Hildebrand, D., Povzner, A., Kuenning, G., and Zadok, E. (2012) Extracting flexible, replayable models from large block traces, in Proceedings of the 10th USENIX Conference on File and Storage Technologies (FAST). Available at http://static.usenix.org/events/fast/tech/.

34 Jin, S. and Bestavros, A. (2001) GISMO: a generator of Internet streaming media objects and workloads. *ACM SIGMETRICS Performance Evaluation Review*, **29** (3), 2–10.

35 Tang, W., Fu, Y., Cherkasova, L., and Vahdat, A. (2003) MediSyn: a synthetic streaming media service workload generator, in Proceedings of the 13th International Workshop on Network and Operating Systems Support for Digital Audio and Video (NOSSDAV), pp. 12–21.

36 Busari, M. and Williamson, C. (2002) ProWGen: a synthetic workload generation tool for simulation evaluation of web proxy caches. *Computer Networks*, **38** (6), 779–794.

37 Ware, P.P., Page, T.W. Jr., and Nelson, B.L. (1998) Automatic modeling of file system workloads using two-level arrival processes. *ACM Transactions on Modeling and Computer Simulation*, **8** (3), 305–330.

38 Hong, B., Madhyastha, T.M., and Zhang, B. (2005) Cluster-based input/output trace synthesis, in Proceedings of the 24th IEEE International Performance, Computing, and Communications Conference, pp. 91–98.

39 Fonseca, R., Almeida, V., Crovella, M., and Abrahao, B. (2003) On the intrinsic locality properties of Web reference streams, in Proceedings of IEEE INFOCOM 2003: 22nd Annual Joint Conference of the IEEE Computer and Communication Societies, vol. 1, pp. 448–458.

40 Chen, Y., Alspaugh, S., and Katz, R. (2012) Interactive analytical processing in big data systems: a cross-industry study of MapReduce workloads. *Proceedings of the VLDB Endowment*, **5** (12), 1802–1813.

41 Abad, C.L., Yuan, M., Cai, C.X., Lu, Y., Roberts, N. and Campbell, R.H. (2013) Generating request streams on Big Data using clustered renewal processes. *Performance Evaluation*, **70** (10), 704–719.

42 Cooper, B.F., Silberstein, A., Tam, E., Ramakrishnan, R., and Sears, R. (2010) Benchmarking cloud serving systems with YCSB, in Proceedings of the 1st ACM Symposium on Cloud Computing (SoCC), pp. 143–154.

43 Anderson, E. (2009) Capture, conversion, and analysis of an intense NFS workload, in Proceedings of the 7th USENIX Conference on File and Storage Technologies (FAST). Available at https://www.usenix.org/legacy/event/fast09/tech/.

44 YouTube, "Statistics." Available at https://www.youtube.com/yt/press/en-GB/statistics.html.

45 Noghabi, S.A., Subramanian, S., Narayanan, P., Narayanan, S., Holla, G., Zadeh, M., Li, T., Gupta, I., and Campbell, R.H. (2016) Ambry: LinkedIn's scalable geo-distributed object store. Proceedings of the International Conference on Management of Data, San Francisco, CA, pp. 253–265.

46 Rosenblum, M. and Ousterhout, J.K. (1992) The design and implementation of a log-structured file system. *ACM Transactions on Computer Systems (TOCS)*, **10** (1), 26–52.

47 Seltzer, M., Bostic, K., McKusick, M.K., and Staelin, C. (1993) An implementation of a log-structured file system for UNIX, in Proceedings of the Winter USENIX, pp. 307–326.

48 Ganger, G.R. and Kaashoek, M.F. (1997) Embedded inodes and explicit grouping: exploiting disk bandwidth for small files, in Proceedings of the USENIX Annual Technical Conference (ATC).

49 Zhang, Z. and Ghose, K. (2007) hFS: a hybrid file system prototype for improving small file and metadata performance, in Proceedings of the 2nd ACM SIGOPS/EuroSys European Conference on Computer Systems, pp. 175–187.

50 Mullender, S.J. and Tanenbaum, A.S. (1984) Immediate files. *Software: Practice and Experience*, **14** (4), 365–368.

51 Sandberg, R., Goldberg, D., Kleiman, S., Walsh, D., and Lyon, B. (1985) Design and implementation of the Sun network file system, in Proceedings of the USENIX Summer Technical Conference, pp. 119–130.

52 Morris, J.H., Satyanarayanan, M., Conner, M.H., Howard, J.H., Rosenthal, D.S., and Smith, F.D. (1986) Andrew: a distributed personal computing environment. *Communications of the ACM (CACM)*, **29** (3), 184–201.

53 Weil, S.A., Brandt, S.A., Miller, E.L., Long, D.D.E., and Maltzahn, C. (2006) Ceph: a scalable, high-performance distributed file system, in Proceedings of the 7th Symposium on Operating Systems Design and Implementation (OSDI), pp. 307–320.

54 Ren, K., Zheng, Q., Patil, S., and Gibson, G. (2014) IndexFS: scaling file system metadata performance with stateless caching and bulk insertion, in Proceedings of the International Conference on High Performance Computing, Networking, Storage and Analysis (SC), pp. 237–248.

55 Lakshman, A. and Malik, P. (2010) Cassandra: a decentralized structured storage system. *ACM SIGOPS Operating Systems Review*, **44** (2), 35–40.

56 Auradkar, A., Botev, C., Das, S., DeMaagd, D., Feinberg, A., Ganti, P., Gao, L., Ghosh, B., Gopalakrishna, K., Harris, B., Koshy, J., Krawez, K., Kreps, J., Lu, S., Nagaraj, S., Narkhede, N., Pachev, S., Perisic, I., Qiao, L., Quiggle, T., Rao, J., Schulman, B., Sebastian, A., Seeliger, O., Silberstein, A., Shkolnik, B., Soman, C., Sumbaly, R., Surlaker, K., Topiwala, S., Tran, C., Varadarajan, B., Westerman, J., White, Z., Zhang, D., and Zhang, J. (2012) Data infrastructure at LinkedIn, in Proceedings of the IEEE 28th International Conference on Data Engineering, pp. 1370–1381.

57 DeCandia, G., Hastorun, D., Jampani, M., Kakulapati, G., Lakshman, A., Pilchin, A., Sivasubramanian, S., Vosshall, P., and Vogels, W. (2007) Dynamo: Amazon's highly available key-value store, in ACM SIGOPS Operating Systems Review, vol. 41, no. 6, pp. 205–220

58 Cooper, B.F., Ramakrishnan, R., Srivastava, U., Silberstein, A., Bohannon, P., Jacobsen, H.-A., Puz, N., Weaver, D., and Yerneni, R. (2008) PNUTS: Yahoo!'s

hosted data serving platform. *Proceedings of the VLDB Endowment,* **1** (2), 1277–1288.

59 Corbett, J.C., Dean, J., Epstein, M., Fikes, A., Frost, C., Furman, J.J., Ghemawat, S., Gubarev, A., Heiser, C., Hochschild, P., Hsieh, W., Kanthak, S., Kogan, E., Li, H., Lloyd, A., Melnik, S., Mwaura, D., Nagle, D., Quinlan, S., Rao, R., Rolig, L., Saito, Y., Szymaniak, M., Taylor, C., Wang, R., and Woodford, D. (2012) Spanner: Google's globally-distributed database, in Proceedings of the 10th USENIX Symposium on Operating Systems Design and Implementation (OSDI). Available at https://www.usenix.org/node/170855.

60 Beaver, D., Kumar, S., Li, H.C., Sobel, J., and Vajgel, P. (2010) Finding a needle in Haystack: Facebook's photo storage, in Proceedings of the 9th USENIX Symposium on Operating Systems Design and Implementation (OSDI). Available at https://www.usenix.org/conference/osdi10/finding-needle-haystack-facebooks-photo-storage.

61 Lee, E.K. and Thekkath, C.A. (1996) Petal: distributed virtual disks, in Proceedings of the 7th International Conference on Architectural Support for Programming Languages and Operating Systems (ASPLOS), pp. 84–92.

62 Muralidhar, S., Lloyd, W., Roy, S., Hill, C., Lin, E., Liu, W., Pan, S., Shankar, S., Sivakumar, V., Tang, L., and Kumar, S. (2014) f4: Facebook's warm BLOB storage system, in Proceedings of the 11th USENIX Symposium on Operating Systems Design and Implementation (OSDI). Available at https://www.usenix.org/conference/osdi14/technical-sessions/presentation/muralidhar.

63 Oracle, Database SecureFiles and large objects developer's guide. Available at https://docs.oracle.com/database/121/ADLOB/toc.htm.

64 Calder, B., Wang, J., Ogus, A., Nilakantan, N., Skjolsvold, A., McKelvie, S., Xu, Y., Srivastav, S., Wu, J., Simitci, H., Haridas, J., Uddaraju, C., Khatri, H., Edwards, A., Bedekar, V., Mainali, S., Abbasi, R., Agarwal, A., ul Haq, M.F., ul Haq, M.I., Bhardwaj, D., Dayanand, S., Adusumilli, A., McNett, M., Sankaran, S., Manivannan, K., and Rigas, L. (2011) Windows Azure storage: a highly available cloud storage service with strong consistency, in Proceedings of the 23rd ACM Symposium on Operating Systems Principles (SOSP), pp. 143–157.

65 Noghabi, S.A., Paramasivam, K., Pan, Y., Ramesh, N., Bringhurst, J., Gupta, I., and Campbell, R.H. (2017) Samza: stateful scalable stream processing at LinkedIn. *Proceedings of the VLDB Endowment,* **10** (12), 1634–1645.

66 TensorFlow.org. An open-source software library for machine intelligence. Available at https://www.tensorflow.org/.

67 Google. Cloud Machine Learning Engine. Available at https://cloud.google.com/ml-engine/.

68 Amazon Web Services. Amazon Machine Learning. Available at https://aws.amazon.com/machine-learning/.

69 Microsoft. Azure Machine Learning. Available at https://azure.microsoft.com/en-us/services/machine-learning/.

6

Resource Management: Performance Assuredness in Distributed Cloud Computing via Online Reconfigurations

Mainak Ghosh, Le Xu, and Indranil Gupta[1]

Department of Computer Science, University of Illinois at Urbana-Champaign, Urbana, IL, USA

Cloud computing relies on software for distributed batch and stream processing, as well as distributed storage. This chapter focuses on an oft-ignored angle of assuredness: performance assuredness. A significant pain point today is the inability to support reconfiguration operations, such as changing of the shard key in a sharded storage/database system, or scaling up (or down) of the number of virtual machines (VMs) being used in a stream or batch processing system. We discuss new techniques to support such reconfiguration operations in an *online* manner, whereby the system does not need to be shut down and the user/client-perceived behavior is indistinguishable regardless of whether a reconfiguration is occurring in the background, that is, the performance continues to be assured in spite of ongoing background reconfiguration. Next, we describe how to scale-out and scale-in (increase or decrease) the number of machines/VMs in cloud computing frameworks like distributed stream processing and distributed graph processing systems, again while offering assured performance to the customer in spite of the reconfigurations occurring in the background. The ultimate performance assuredness is the ability to support SLAs/ SLOs (service-level agreements/objectives) such as deadlines. We present a new real-time scheduler that supports priorities and hard deadlines for Hadoop jobs. We implemented our reconfiguration systems as patches to several popular and open-source cloud computing systems, including MongoDB and Cassandra (storage), Storm (stream processing), LFGraph (graph processing), and Hadoop (batch processing).

---

1 This material is based on research sponsored by the Air Force Research Laboratory and the Air Force Office of Scientific Research under agreement number FA8750-11-2-0084, as well as by the National Science Foundation under grants NSF CNS 1319527 and NSF CCF 0964471. The U.S. government is authorized to reproduce and distribute reprints for governmental purposes notwithstanding any copyright notation thereon. Any opinions, findings, and conclusions or recommendations expressed in this material are those of the authors and do not necessarily reflect the views of the National Science Foundation. This chapter is based on work done in collaboration with Mayank Pundir, Luke Leslie, Boyang Peng, Yosub Shin, Wenting Wang, Gopalakrishna Holla, Muntasir Rahman, Tej Chajed, and Roy H. Campbell.

Assured Cloud Computing, First Edition. Edited by Roy H. Campbell, Charles A. Kamhoua, and Kevin A. Kwiat.
© 2018 the IEEE Computer Society. Published 2018 by John Wiley & Sons, Inc.

6.1 Introduction

Cloud computing relies on distributed systems for storing (and querying) large amounts of data, for batch computation of these data and for streaming computation of big data, incoming at high velocity. Distributed storage systems include NoSQL databases such as MongoDB [1] and key-value stores such as Cassandra [2], which have proliferated for workloads that need fast and weakly consistent access to data. A large amount of batch computation today relies on Apache Hadoop [3], which uses the MapReduce paradigm. Apache Storm [4] is a prominent stream processing system used by a variety of companies; others include Heron [5] (used by Twitter) and Samza [6] (used by LinkedIn).

Design of the first generation of these systems focused largely on scale and optimizing performance, but these systems have reached a level of maturity that now calls for performance assuredness. In other words, in spite of changes made to a distributed cloud service (such as the ones already listed), the service should run uninterrupted and offer performance (latency or throughput) and availability indistinguishable from those in the common-case operations. In general, the requirement is that clients (front-end machines) and eventually user devices should be unaware of any reconfiguration changes taking place in the background.

Examples of such reconfiguration operations include (i) changing the layout of data in a sharded cloud database, for example, by changing the shard key; (ii) scaling-out or scaling-in the number of machines/VMs involved in a distributed computation or database; and (iii) eventually, supporting predictable service level agreements/objectives (SLAs/SLOs) for jobs. Support for those operations will make the performance of each job predictable, push systems toward supporting multitenant scenarios (in which multiple jobs share the same cluster), and allow developers to specify requirements (in the form of SLAs/SLOs); the scheduler would then be able to manage resources automatically to satisfy these SLAs/SLOs without human involvement.

This chapter is not focused on security challenges. However, even meeting performance assuredness goals is a significant undertaking today, and it entails several challenges. First, the sheer scale of the data being stored or processed, and the nontrivial number of machines/VMs involved, means that decision to migrate data or computation need to be made *wisely*. Second, the de facto approach used today in industry systems is to shut down services before reconfiguration; this is unacceptable, as it negatively affects availability and latency (in storage/databases), drops throughput to zero (in stream processing systems), and wastes work (in batch processing systems). What is needed is a way to migrate components in the *background*, without affecting the foreground performance perceived by clients and users. Third, the mechanisms to scale-out and scale-in can be used *either manually or in an adaptive fashion*, such that they are driven by logic that seeks to satisfy SLAs/SLOs.

In this chapter, we describe multiple new approaches to attacking different pieces of the broad problem already described. We incorporate performance assuredness into cloud storage/database systems, stream processing systems, and batch processing systems. Specifically, we describe the following innovations:

1) **Reconfigurations in NoSQL distributed storage/database systems:**
    a) The *Morphus* system supports reconfiguration operations such as shard key change in (flexibly) sharded databases such as MongoDB (in which we implemented our techniques). Morphus is described in Section 6.4.2.
       - *Key idea:* Our approach is to make optimal decisions on where to place new chunks based on bipartite graph matching (which minimizes network volume transferred), to support concurrent migration of data and processing of foreground queries in a careful and staged manner, and to replay logged writes received while the reconfiguration was in progress.
    b) The *Parqua* system extends our Morphus design to ring-based key-value stores. Our implementation is in Apache Cassandra, the most popular key-value store in industry today. Parqua is described in Section 6.4.3.
2) **Scale-out/scale-in in distributed stream and distributed graph processing systems:**
    a) The *Stela* system supports automated scale-out/scale-in in distributed stream processing systems. Our implementation is in Apache Storm, the most popular stream processing system in industry today. Stela is described in Section 6.5.1.
       - *Key idea:* We identify the congestion level of operators in the stream processing job (which is a DAG of operators). For scale-out, we give more resources to the most congested operators, while for scale-in we take resources away from the least congested operators. We do so in the background without affecting ongoing processing at other operators in the job.
    b) For graph processing elasticity, we support scale-out/scale-in in distributed graph processing systems; our implementation is on our home-baked system called LFGraph, which is described in Section 6.5.2.
       - *Key idea:* We optimally decide how to repartition the vertices of the graphs (e.g., a Web graph or Facebook-style social network) among the remaining servers, so as to minimize the amount of data moved and thus the time to reconfigure. We perform the migration carefully and in the background, allowing the iteration-based computation to proceed normally.
3) **Priorities and deadlines in batch processing systems:** The Natjam system incorporates support for job priorities and deadlines (i.e., SLAs/SLOs) in Apache YARN. Our implementation is in Apache Hadoop, the most popular distributed batch processing system in industry today. Natjam is described in Section 6.6.
    - *Key idea:* We quickly checkpoint individual tasks, so that if one is preempted by a task of a higher priority (or lower deadline) job, then it can resume from where it left off (thus avoiding wasted work). We

developed policies for both job-level and task-level evictions that allow the scheduler to decide which running components will be victimized when a more important job arrives.

In this chapter, we present the key design techniques and algorithms, outline design and implementation details, and touch on key experimental results. We performed all experiments by deploying both the original system(s) and our modified system on real clusters and subjecting them to real workloads.

6.2 Vision: Using Cloud Technology in Missions

Missions that rely on the cloud, by definition, have strong time criticality. This implies two requirements:

1) Any changes that the mission team needs to make must be made on an *already-running* system in the cloud (e.g., a database or a processing system), without shutting it down and without affecting the ongoing foreground performance of the system. This capability will maximize the flexibility with which the team can change configuration settings on the fly, including the data layout (affecting query time in a database) and the number of machines/ VMs running a computation.
2) The team must be able to specify *priorities* or *deadlines* for certain jobs. A mission typically consists of a diverse set of jobs running simultaneously on the cloud. For instance, a Hadoop job tracking the flight of a fighter aircraft must be run at a higher priority than a Hadoop job that is computing the overnight logs of yesterday's missions (and perhaps must be run with a deadline).

The research contributions described in this chapter fulfill both of these requirements. First, our Morphus and Parqua systems make it possible to change configuration parameters (such as the shard key) that affect the layout of data in a database, without affecting query latency, and while minimizing total transfer time. That gives the team flexibility to make such configuration changes at any time. Our Stela system allows the team to scale-out or scale-in, at will, the number of machines or VMs associated with a stream processing job. For instance, consider a stream processing job tracking the camera feed from an aircraft fleet (e.g., in order to identify enemy positions). When the volume of such data increases, our Stela system allows the team to scale-out the number of machines quickly to cope with the increased workload, while still continuing to process the incoming stream at a high rate. Similar flexibility extends to our scale-out/scale-in techniques for distributed graph processing.

Second, our Natjam system allows Hadoop jobs (running in a YARN cluster) to be associated with a priority or a deadline. The Natjam scheduler automatically ensures that higher priority (or earlier deadline) jobs preempt other jobs, run sooner, and thus finish sooner – an ability that is much needed in critical missions that rely on the cloud.

6.3 State of the Art

Here, we summarize the state of the art for each of the individual challenges and systems covered in this chapter.

6.3.1 State of the Art: Reconfigurations in Sharded Databases/Storage

6.3.1.1 Database Reconfigurations
Research in distributed databases has focused on query optimization and load balancing [7], and orthogonally on using group communication for online reconfiguration [8]; however, that work has not solved the core algorithmic problems for efficient reconfiguration. Online schema change was targeted in Google [9], but the resultant availabilities were lower than those provided by Morphus and Parqua. In a parallel work, Elmore *et al.* [10] looked into the reconfiguration problem for a partitioned main memory database such as H-Store. Data placement in parallel databases has used hash-based and range-based partitioning [11,12], but optimality for reconfiguration has not been a goal.

6.3.1.2 Live Migration
The problem of live migration has been looked into in the context of databases. Albatross [13], Zephyr [14], and ShuttleDB [15] address live migration in multi-tenant transactional databases. Albatross and ShuttleDB use iterative operation replay, such as Morphus, while Zephyr routes updates based on current data locations. Data migration in these systems happens between two different sets of servers, while Morphus and Parqua achieve data migration within a replica set. Also, these papers do not propose optimal solutions for any reconfiguration operation. Opportunistic lazy migration, explored in the Relational Cloud [16], entails longer completion times. Tuba [17] addressed the problem of migration in a geo-replicated setting. The authors avoided write throttle by having multiple masters at the same time, which is not supported by MongoDB and Cassandra.

Morphus's techniques naturally bear some similarities to live VM migration. Precopy techniques migrate a VM without stopping the OS, and if this fails, then the OS is stopped [18]. Like precopy, Morphus also replays operations that occurred during the migration. Precopy systems also use write throttling [19], and precopy has been used in database migration [20].

6.3.1.3 Network Flow Scheduling
For network flow scheduling, Chowdhury *et al.* [21] proposed a weighted flow-scheduling approach that allocates multiple TCP connections to each flow to minimize migration time. Our WFS approach improves upon their approach by considering network latencies as well. Morphus's performance will likely improve further if we also consider bandwidth. Hedera [22] also provides a dynamic flow scheduling algorithm for multirooted network topology. While these techniques

may improve reconfiguration time, Morphus's approach is end-to-end and is less likely to disrupt normal reads and writes that use the same network links.

6.3.2 State of the Art: Scale-Out/Scale-In in Distributed Stream Processing Systems

6.3.2.1 Real-Time Reconfigurations

Based on its predecessor Aurora [23], Borealis [24] is a stream processing engine that enables modification of queries on the fly. Borealis focuses on load balancing on individual machines and distributes load shedding in a static environment. Borealis also uses ROD (resilient operator distribution) to determine the best operator distribution plan that is closest to an "ideal" feasible set: a maximum set of machines that are underloaded. Borealis does not explicitly support elasticity. Twitter's Heron [5] improves on Storm's congestion-handling mechanism by using back pressure; however, elasticity is not explicitly addressed.

6.3.2.2 Live Migration

Stormy [25] uses a logical ring and consistent hashing to place new nodes upon a scale-out. Unlike Stela, it does not take congestion into account. Stream-Cloud [26] builds elasticity into the Borealis Stream Processing Engine [27]. StreamCloud modifies the parallelism level by splitting queries into subqueries and uses rebalancing to adjust resource usage. Stela does not change running topologies, because we believe that would be intrusive to the applications.

6.3.2.3 Real-Time Elasticity

SEEP [28] uses an approach to elasticity that focuses mainly on an operator's state management. It proposes mechanisms to back up, restore, and partition operators' states in order to achieve short recovery time. Several papers have focused on elasticity for stateful stream processing systems. Work [29,30] from IBM enabled elasticity for both IBM System S [31–33] and SPADE [34] by increasing the parallelism of processing operators. It applied networking concepts such as congestion control to expand and contract the parallelism of a processing operator by constantly monitoring the throughput of its links. They do not assume that a fixed number of machines is provided (or taken away) by the users. Our system aims at intelligent prioritization of target operators to parallelize more toward (or migrate further from) the user-determined number of machines that join (or are taken away from) the cluster, with a mechanism for optimizing throughput.

Recent work [35] proposes an elasticity model that provides latency guarantees by tuning the task-wise parallelism level in a fixed-size cluster. Meanwhile, another recent effort [36] implemented stream processing system elasticity; however, it focused on both latency (not throughput) and policy (not mechanisms). Nevertheless, Stela's mechanisms can be used as a black box inside this system.

Some of these recent efforts have looked at *policies* for adaptivity [36], while others [25,26,29,30] focus on the *mechanisms* for elasticity. These are important building blocks for adaptivity. To the best of our knowledge, Ref. [37] describes the only prior mechanism for elasticity in stream processing systems; in Sections 6.5.1.9–6.5.1.12, we compare it to Stela.

6.3.3 State of the Art: Scale-Out/Scale-In in Distributed Graph Processing Systems

To the best of our knowledge, we are the first to explore elasticity for distributed graph computations. However, elasticity has been explored in many other areas.

6.3.3.1 Data Centers

AutoScale [38] enables elastic capacity management for data centers. Its goal is to reduce power wastage by maintaining just enough server capacity for the current workload. Zhang et al. [39] proposes to solve the same problem using a Model Predictive Control framework.

6.3.3.2 Cloud and Storage Systems

The creators of CloudScale [40] and Jiang et al. [41] propose mechanisms for scaling up VM resources based on predicted application needs in an Infrastructure as a Service (IaaS) environment. AGILE [42] proposes mechanisms to scale-out VM resources based on predicted application needs.

Pujol et al. [43] propose a social partitioning and replication middleware to enable efficient storage of social network graphs. The technique enables the storage system to scale-out/scale-in based on need. Albatross [13] enables scaling of multitenant databases by using live migration. TIRAMOLA [44] allows NoSQL storage clusters to scale-out or scale-in by using user-provided policies to apply a Markov decision process on the workload. The creators of Transactional Auto Scaler [45] propose another storage resource scaling mechanism, which uses analytical modeling and machine learning to predict workload behavior.

6.3.3.3 Data Processing Frameworks

Starfish [46] performs tuning of Hadoop parameters at job, workflow, and workload levels. Herodotou et al. [47] propose Elastisizer, which has the ability to predict cluster size for Hadoop workloads.

6.3.3.4 Partitioning in Graph Processing

PowerGraph [48] performs vertex assignment using balanced partitioning of edges. The aim is to limit the number of servers spanned by the vertices. Distributed GraphLab [49] involves a two-phase assignment of vertices to servers; the first phase creates partitions that are more numerous than the servers. In the second phase, the servers load their respective partitions based on a balanced partitioning. That makes it possible to load the graph in a distributed manner, and to change the number of servers, without affecting the initial load phase. Stanton et al. [50] discuss partitioning strategies of streaming graph data. However, they do not explore elasticity. Use of partitioning during a scale-out/scale-in operation is not viable, as it would partition the entire graph; instead, we do incremental repartitioning.

6.3.3.5 Dynamic Repartitioning in Graph Processing

Vaquero et al. [51] proposes a vertex migration heuristic between partitions, to maintain balance and reduce communication cost. GPS [52] involves dynamic repartition of vertices by colocating the vertices that exchange larger numbers of messages with each other. While those efforts did not explicitly explore on-demand elasticity, our techniques are orthogonal and can be applied to such systems. More specifically, one can use our vertex repartitioning technique in such systems by treating each partition as the set of vertices currently assigned to that server.

6.3.4 State of the Art: Priorities and Deadlines in Batch Processing Systems

6.3.4.1 OS Mechanisms

Sharing finite resources among applications is a fundamental issue in operating systems [53]. Not surprisingly, Natjam's eviction policies are analogous to multiprocessor scheduling techniques (e.g., shortest task first) and to eviction policies for caches and paged OSes. However, our results are different, because MapReduce jobs need to have all tasks finished. PACMan [54] looks at eviction policies for caches in MapReduce clusters, and it can be used orthogonally with Natjam.

6.3.4.2 Preemption

Amoeba, a system built in parallel with ours, provides instantaneous fairness with elastic queues and uses a checkpointing mechanism [55]. The main differences between Natjam and Amoeba are that (i) Natjam focuses on job and task eviction policies, (ii) Natjam focuses on jobs with hard deadlines, and (iii) our implementation of it works directly with Hadoop 0.23, while Amoeba requires use of the prototype Sailfish system [56]. Further, Sailfish was built on Hadoop 0.20, and Hadoop 0.23 later addressed many relevant bottlenecks, for example, the use of read-ahead seeks and the use of Netty [57] to speed up shuffle. Finally, our eviction policies and scheduling can be implemented orthogonally in Amoeba.

Delay scheduling [58] avoids killing map tasks while achieving data locality. In comparison, Natjam focuses on reduce tasks, as they are longer than maps and release resources more slowly, making our problem more challenging. Global preemption [59] selects tasks to kill across all jobs, which is a suboptimal solution.

A recently started Hadoop JIRA issue [60] also looks at checkpointing and preemption of reduce tasks. Such checkpointing can be used orthogonally with our eviction policies, which comprise the primary contribution of this part of our work. Finally, Piccolo [61] is a data-processing framework that uses checkpointing based on consistent global snapshots; in comparison, Natjam's checkpoints are local.

6.3.4.3 Real-Time Scheduling

ARIA [62] and Conductor [63] estimate how a Hadoop job needs to be scaled up to meet its deadline, for example, based on past execution profiles or a constraint satisfaction problem. They do not target clusters with finite resources. Real-time constraint satisfaction problems have been solved analytically [64], and Jockey [65] addressed DAGs of data-parallel jobs; however, eviction policies and Hadoop integration were not fleshed out in that work. Statistics-driven approaches have been used for cluster management [66] and for Hadoop [67]. Much work has also been done in speeding up MapReduce environments by tackling stragglers, for example [68,69], but those efforts do not support job priorities.

Dynamic proportional share scheduling [70] allows applications to bid for resources but is driven by economic metrics rather than priorities or deadlines. The network can prioritize data for time-sensitive jobs [21], and Natjam can be used orthogonally.

Natjam focuses on batch jobs rather than stream processing or interactive queries. Stream processing in the cloud has been looked at intensively, for example, in the work on Hadoop Online [71], Spark [72], Storm [73], Timestream [74], and Infosphere [75]. BlinkDB [76] and MeT [77] optimize interactive queries for SQL and NoSQL systems.

Finally, classical work on real-time systems has proposed a variety of scheduling approaches, including classical EDF and rate monotonic scheduling [78,79], priority-based scheduling of periodic tasks [80], laxity-based approaches [81], and handling of task DAGs [82]. Natjam is different in its focus on MapReduce workloads.

6.3.4.4 Fairness

Providing fairness across jobs has been a recent focus in cloud computing engines. Outcomes of such work include Hadoop's Capacity Scheduler [83] and Fair Scheduler [84], which provide fairness by allowing an administrator to configure queue capacities and job priorities. They do not, however, allow resource preemption [85]. Quincy [86] solves an optimization problem to provide fairness in DryadLINQ [87]. Quincy does consider preemption but proposes neither eviction policies nor checkpointing mechanisms. Finally, there has been a recent focus on satisfying SLAs [88] and satisfying real-time QoS [89], but such efforts have not targeted MapReduce clusters.

6.3.4.5 Cluster Management with SLOs

Recent cluster management systems have targeted SLOs, for example, Omega [90], Cake [91], Azure [92], Centrifuge [93], and Albatross [13]. Mesos [94] uses dominant resource fairness across applications that share a cluster, and Pisces [95] looks at multitenant fairness in key-value stores.

6.4 Reconfigurations in NoSQL and Key-Value Storage/Databases

In this section, we describe how to perform reconfiguration operations in both sharded NoSQL databases (e.g., MongoDB) and key-value stores that use hash-based partitioning (e.g., Cassandra). We first discuss motivations in Section 6.4.1, then our Morphus system (integrated into MongoDB) in Section 6.4.2, and finally our Parqua system (integrated into Cassandra) in Section 6.4.3. (We discussed related work in Section 6.3.1.)

6.4.1 Motivation

Distributed NoSQL storage systems comprise one of the core technologies in today's cloud computing revolution. These systems are attractive because they offer high availability and fast read/write operations for data. They are used in production deployments for content management, archiving, e-commerce, education, finance, gaming, e-mail, and health care. The NoSQL market is expected to earn $14 billion in revenue during 2015–2020 and become a $3.4 billion market by 2020 [96].

In today's NoSQL deployments [1,2,97,98], data-centric[2] global reconfiguration operations are quite inefficient. The reason is that executing them relies on ad hoc mechanisms rather than solution of the core underlying algorithmic and system design problems. The most common solution involves first saving a table or the entire database, and then reimporting all the data into the new configuration [99]. This approach leads to a significant period of unavailability. A second option may be to create a new cluster of servers with the new database configuration and then populate it with data from the old cluster [13–15]. This approach does not support concurrent reads and writes during migration, a feature we would like to provide.

Consider an admin who wishes to change the shard key inside a sharded NoSQL store such as MongoDB [99]. The shard key is used to split the database into blocks, where each block stores values for a contiguous range of shard keys. Queries are answered much faster if they include the shard key as a query parameter (because otherwise the query needs to be multicast). For today's systems, it is strongly recommended that the admin choose the shard key at database creation time and not change it afterward. However, that approach is challenging because it is hard to guess how the workload will evolve in the future. In reality, there are many reasons why admins might need to change the shard key, such as changes in the nature of the data being received, evolving business logic, the need to perform operations such as joins with other tables, and the discovery in hindsight that prior design choices were suboptimal. As a result, the

---

2 Data-centric reconfiguration operations deal only with migration of data residing in database tables. Non-data-centric reconfigurations, for example, software updates or configuration table changes, are beyond our scope.

reconfiguration problem has been a fervent point of discussion in the community for many years [100,101].

In this work, we present two systems that support automated reconfiguration. Our systems, called *Morphus* and *Parqua*, allow reconfiguration changes to happen in an online manner, that is, by concurrently supporting reads and writes on the database table while its data are being reconfigured.

6.4.2 Morphus: Reconfigurations in Sharded Databases/Storage

This section is based on Ref. [102], and we encourage the reader to refer to it for further details on design, implementation, and experiments.

6.4.2.1 Assumptions

Morphus assumes that the NoSQL system features master–slave replication, range-based (as opposed to hash-based) sharding,[3] and flexibility in data assignment.[4] Several databases satisfy these assumptions, for example, MongoDB [1], RethinkDB [103], and CouchDB [104]. To integrate our Morphus system, we chose MongoDB because of its clean documentation, strong user base, and significant development activity. To simplify discussion, we assume a single data center, but our paper [105] present results for geo-distributed experiments. Finally, we focus on NoSQL rather than ACID databases because the simplified CRUD (Create, Read, Update, Delete) operations allow us to focus on the reconfiguration problem. Addressing of ACID transactions is an exciting avenue that our work opens up.

Morphus solves three major challenges: (i) In order to be fast, data migration across servers must incur the least traffic volume. (ii) Degradation of read and write latencies during reconfiguration must be small compared to operation latencies when there is no reconfiguration. (iii) Data migration traffic must adapt itself to the data center's network topology.

6.4.2.2 MongoDB System Model

We have chosen to incorporate Morphus into a popular sharded key-value store, MongoDB v2.4 [1]. As noted earlier, our choice of MongoDB was driven not just by its popularity but also by its clean documentation, its strong user base, and the significant development and discussion around it.

A MongoDB deployment consists of three types of servers. The *mongod* servers store the data chunks themselves, which typically are grouped into disjoint *replica sets*. Each replica set contains the same number of servers

---

3 Most systems that hash keys use range-based sharding of the hashed keys, so our system applies there as well. Our system also works with pure hash sharded systems, though it is less effective.

4 This flexibility allows us to innovate on data placement strategies. Inflexibility in consistent hashed systems such as Cassandra [2] requires a different solution and was investigated in Parqua.

(typically 3), which are exact replicas of each other, with one of the servers marked as a primary (master) and the others acting as secondaries (slaves). The configuration parameters of the database are stored at the *config servers*. Clients send CRUD queries to a set of front-end servers, also called *mongos*. The mongos servers also cache some of the configuration information from the config servers, for example, in order to route queries, they cache mappings from each chunk range to a replica set.

A single database table in MongoDB is called a *collection*. Thus, a single MongoDB deployment consists of several collections.

6.4.2.3 Reconfiguration Phases in Morphus

Morphus allows a reconfiguration operation to be initiated by a system administrator on any collection. Morphus executes the reconfiguration via five sequential phases, as shown in Figure 6.1.

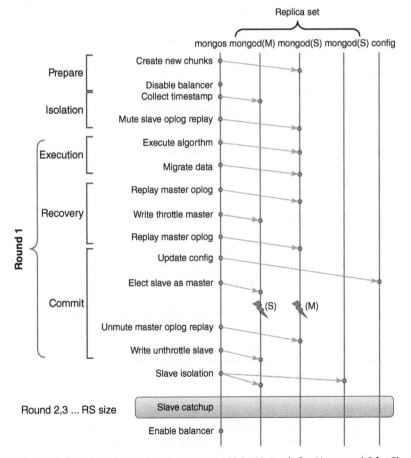

Figure 6.1 Morphus phases. Arrows represent RPCs. *M* stands for *Master* and *S* for *Slave*.

First, in the prepare phase, Morphus prepares for the reconfiguration by creating partitions (with empty new chunks) by using the new shard key (prepare phase). Second, in the isolation phase, Morphus isolates one secondary server from each replica set. Third, in the execution phase, these secondaries exchange data based on the placement plan chosen by the mongos. In the meantime, further operations may have arrived at the primary servers, and they are now replayed at the secondaries in the fourth phase – recovery phase. When the reconfigured secondaries have caught up, they swap places with their primaries in the fifth phase – commit phase.

At that point, the database has been reconfigured and can start serving queries with the new configuration. However, other secondaries in all replica sets need to reconfigure as well. This slave catchup is done in multiple rounds, with the number of rounds equal to the size of the replica set.

We discuss the individual phases in detail in our paper [105].

Read–Write Behavior

The end of the commit phase marks the switch to the new shard key. Until this point, all queries with the old shard key were routed to the mapped server and all queries with the new shard key were multicast to all the servers (which is normal MongoDB behavior). After the commit phase, a query with the new shard key is routed to the appropriate server (the new primary). Queries that do not use the new shard key are handled with a multicast, which again is normal MongoDB behavior.

Reads in MongoDB offer per-key sequential consistency. Morphus is designed so that it continues to offer the same consistency model for data undergoing migration.

6.4.2.4 Algorithms for Efficient Shard Key Reconfigurations

In a reconfiguration operation, the data present in shards across multiple servers are resharded. The new shards need to be placed at the servers in such a way as to reduce the total network transfer volume during reconfiguration and achieve load balance. This section presents optimal algorithms for this planning problem.

We present two algorithms for placement of the new chunks in the cluster. Our first algorithm is greedy and is optimal in the total network transfer volume. However, it may create bottlenecks by clustering many new chunks at a few servers. Our second algorithm, based on bipartite matching, is optimal in network transfer volume among all those strategies that ensure load balance.

Greedy Assignment

The greedy approach considers each new chunk independently. For each new chunk NC_i, the approach evaluates all the N servers. For each server S_j, it calculates the number of data items W_{NC_i,S_j} of chunk NC_i that are already present in old chunks at server S_j. The approach then allocates each new chunk NC_i to the server S_j that has the maximum value of W_{NC_i,S_j}, that is, $\text{argmax}_{S_*}\left(W_{NC_i,S_j}\right)$. As chunks are considered independently, the algorithm produces the same output irrespective of the order in which chunks are considered by it.

Lemma 6.1   The greedy algorithm is optimal in total network transfer volume.

To illustrate the greedy scheme in action, Figure 6.2 provides two examples for the shard key change operation. In each example, the database has three old chunks, OC_1–OC_3, each of which contains three data items. For each data item, we show the old shard key K_o and the new shard key K_n (both in the range 1–9). The new configuration splits the new key range evenly across the three chunks, shown as NC_1–NC_3.

In Figure 6.2a, the old chunks are spread evenly across servers S_1–S_3. The edge weights in the bipartite graph show the number of data items of NC_i that are local at S_j, that is, W_{NC_i,S_j} values. Thick lines show the greedy assignment.

However, the greedy approach may produce an unbalanced chunk assignment for skewed bipartite graphs, as in Figure 6.2b. While the greedy approach minimizes network transfer volume, it assigns new chunks NC_2 and NC_3 to server S_1, while leaving server S_3 empty.

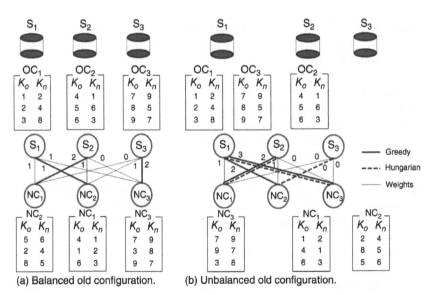

(a) Balanced old configuration.   (b) Unbalanced old configuration.

Figure 6.2 Greedy and Hungarian strategies for shard key change using (a) balanced, (b) unbalanced old chunk configuration. *S1–S3* represent servers. OC_1–OC_3 and NC_1–NC_3 are old and new chunks, respectively. K_o and K_n are old and new shard keys, respectively. Edges are annotated with W_{NC_i,S_j} weights.

Load Balance via Bipartite Matching

Load balancing of chunks across servers is important for several reasons. First, it improves read/write latencies for clients by spreading data and queries across more servers. Second, it reduces read/write bottlenecks. Finally, it reduces the tail of the reconfiguration time by preventing the allocation of too many chunks to any one server.

Our second strategy achieves load balance by capping the number of new chunks allocated to each server. With m new chunks, this per-server cap is $\lceil m/N \rceil$ chunks. We then create a bipartite graph with two sets of vertices, top and bottom. The top set consists of $\lceil m/N \rceil$ vertices for each of the N servers in the system; the vertices for server S_j are denoted by $S_j^1 - S_j^{\lceil m/N \rceil}$. The bottom set of vertices consists of the new chunks. All edges between a top vertex S_j^k and a bottom vertex NC_i have an edge cost equal to $|NC_i| - W_{NC_i, S_j}$, that is, the number of data items that will move to server S_j if new chunk NC_i is allocated to it.

Assigning new chunks to servers in order to minimize data transfer volume now becomes a bipartite matching problem. Thus, we find the minimum weight matching by using the classical Hungarian algorithm [106]. The complexity of this algorithm is $O((N \cdot V + m) \cdot N \cdot V \cdot m)$, where $V = \lceil m/N \rceil$ chunks. This reduces to $O(m^3)$. The greedy strategy becomes a special case of this algorithm with $V = m$.

Lemma 6.2 Among all load-balanced strategies that assign at most $V = \lceil m/N \rceil$ new chunks to any server, the Hungarian algorithm is optimal in total network transfer volume.

Figure 6.2b shows the outcome of the bipartite matching algorithm with dotted lines in the graph. While it incurs the same overall cost as the greedy approach, it provides the benefit of a load-balanced new configuration, wherein each server is allocated exactly one new chunk.

While we focus on the shard key change, this technique can also be used for other reconfigurations, such as changing shard size, or cluster scale-out and scale-in. The bipartite graph would be drawn appropriately (depending on the reconfiguration operation) and the same matching algorithm used. For the purpose of concreteness, the rest of this section of the chapter focuses on shard key change.

Finally, we have used data size (number of key-value pairs) as the main cost metric. Instead, we could use traffic to key-value pairs as the cost metric and derive edge weights in the bipartite graph (Figure 6.2) from these traffic estimates. The Hungarian approach on this new graph would balance out traffic load, while trading off optimality. Further exploration of this variant is beyond our scope here.

6.4.2.5 Network Awareness

Data centers use a wide variety of topologies, the most popular being hierarchical; for example, a typical two-level topology consists of a core switch and multiple rack switches. Others that are commonly used in practice include fat-trees [107], CLOS [108], and butterfly topologies [109].

Our first-cut data migration strategy, discussed in Section 6.4.2.3, was *chunk based*: It assigned as many sockets (TCP streams) to a new chunk C at its destination server as there were source servers for C, that is, it assigned one TCP stream per server pair. Using multiple TCP streams per server pair has been shown to better utilize the available network bandwidth [21]. Further, the chunk-based approach also results in *stragglers* in the execution phase. In particular, we observe that 60% of the chunks finish quickly, followed by a 40% cluster of chunks that finish late.

To address these two issues, we propose a weighted fair sharing (WFS) scheme that takes both data transfer size and network latency into account. Consider a pair of servers i and j, where i is sending some data to j during the reconfiguration. Let $D_{i,j}$ denote the total amount of data that i needs to transfer to j, and $L_{i,j}$ denote the latency in the shortest network path from i to j. Then, we set $X_{i,j}$, the weight for the flow from server i to j, as follows:

$$X_{i,j} \propto D_{i,j} \times L_{i,j}.$$

In our implementation, the weights determine the number of sockets that we assign to each flow. We assign each destination server j a total number of sockets $X_j = K \times \frac{\sum_i D_{i,j}}{\sum_{i,j} D_{i,j}}$, where K is the total number of sockets throughout the system. Thereafter, each destination server j assigns each source server i a number of sockets, $X_{i,j} = X_j \times \frac{C_{i,j}}{\sum_i C_{i,j}}$.

However, $X_{i,j}$ may be different from the number of new chunks that j needs to fetch from i. If $X_{i,j}$ is larger, we treat each new chunk as a data slice, and iteratively split the largest slice into smaller slices until $X_{i,j}$ equals the total number of slices. Similarly, if $X_{i,j}$ is smaller, we use iterative merging of the smallest slices. Finally, each slice is assigned a socket for data transfer. Splitting or merging of slices is done only for the purpose of socket assignment and to speed up data transfer; it does not affect the final chunk configuration that was computed in the prepare phase.

Our approach above could have used estimates of available bandwidth instead of latency estimates. We chose the latter because (i) they can be measured with a lower overhead, (ii) they are more predictable over time, and (iii) they are correlated to the effective bandwidth.

6.4.2.6 Evaluation

Setup

We used the data set of Amazon reviews as our default collection [110]. Each data item had 10 fields. We chose *product ID* as the old shard key and *userID* as the new shard key; update operations used these two fields and a *price* field. Our default database size was 1 GB. (We later show scalability with data size.)

The default Morphus cluster used 10 machines, which included one mongos (front end) and three replica sets, each containing a primary and two secondaries. There were three config servers, each of which was colocated on a physical machine with a replica set primary; this is an allowed MongoDB installation. Each physical machine was a d710 Emulab node [111] with a 2.4 GHz processor, four cores, 12 GB RAM, two hard disks of 250 GB and 500 GB, and 64-bit CentOS 5.5, connected to a 100 Mbps LAN switch.

We implemented a custom workload generator that injects YCSB-like workloads via MongoDB's *pymongo* interface. Our default injection rate was 100 ops/s with 40% reads, 40% updates, and 20% inserts. To model realistic key access patterns, we selected keys for each operation via one of three YCSB-like [112] distributions: (i) Uniform (default), (ii) Zipf, and (iii) Latest. For the Zipf and Latest distributions, we employed a shape parameter $\alpha = 1.5$. The workload generator ran on a dedicated pc3000 node in Emulab that ran a 3 GHz processor, 2 GB RAM, two 146 GB SCSI disks, and 64-bit Ubuntu 12.04 LTS.

Morphus was implemented in about 4000 lines of C++ code, which is publicly available at http://dprg.cs.uiuc.edu/downloads. Each plotted data point is an average of at least three experimental trials, shown along with standard deviation bars. Section 6.4.2.4 outlined two algorithms for the shard key change reconfiguration: Hungarian and greedy. We implemented both in Morphus, and call them variants *Morphus-H* and *Morphus-G*, respectively.

While we present only selected experimental results in this chapter, we refer the reader to Ref. [102] for extensive information on experiments and evaluation of our system.

Effect on Reads and Writes

A key goal of Morphus is to ensure the availability of the database during reconfiguration. To evaluate its success, we generated read and write requests and measured their latency while a reconfiguration was in progress. We used Morphus-G with a chunk-based migration scheme. We ran separate experiments for all the key access distributions and also for a read-only workload.

Table 6.1 lists the percentages of read and write requests that succeeded during reconfiguration. The number of writes that failed is low: For the Uniform and Zipf workloads, fewer than 2% of writes failed. We observe that many of the failed writes occurred during one of the write-throttling periods. Recall from Section 6.4.2.3 that the number of write-throttling periods is the same as the replica set size, with one throttle period at the end of each reconfiguration round. The Latest workload has a slightly higher failure rate, since if an attempt has been made to write a particular key, that increases the likelihood that in the near future there will be another attempt to write or read that same key. Still, the write failure rate of 3.2% and the read failure rate of 2.8% are reasonably low.

Table 6.1 Percentages of reads and writes that succeeded under reconfiguration.

	Read	Write
Read only	99.9	—
Uniform	99.9	98.5
Latest	97.2	96.8
Zipf	99.9	98.3

Overall, the availability numbers are higher, at 99% to 99.9% for Uniform and Zipf workloads, which is comparable to the numbers for a scenario with no insertions. We conclude that unless there is temporal and spatial (key-wise) correlation between writes and reads (i.e., Latest workloads), the read latency is not affected much by concurrent writes. When there is correlation, Morphus mildly reduces the offered availability.

Going further, we plot in Figure 6.3a the CDF of read latencies for the four settings, and for a situation in which there was no reconfiguration (Uniform workload). Note that the horizontal axis is logarithmic scale. We consider latencies only for successful reads. We observe that the 96th percentile latencies for all workloads are within a range of 2 ms. The median (50th percentile) latency for No Reconfiguration is 1.4 ms, and this median holds for both the Read only (No Write) and Uniform workloads. The medians for the Zipf and Latest workloads are lower, at 0.95 ms. This lowered latency has two causes: caching at the mongod servers for the frequently accessed keys, and, in the case of Latest, the lower percentage of successful reads. In Figure 6.3b, we plot the corresponding CDF for write latencies. The median for writes when there is no reconfiguration (Uniform workload) is similar to that of the other distributions.

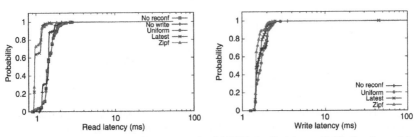

(a) CDF of read latency distribution (log axis). (b) CDF of write latency distribution (log axis).

Figure 6.3 CDF of (a) read and (b) write latency distribution for no reconfiguration (No Reconf) and three under-reconfiguration workloads.

We conclude that under reconfiguration, the read and write availability provided by Morphus is high (close to 99%), while latencies of successful writes degrade only mildly compared to those observed when there is no reconfiguration in progress.

Effect of Network Awareness

First, Figure 6.4a shows the length of the execution phase (when a 500 MB Amazon collection was used) for two hierarchical topologies and five migration strategies. The topologies were (i) homogeneous, in which nine servers were distributed evenly across three racks; and (ii) heterogeneous, in which three racks contained six, two, and one servers, respectively. The switches were Emulab pc3000 nodes and all links were 100 Mbps. The inter-rack and intra-rack latencies were 2 and 1 ms, respectively. The five strategies were (i) fixed sharing, with one socket assigned to each destination node; (ii) a chunk-based approach (see Section 6.4.2.3); (iii) Orchestra [21] with $K = 21$; (iv) WFS with $K = 21$ (see Section 6.4.2.5); and (v) WFS with $K = 28$.

We observed that in the homogeneous clusters, the WFS strategy with $K = 28$ was 30% faster than fixed sharing and 20% faster than the chunk-based strategy. Compared to Orchestra, which weights flow only by their data size, WFS with $K = 21$ does 9% better, because it takes the network into account as well. Increasing K from 21 to 28 improves completion time in the homogeneous cluster, but causes degradation in the heterogeneous cluster. The reason is that a higher K results in more TCP connections, and at $K = 28$, this begins to cause congestion at the rack switch of six servers.

Second, Figure 6.4b shows that Morphus's network-aware WFS strategy has a shorter tail and finishes earlier. Network awareness lowers the median chunk finish time by around 20% in both the homogeneous and heterogeneous networks.

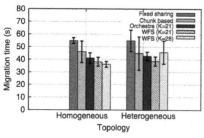

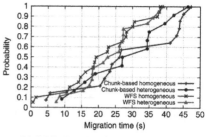

(a) Execution phase migration time for five strategies: (i) fixed sharing, (ii) a chunk-based strategy, (iii) Orchestra with $K = 21$, (iv) WFS with $K = 21$, and (v) WFS with $K = 28$.

(b) CDF of total reconfiguration time in chunk-based strategy versus WFS with $K = 28$.

Figure 6.4 Network-aware evaluation.

We conclude that the WFS strategy improves performance compared to existing approaches, and K should be chosen to be as high as possible without leading to congestion.

Large-Scale Experiment

In this experiment, we increased data and cluster size simultaneously such that the amount of data per replica set was constant. We ran this experiment on Google Cloud [113]. We used n1-standard-4 VMs, each with four virtual CPUs and 15 GB of memory. The disk capacity was 1 GB, and the VMs were running Debian 7. We generated a synthetic data set by randomly dispersing data items among new chunks. Morphus-H was used for reconfiguration with the WFS migration scheme, and $K =$ the number of old chunks.

Figure 6.5 shows a sublinear increase in reconfiguration time as data size and cluster size increased. Note that the x-axis uses a log scale. In the execution phase, all replica sets communicated among themselves to migrate data. As the number of replica sets increased with cluster size, the total number of connections increased, leading to network congestion. Thus, the execution phase took longer.

The amount of data per replica set affects reconfiguration time superlinearly. On the other hand, cluster size has a sublinear impact. In this experiment, the latter dominated, as the amount of data per replica set was constant.

6.4.3 Parqua: Reconfigurations in Distributed Key-Value Stores

This section is based on Ref. [114], and we refer the reader to that publication for further details on design, implementation, and experiments.

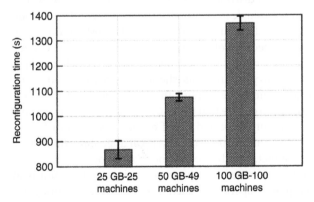

Figure 6.5 Running Morphus-H with WFS ($K =$ number of old chunks) for reconfiguring databases with sizes of 25, 50, and 100 GB running on clusters of 25 machines (8 replica sets * 3 + 1 mongos), 49 machines (16 replica sets), and 100 machines (33 replica sets).

In this section, we describe how to perform reconfigurations in *ring-based* key-value/NoSQL stores such as Cassandra [2], Riak [115], Dynamo [116], and Voldemort [117].

The techniques described for Morphus in Section 6.4.2 cannot be applied directly, for two reasons. First, ring-based systems place data strictly in a deterministic fashion around the ring (e.g., using *consistent hashing*), which determines which keys can be placed where. Thus, our optimal placement strategies from Morphus do not apply to ring-based systems. Second, unlike sharded systems (e.g., MongoDB), ring-based systems do not allow isolation of a set of servers for reconfiguration (a fact that Morphus leveraged). In sharded databases, each participating server exclusively owns a range of data (as master or slave). In ring-based stores, however, ranges of keys overlap across multiple servers in a chained manner (because a node and its successors on the ring are replicas), and this makes full isolation impossible.

That motivated us to build a new reconfiguration system oriented toward ring-based key-value/NoSQL stores. Our system, named *Parqua*,[5] enables online and efficient reconfigurations in virtual ring-based key-value/NoSQL systems. Parqua suffers no overhead when the system is not undergoing reconfiguration. During reconfiguration, Parqua minimizes the impact on read and write latency by performing reconfiguration in the background while responding to reads and writes in the foreground. It keeps the availability of data high during the reconfiguration and migrates to the new reconfiguration at an atomic switch point. Parqua is fault-tolerant, and its performance improves as cluster size increases. We have integrated Parqua into Apache Cassandra.

6.4.3.1 System Model

Parqua is applicable to any key-value/NoSQL store that satisfies the following assumptions. First, we assume a distributed key-value store that is fully decentralized, without the notion of a single master node or replica. Second, each node in the cluster must be able to deterministically choose the destination of the entries that are being moved because of the reconfiguration. This is necessary because there is no notion of a master in a fully decentralized distributed key-value store, and for each entry, all replicas should be preserved after the reconfiguration is finished. Third, we require the key-value store to utilize *SSTable (Sorted String Table)* to ensure that the entries persist permanently. An SSTable is essentially an immutable sorted list of entries stored on disk [98]. Fourth, each write operation accompanies a timestamp or a version number that can be used to resolve a conflict. Finally, we assume that the operations issued are idempotent. Therefore, supported operations are insert, update, and read operations, and nonidempotent operations such as counter-incrementing are not supported.

---

5 The Korean word for "change."

6.4.3.2 System Design and Implementation

Parqua runs reconfiguration in four phases. A graphical overview of the Parqua phases is given in Figure 6.6. Next, we discuss these individual phases in detail.

Isolate Phase

In this phase, the initiator node, in which the reconfiguration command is run, creates a new (and empty) column family (i.e., database table), denoted by *Reconfigured CF* (*column family*). It does so using a schema derived from the Original CF, except it uses the desired key as the new primary key. The Reconfigured CF enables reconfiguration to happen in the background while the Original CF continues to serve reads and writes using the old reconfiguration. We also record the timestamp of the last operation before the Reconfigured CF is created so that all operations that arrive while the execute phase is running can be applied later in the recovery phase.

Execute Phase

The initiator node notifies all other nodes to start copying data from the Original CF to the Reconfigured CF. Read and write requests from clients continue to be served normally during this phase. At each node, Parqua iterates through all entries for which it is responsible and sends them to the appropriate new destination nodes. The destination node for an entry is determined by (i) hashing the new primary key-value on the hash ring, and (ii) using the replica number associated with the entry. Key-value pairs are transferred between corresponding nodes that have matching replica numbers in the old configuration and the new configuration.

For example, in the execute phase of Figure 6.6, the entry with the old primary key "1" and the new primary key "10" has a replica number of 1 at node A, 2 at B, and 3 at C. In this example, after the primary key is changed, the new position of the entry on the ring is between nodes C and D, where nodes D, E, and F are replica numbers 1, 2, and 3, respectively. Thus, in the execute phase, the said entry in node A is sent to node D, and similarly the entry in B is sent to E, and from C to F.

Commit Phase

After the execute phase, the Reconfigured CF has the new configuration, and the entries from the Original CF have been copied to the Reconfigured CF. Now, Parqua atomically swaps both the schema and the SSTables between the Original CF and the Reconfigured CF. The write requests are locked in this phase, while reads still continue to be served. To implement the SSTable swap, we leverage the fact that SSTables are maintained as files on disk, stored in a directory named after the column family. Therefore, we move SSTable files from one directory to another. This does not cause disk I/O, as we update the inodes only when moving files.

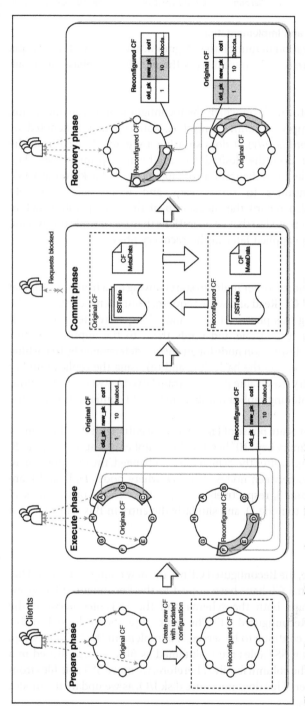

Figure 6.6 Overview of Parqua phases. The gray solid lines represent internal entry transfers and the gray dashed lines are client requests. The phases progress from left to right.

At the end of the commit phase, the write lock is released at each node. At this point, all client-facing requests are processed according to the new configuration. In our case, the new primary key is now in effect, and the read requests must use the new primary key.

Recovery Phase

During this phase, the system catches up with the recent writes that were not transferred to the Reconfigured CF in the execute phase. Read/write requests are processed normally, with the difference that until the recovery is done, the read requests may return stale results.[6] At each node, Parqua iterates through the SSTables of the Original CF to recover the entries that were written during the reconfiguration. We limit the amount of disk accesses required for recovery by iterating only the SSTables that were created after the reconfiguration started. The iterated entries are routed to appropriate destinations in the same way as in the execute phase.

Since all writes in Cassandra carry a timestamp [118], Parqua can ensure that the recovery of an entry does not overshadow newer updates, thus guaranteeing the eventual consistency.

6.4.3.3 Experimental Evaluation

Setup

We used the Yahoo! Cloud Service Benchmark (YCSB) [112] to generate the data set and used the Uniform, Zipfian, and Latest key access distributions to generate CRUD workloads. Our default database size was 10 GB in all experiments. The operations consisted of 40% reads, 40% updates, and 20% inserts. Again, we present only selected experimental results in this chapter, and refer the reader to Ref. [114] for extensive information on experiments and evaluation of our system.

Availability

In this experiment, we measured the availability of our system during reconfiguration, shown in Table 6.2. The slight degradation in availability was due to the rejection of writes in the commit phase. The total duration of the unavailability was only a few seconds, which is orders of magnitude better than the current state of the art.

The lowest availability was observed for the Latest distribution. The reason is that YCSB does not wait for the database to acknowledge an insert of a key. Because of the temporal nature of the distribution, as keys are further read and updated, the operations fail because the inserts are still in progress.

---

6 This is acceptable as Cassandra only guarantees eventual consistency.

Table 6.2 Percentages of reads and writes that succeeded during reconfiguration.

	Read (%)	Write (%)
Read only	99.17	—
Uniform	99.27	99.01
Latest	96.07	98.92
Zipfian	99.02	98.92

Effect on Read Latency

Figure 6.7a shows the CDF of read latencies under various workloads while reconfiguration is being executed. As a baseline, we also plot the CDF of read latency when no reconfiguration is being run using the Uniform key access distribution. We plot the latencies of successful reads only.

The median (50th percentile) latencies for the read-only workload and the baseline are similar because they both use Uniform distribution. Under

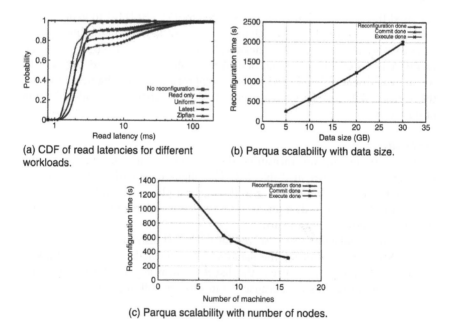

(a) CDF of read latencies for different workloads.

(b) Parqua scalability with data size.

(c) Parqua scalability with number of nodes.

Figure 6.7 Parqua experiments.

reconfiguration, 20% of the reads take longer. With writes in the workload, the observed latencies for the Uniform curve are higher overall.

Compared to other workloads, Latest had the smallest median latency. Because of that workload's temporal nature, recently inserted keys were present in Memtables, which is a data structure maintained in memory. As a result, reads were faster than in other distributions that require disk accesses.

Overall, Parqua affects median read latency minimally across all the distributions. Our observations for write latency are similar. We refer the reader to our technical report for more detail in Ref. [119].

Scalability

Next, we measured how well Parqua scales with (i) database size, (ii) cluster size, (iii) operation injection rate, and (iv) replication factor. For lack of space, we omit the plots for the last two experiments and refer the reader our technical report for them [119]. To evaluate our system's scalability, we measured the total reconfiguration times along with the time spent in each phase. We did not inject operations for the experiments presented next.

Figure 6.7b depicts the reconfiguration times as the database size is increased up to 30 GB. Since we used a replication factor (number of copies of the same entry across the cluster) of 3 for fault tolerance, 30 GB of data implies 90 GB of total data in the database. In this plot, we observe that the total reconfiguration time scales linearly with database size. The bulk of the reconfiguration time is spent in transferring data in the execute phase.

In Figure 6.7c, we observe that the reconfiguration time *decreases* as the number of Cassandra peers increases. The decrease occurs because as the number of machines increases, the same amount of data, divided into smaller chunks, gets transferred by a larger number of peers. Again, the execute phase dominated the reconfiguration time.

6.5 Scale-Out and Scale-In Operations

Next, we describe a transparent way to scale-out and scale-in cloud computing applications. *Scaling-out* means increasing the number of machines (or VMs) running the application, and *scaling-in* means reducing the number. First, we tackle distributed stream processing systems in Section 6.5.1, where we describe our Stela system, which supports scale-out/scale-in and is implemented in Apache Storm [4]. Then, in Section 6.5.2, we address distributed graph processing systems, describing how to support scale-out/scale-in; our implementation is in LFGraph.

6.5.1 Stela: Scale-Out/Scale-In in Distributed Stream Processing Systems

This section is based on Ref. [120], and we refer the reader to that paper for further details on design, implementation, and experiments.

6.5.1.1 Motivation

The volume and velocity of real-time data require frameworks that can process large dynamic streams of data on the fly and serve results with high throughput. To meet this demand, several new stream processing engines have recently been developed that are now widely in use in industry, for example, Storm [4], System S [33], and Spark Streaming [121], among others [25,27,31]. Apache Storm is one of the most popular.

Unfortunately, these new stream processing systems used in industry lack an ability to scale the number of servers *seamlessly* and *efficiently* in an *on-demand* manner. *On-demand* means that the scaling is performed when the user (or some adaptive program) requests an increase or decrease in the number of servers in the application. Today, Storm supports an on-demand scaling request by simply unassigning all processing operators and then reassigning them in a round-robin fashion to the new set of machines. This approach is not seamless, as it interrupts the ongoing computation for a long time. It is not efficient either, as it results in suboptimal throughput after the scaling is completed (as our experiments showed, as we will discuss later).

Scaling-out and scaling-in are critical tools for customers. For instance, a user might start running a stream processing application with a given number of servers, but if the incoming data rate rises or if there is a need to increase the processing throughput, the user may wish to add a few more servers (scale-out) to the stream processing application. On the other hand, if the application is currently underutilizing servers, then the user may want to remove some servers (scale-in) in order to reduce the dollar cost (e.g., if the servers are VMs in AWS [122]).

On-demand scaling operations should meet two goals: (i) The post-scaling throughput (tuples per second) should be optimized, and (ii) the interruption to the ongoing computation (while the scaling operation is being carried out) should be minimized. We have created a new system, named *Stela* (*STream processing ELAsticity*), that meets those two goals. For scale-out, Stela carefully selects which operators (inside the application) are given more resources, and does so with minimal intrusion. Similarly, for scale-in, Stela carefully selects which machine(s) to remove in a way that minimizes the overall detriment to the application's performance.

To select the best operators to give more resources when scaling out, Stela uses a new metric called ETP (*effective throughput percentage*). The key intuition behind ETP is that it is used to capture those operators (e.g., bolts and spouts in Storm) that both (i) are congested (i.e., are being overburdened with incoming tuples), and (ii) affect throughput the most because they reach a large number of sink operators. For scale-in, we also use an ETP-based approach to decide which machine(s) to remove and where to migrate operator(s).

The ETP metric is both hardware- and application-agnostic. Thus, Stela needs neither hardware profiling (which can be intrusive and inaccurate) nor knowledge of application code.

The design of Stela is generic to any data flow system (see Section 6.5.1.2). For concreteness, we integrated Stela into Apache Storm. We compare Stela against the most closely related elasticity techniques in the literature [37]. We generated experimental results by using microbenchmark Storm applications, as well as production applications from industry (Yahoo! Inc. and IBM [29]). We believe our metric can be applied to other systems as well.

Our main contributions in this are (i) development of the novel metric, ETP, that captures the "importance" of an operator; (ii) to the best of our knowledge, the first description and implementation of on-demand elasticity within Storm; and (iii) the evaluation of our system both on microbenchmark applications and on applications used in production.

6.5.1.2 Data Stream Processing Model and Assumptions

We target distributed data stream processing systems that represent each application as a directed acyclic graph (DAG) of operators. An *operator* is a user-defined logical processing unit that receives one or more streams of *tuples*, processes each tuple, and outputs one or more streams of tuples. We assume operators are stateless, and that tuple sizes and processing rates follow an ergodic distribution. These assumptions hold true for most Storm topologies used in industry. Operators that have no parents are *sources* of data injection. They may read from a Web crawler. Operators with no children are *sinks*. The intermediate operators perform processing of tuples. Each sink outputs data to a GUI or database, and the application throughput is the sum of the throughputs of all sinks in the application. An application may have multiple sources and sinks.

An *instance* of an operator is an instantiation of the operator's processing logic and is the physical entity that executes the operator's logic. The number of instances is correlated with the operator's parallelism level. For example, in Storm, these instances are called "executors" [4].

6.5.1.3 Stela: Scale-Out Overview

In this section, we give an overview of how Stela supports scale-out. When a user requests a scale-out with a given number of new machines, Stela needs to choose the operators to which it will give more resources by increasing their parallelism.

Stela first identifies operators that are congested based on their input and output rates. It identifies all congested operators in the graph by continuously sampling the input rate and processing rate of each operator. When the ratio of input to processing exceeds a threshold *CongestionRate*, we consider that operator to be congested. The *CongestionRate* parameter can be tuned as needed and controls the sensitivity of the algorithm. For our Stela experiments, we set *CongestionRate* to be 1.2.

After an operator is identified as congested, Stela calculates a per-operator metric called the ETP. ETP takes the topology into account: It captures the percentage of total application throughput (across all sinks) on which the operator has direct impact but ignores all down-stream paths in the topology that are already congested; it selects the next operator to increase its parallelism, and iterates this process. To ensure load balance, the total number of such iterations equals the number of new machines added times the average number of instances per machine prescale. We determine the number of instances to allocate a new machine with $N_{instances} = (total\ number\ of\ instances)/(number\ of\ machines)$; in other words, $N_{instances}$ is the average number of instances per machine prior to scale-out. This ensures load balance post-scale-out. The schedule of operators on existing machines is left unchanged.

6.5.1.4 Effective Throughput Percentage (ETP)

Effective Throughput Percentage

To estimate the impact of each operator on the application throughput, Stela uses a new metric we developed called ETP. An operator's ETP is defined as the percentage of the final throughput that would be affected if the operator's processing speed were changed.

The ETP of an operator o is computed as

$$ETP_o = \frac{Throughput_{EffectiveReachableSinks}}{Throughput_{workflow}}$$

Here, $Throughput_{EffectiveReachableSinks}$ denotes the sum of the throughputs of all sinks reachable from o by at least one uncongested path, that is, a path consisting only of operators that are not classified as congested. $Throughput_{workflow}$ denotes the sum of the throughputs of all sink operators of the entire application. The algorithm that calculates ETPs does a depth-first search throughout the application DAG and calculates ETPs via a postorder traversal. Processing-RateMap stores the processing rates of all operators. Note that if an operator o has multiple parents, then the effect of o's ETP is the same at each of its parents (i.e., it is replicated, not split).

ETP Calculation Example and Intuition

In Figure 6.8, we illustrate the ETP calculation with an example application.[7] The processing rate of each operator is shown. The congested operators, that is, operators 1, 3, 4, and 6, are shaded. The total throughput of the workflow is calculated with $Throughput_{workflow} = 5000$ tuples/s as the sum of throughputs of sink operators 4, 7, 8, 9, and 10.

Let us calculate the ETP of operator 3. Its reachable sink operators are 7, 8, 9, and 10. Of these, only 7 and 8 are considered to be the "effectively" reachable sink operators, as they are both reachable via an uncongested path. Thus, increasing the speed of operator 3 will improve the throughput of operators 7 and 8. However, operator 6 is not effectively reachable from operator 3, because operator 6 is already congested; thus, increasing operator 3's resources will only increase operator 6's input rate and make operator 6 further congested, without improving its processing rate. Thus, we ignore the subtree of operator 6 when calculating 3's ETP. The ETP of operator 3 is $ETP_3 = (1000 + 1000)/5000 = 40\%$.

Similarly, for operator 1, the sink operators 4, 7, 8, 9, and 10 are reachable, but none of them are reachable via an uncongested path. Thus, the ETP of operator 1

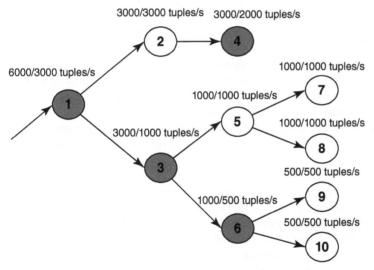

Figure 6.8 A sliver of a stream processing application. Each operator is labeled with its input/execution speed. Shaded operators are congested. *CongestionRate* = 1.

---

7 This corrects a slightly erroneous version of our example in Ref. [120].

is 0. Likewise, we can calculate the ETP of operator 4 as 40% and the ETP of operator 6 as 20%. Therefore, the priority order in which Stela will assign resources to these operators is 3, 4, 6, 1.

6.5.1.5 Iterative Assignment and Intuition

During each iteration, Stela calculates the ETP for all congested operators. Stela targets the operator with the highest ETP and increases the parallelism of that operator by assigning a new instance of it at the newly added machine. If multiple machines are being added, then the target machine is chosen in a round-robin manner. Overall, this algorithm runs $N_{instances}$ iterations to select $N_{instances}$ target operators. (Section 6.5.1.3 showed how to calculate $N_{instances}$.)

In each iteration, Stela constructs a *CongestedMap* to store all congested operators. If there are no congested operators in the application, Stela chooses a source operator as a target in order to increase the input rate of the entire application. If congested operators do exist, for each one, Stela finds its ETP using the algorithm discussed in Section 6.5.1.4. The result is sorted in *ETPMap*. Stela chooses the operator that has the highest ETP value from *ETPMap* as a target for the current iteration. It increases the parallelism of this operator by assigning one additional random instance to it on one of the new machines in a round-robin manner.

For the next iteration, Stela estimates the processing rate of the previously targeted operator o proportionally, that is, if the o previously had an output rate E and k instances, then o's new *projected* processing rate is $E \cdot \frac{k+1}{k}$. This is a reasonable approach since all machines have the same number of instances and thus proportionality holds. (This approach may not be accurate, but we find that it works in practice.) Then Stela uses the projected processing rate to update the output rate for o, and the input rates for o's children. (The processing rates of o's children, and indeed o's grand-descendants, do not need updates, as their resources remain unchanged.) Stela updates the emit rate of the target operator in the same manner to ensure that the estimated operator submission rate can be applied.

Once it has done that, Stela recalculates the ETPs of all operators again using the same algorithm. We call these new ETPs *projected ETPs*, or *PETPs*, because they are based on estimates. The PETPs are used as ETPs for the next iteration. Iterations are performed until all available instance slots at the new machines are filled. Once that procedure is complete, the schedule is committed through starting of the appropriate executors on new instances.

The algorithm involves searching for all reachable sinks for every congested operator; as a result, each iteration of Stela has a running time complexity of $O(n^2)$, where n is the number of operators in the workflow. The entire algorithm has a running time complexity of $O(m \cdot n^2)$, where m is the number of new instance slots at the new workers.

6.5.1.6 Stela: Scale-In

For scale-in, we assume that the user specifies only the number of machines to be removed and Stela picks the "best" machines from the cluster to remove. (If the user specifies the exact machines to remove, the problem is no longer challenging.) We describe how techniques used for scale-out, particularly the ETP metric, can also be used for scale-in. For scale-in, we will calculate the ETP not merely per operator but instead per *machine* in the cluster. That is, we first calculate the *ETPSum* for each machine, as follows:

$$ETPSum(machine_k) = \sum_{i=1}^{n} FindETP(FindComp(\tau_i)).$$

The scale-in procedure is called iteratively, as many times as the number of machines the user asked Stela to remove. The procedure calculates the *ETPSum* for every machine in the cluster and puts the machine and its corresponding *ETPSum* into the ETPMachineMap. The *ETPSum* for a machine is the sum of all the ETPs of instances of all operators that currently reside on the machine. Thus, for every instance τ_i, we first find the operator of which τ_i is an instance (e.g., *operator$_o$*), and then find the ETP of that *operator$_o$*. Then, we sum all of these ETPs. The *ETPSum* of a machine is thus an indication of how much the instances executing on that machine contribute to the overall throughput.

The ETPMachineMap is sorted by increasing order of *ETPSum* values. The machine with the lowest *ETPSum* will be the target machine to be removed in this round of scale-in. Operator migration to machines with lower *ETPSum*s will have less of an effect on the overall performance, since machines with lower *ETPSum*s contribute less to the overall performance. This approach also helps shorten the amount of downtime the application experiences because of the rescheduling. Operators from the machine that is chosen to be removed are reassigned to the remaining machines in the cluster in a round-robin fashion in increasing order of their *ETPSum*s.

After the schedule is created, Stela commits it by migrating operators from the selected machines, and then releases these machines. The scale-in algorithm involves sorting of *ETPSum*, which results in a running time complexity of $O(nlog(n))$.

6.5.1.7 Core Architecture

Stela runs as a custom scheduler in a Java class that implements a predefined IScheduler interface in Storm. A user can specify which scheduler to use in a YAML-formatted configuration file called *storm.yaml*. Our scheduler runs as part of the Storm Nimbus daemon. The architecture of Stela's implementation in Storm is presented in Figure 6.9. It consists of the following three modules:

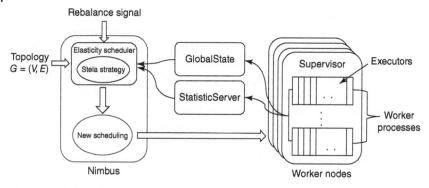

Figure 6.9 Stela architecture.

- *StatisticServer:* This module is responsible for collecting statistics in the Storm cluster, for example, the throughput at each task, at each bolt, and for the topology. These data are used as input to the congestion detection.
- *GlobalState:* This module stores important state information regarding the scheduling and performance of a Storm cluster. It holds information about where each task is placed in the cluster. It also stores statistics such as sampled throughputs of incoming and outgoing traffic for each bolt for a specific duration, and these statistics are used to identify congested operators, as mentioned in Section 6.5.1.3.
- *ElasticityScheduler:* This module is the custom scheduler that implements the IScheduler interface. This class starts the StatisticServer and GlobalState modules and invokes the Strategy module when needed.
  - *Strategy:* Contained inside ElasticityScheduler, this provides an interface for implementation of scale-out strategies so that different strategies can be easily swapped in and out for evaluation purposes. This module calculates a new schedule based on the scale-in or scale-out strategy in use and uses information from the Statistics and GlobalState modules. The core Stela policy (Sections 6.5.1.3 and 6.5.1.6) and alternative strategies [37] are implemented here.

When a scale-in or scale-out signal is sent by the user to the ElasticityScheduler, a procedure is invoked that detects newly joined machines based on previous membership. The ElasticityScheduler invokes the Strategy module, which calculates the entire new scheduling; for example, for scale-out, it decides on all the newly created executors that need to be assigned to newly joined machines. The new scheduling is then returned to the ElasticityScheduler, which atomically (at the commit point) changes the current scheduling in the cluster. Computation is thereafter resumed.

Fault Tolerance

When no scaling is occurring, failures are handled the same way as in Storm, that is, Stela inherits Storm's fault tolerance. If a failure occurs during a scaling operation, Stela's scaling will need to be aborted and restarted. If the scaling is already committed, failures are handled as in Storm.

6.5.1.8 Evaluation

Our evaluation is two-pronged, and includes use of both microbenchmark topologies and real topologies (including two from Yahoo!). We adopted this approach because of the absence of standard benchmark suites (e.g., TPC-H or YCSB) for stream processing systems. Our microbenchmarks include small topologies such as star, linear, and diamond, because we believe that most realistic topologies will be a combination of these. We also use two topologies from Yahoo! Inc., which we call the Page Load topology and Processing topology, as well as a Network Monitoring topology [29]. In addition, we present a comparison among Stela, the Link Load Strategy [37], and Storm's default scheduler (which is state of the art). We present only selected experimental results in this chapter and refer the reader to Ref. [120] for extensive information on the experiments and evaluation of our system.

6.5.1.9 Experimental Setup

For our evaluation, we used two types of machines from the Emulab [111] test bed to perform our experiments. Our typical Emulab setup consisted of a number of machines running Ubuntu 12.04 LTS images, connected via a 100 Mpbs VLAN. A type 1 machine had one 3 GHz processor, 2 GB of memory, and 10,000 RPM 146 GB SCSI disks. A type 2 machine had one 2.4 GHz quad-core processor, 12 GB of memory, and 750 GB SATA disks. The settings for all topologies tested are listed in Table 6.3. For each topology, the same scaling operations were applied to all strategies.

6.5.1.10 Yahoo! Storm Topologies and Network Monitoring Topology

We obtained the layout of a topology in use at Yahoo! Inc. We refer to this topology as the *Page Load* topology (which is not its original name). The layout of the Page Load topology is displayed in Figure 6.10a and the layout of the Network Monitoring topology, which we derived from Ref. [29], is displayed in Figure 6.10b.

We examine the performance of three scale-out strategies: default, Link-based [37], and Stela. The throughput results are shown in Figure 6.11. Recall that link-load-based strategies reduce the network latency of the workflow by colocating communicating tasks on the same machine.

From Figure 6.11, we observe that Stela improves the throughput by 80% after a scale-out of the Page Load topology. In comparison, the Least Link Load

Table 6.3 Experiment settings and configurations.

Topology type	Number of tasks per component	Initial number of executors per component	Number of worker processes	Initial cluster size	Cluster size after scaling	Machine type
Page Load	8	4	28	7	8	1
Network	8	4	32	8	9	2
Page Load Scale-in	15	15	32	8	4	1

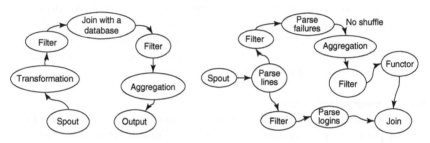

(a) Layout of page load topology.　　　(b) Layout of network monitoring topology [60].

Figure 6.10 Yahoo! topology and a Network Monitoring topology derived from Ref. [29].

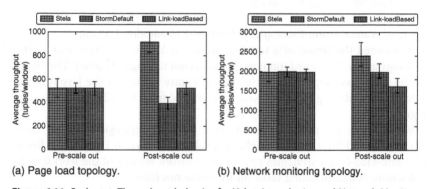

(a) Page load topology.　　　(b) Network monitoring topology.

Figure 6.11 Scale-out: Throughput behavior for Yahoo! topologies and Network Monitoring topology (Window size of 10 s).

strategy barely improves the throughput after a scale-out, because migration of tasks that are not resource-constrained will not significantly improve performance. The default scheduler actually decreases the throughput after the scale-out, since it simply unassigns all executors and reassigns them in a round-robin fashion to all machines, including the new ones. That may cause machines with "heavier" bolts to be overloaded, thus creating newer bottlenecks that damage performance, especially for topologies with a linear structure. In comparison, Stela's postscaling throughput is about 80% better than Storm's for both the Page Load and Network Monitoring topologies, indicating that Stela is able to find the most congested bolts and paths and give them more resources.

In addition to the Page Load and Network Monitoring topologies, we also looked at a published application from IBM [29], and we wrote from scratch a similar Storm topology (shown in Fig. 6.10b). Because we increased the cluster size from 8 to 9, our experimental result (Figure 6.11b) shows that Stela improves the throughput by 21% by choosing to parallelize the congested operator closest to the sink. The Storm default scheduler does not improve postscale throughput, and the Least Link Load strategy decreases system throughput.

6.5.1.11 Convergence Time

We measured interruptions to ongoing computation by measuring the convergence time. A *convergence time* is the duration of time between the start of a scale-out operation and the stabilization of the overall throughput of the Storm topology. More specifically, the convergence time duration stopping criteria are (i) the throughput oscillates twice above and twice below the average of postscale-out throughput, and (ii) the oscillation is within a small standard deviation of 5%. Thus, a lower convergence time means that the system is less intrusive during the scale-out operation, and it can resume meaningful work earlier.

Figure 6.12a shows the convergence time for the Yahoo! topology. We observe that Stela is far less intrusive than Storm (with an 88% lower convergence time)

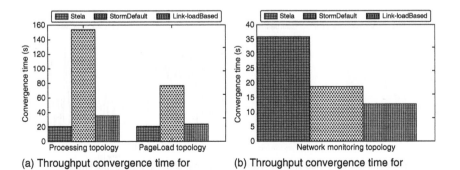

(a) Throughput convergence time for Yahoo! topology.

(b) Throughput convergence time for network monitoring topology.

Figure 6.12 Scale-out: Convergence time comparison (in seconds).

when scaling out. The main reason why Stela has a better convergence time than either Storm's default scheduler or the Least Link Load strategy [37] is that Stela (unlike the other two) does not change the current scheduling at existing machines, instead choosing to schedule operators at the new machines only.

In the Network Monitoring topology, Stela experiences longer convergence times than Storm's default scheduler and the Least Link Load strategy, because of re-parallelization during the scale-out operation (Figure 6.12b). On the other hand, Stela provides the benefit of higher post-scale throughput, as shown in Figure 6.11b.

6.5.1.12 Scale-In Experiments

We examined the performance of Stela scale-in by running Yahoo's Page Load topology. The initial cluster size was 8, and Figure 6.13a shows how the throughput changed after the cluster size shrank to four machines. (We initialized the operator allocation so that each machine could be occupied by tasks from fewer than two operators (bolts and spouts).) We compared the performance against that of a round-robin scheduler (the same as Storm's default scheduler), using two alternative groups of randomly selected machines.

We observe that Stela preserved throughput after scale-in, while the two Storm groups experienced 80% and 40% decreases in throughput, respectively. Thus, Stela's post-scale-in throughput is 25× higher than that obtained when the machines to remove are chosen randomly. Stela also achieved 87.5% and 75% less downtime (time during which the throughput is zero) than group 1 and group 2, respectively (see Fig. 6.13b). The main reason is that Stela's migration of operators with low ETPs will intrude less on the application, which will allow downstream congested components to digest tuples in their queues and continue producing output. In the Page Load topology, the two machines with the lowest ETPs were chosen for redistribution by Stela, and that resulted in less intrusion for the application and, thus, significantly better performance than Storm's default scheduler.

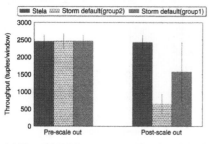

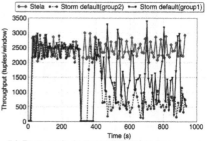

(a) Throughput convergence time for Yahoo! topologies.   (b) Post-scale-in throughput timeline.

Figure 6.13 Scale-in experiments. (Window size is 10 s).

Therefore, Stela is intelligent at picking the best machines to remove (via *ETPSum*). In comparison, Storm has to be lucky. In the above scenario, two out of the eight machines were the "best." The probability that Storm would have been lucky enough to pick both (when it picks 4 at random) $= \binom{6}{2}/\binom{8}{4} = 0.21$, which is low.

6.5.2 Scale-Out/Scale-In in Distributed Graph Processing Systems

This section is based on Ref. [123], and we refer the reader to that publication for further details on design, implementation, and experiments.

6.5.2.1 Motivation

Large graphs are increasingly common; examples include online social networks such as Twitter and Facebook, Web graphs, Internet graphs, biological networks, and many others. Processing and storing these graphs in a single machine is not feasible. Google's Pregel [124] and GraphLab [125] were the first attempts at processing these graphs in a distributed way. Subsequently, the research community has developed more efficient engines that adopt the vertex-centric approach for graph processing, such as LFGraph [126], PowerGraph [48], and GPS [52].

Today's graph processing frameworks operate on statically allocated resources; the user must decide on resource requirements before an application starts. However, partway through computation, the user may wish to scale-out (e.g., to speed up computation) or scale-in (e.g., to reduce hourly costs). The capability to scale-out/scale-in when required by the user is called *on-demand elasticity*. Alternatively, an adaptive policy may request scale-out or scale-in.[8] Such a concept has been explored for data centers [38,39], cloud systems [40–42], storage systems [13,43–45,127], and data processing frameworks such as Hadoop [46,47] and Storm [37]. However, on-demand elasticity remains relatively unexplored in batch-distributed graph processing systems.

Partitioning techniques have been proposed to optimize computation and communication [48], but they partition the entire graph across servers and are thus applicable only at the start of the graph computation. On-demand elasticity requires an *incremental* approach to (re-)partitioning vertices on demand. Solving the problem of on-demand elasticity is also the first step toward adaptive elasticity (e.g., satisfying an SLA in a graph computation), for which our techniques may be employed as black boxes.

---

8 This chapter does not deal with the details of adaptive policies or triggers. We focus instead on the on-demand mechanisms, which are an important building block for such policies.

A distributed graph processing system that supports on-demand scale-out/scale-in must overcome three challenges:

1) *The need to perform scale-out/scale-in without interrupting graph computation.* A scale-out/scale-in operation requires a reassignment of vertices among servers. During scale-out, new servers must obtain some vertices (and their values) from existing servers. Similarly, during scale-in, vertices from the departing servers must be reassigned to the remaining servers. These transfers must be done while minimally affecting ongoing computation times.

2) *The need to minimize the background network overhead involved in the scale-out/scale-in.* To reduce the impact of the vertex transfer on computation time, we wish to minimize the total amount of vertex data transferred during scale-out/scale-in.

3) *The need to mitigate stragglers by maintaining load balance across servers.* Graph processing proceeds in iterations, and stragglers will slow an entire iteration down. Thus, while reassigning vertices at the scale-out/scale-in point, we aim to achieve load balance in order to mitigate stragglers and keep computation time low.

Our approach to solving the problem of on-demand elasticity and overcoming the above challenges is motivated by two critical questions:

1) *What should be migrated, and how?* Which vertices from which servers should be migrated in order to reduce the network transfer volume and maintain load balance?

2) *When should they be migrated?* At what points during computation should migration begin and end?

To answer the first question, we created and analyzed two techniques. The first, called contiguous vertex repartitioning (CVR), achieves load balance across servers. However, it may result in high overhead during the scale-out/scale-in operation. Thus, we developed a second technique, called ring-based vertex repartitioning (RVR), that relies on ring-based hashing to lower the overhead. To address the second question, of when to migrate, we integrated our techniques into the LFGraph graph processing system [126], and used our implementation to carefully decide when to begin and end background migration, and when to migrate static versus dynamic data. We also use our implementation to explore system optimizations that make migration more efficient.

We performed experiments with multiple graph benchmark applications on a real Twitter graph with 41.65 million vertices and 1.47 billion edges. Our results indicate that our techniques are within 9% of an optimal mechanism for scale-out operations and within 21% for scale-in operations.

6.5.2.2 What to Migrate, and How?

In this section, we address the question of which vertices to migrate when the user requests a scale-out/scale-in operation.

Contiguous Vertex Repartitioning (CVR)

Our first technique assumes that the hashed vertex space is divided into as many partitions as there are servers, and each server is assigned one partition. Partitions are equisized in order to accomplish load balancing. The top of Figure 6.14a shows an example graph containing 100 vertices, split across 4 servers. The vertex sequence (i.e., V_i) is random but consistent due to our use of consistent hashing and is split into four equisized partitions, which are then assigned to servers S_1–S_4 sequentially.

Upon a scale-out/scale-in operation, the key problem we need to solve is, how do we assign the (new) equisized partitions to servers (one partition per server), such that network traffic volume is minimized? For instance, the bottom of Figure 6.14 shows the problem when scaling out from four to five servers. To solve this problem, we now (i) show how to reduce the problem to one of graph matching, and (ii) propose an efficient heuristic.

When we scale-out/scale-in, we repartition the vertex sequence into equisized partitions. Assigning these new partitions in an arbitrary fashion to servers may be suboptimal and involve transfer of large amounts of vertex data across the network. For instance, in the bottom of Figure 6.14a, we scale-out by adding one server, resulting in five new partitions. Merely adding the new server to the end of the server sequence and assigning partitions to servers in that order results in movement of 50 total vertices. On the other hand, Figure 6.14b shows the optimal solution for this example, wherein adding the new server in the middle of the partition sequence results in movement of only 30 vertices.

To achieve the optimal solution, we consider the scale-out problem formally. (The solution for scale-in is analogous and excluded for brevity.) Let the cluster initially have N servers $S_1, \ldots, S_N$. With a graph of V vertices, the initial size of each partition P_i^{old} is $M^{old} = \frac{V}{N}$, where $1 \leq i \leq N$. Each jth vertex ID is hashed, and

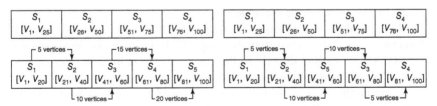

(a) Suboptimal partition assignment with total vertex transfer = 50.

(b) Optimal partition assignment with total vertex transfer = 30.

Figure 6.14 Scale-out from four (top) to five (bottom) servers using CVR. Fewer vertices are transferred in the optimal partition assignment (30 versus 50).

then the resulting value V_j is used to assign the vertex to partition P_i^{old}, where $i = \lceil * \rceil \frac{V_j}{M^{old}}$. If we add k servers to this cluster, the size of each new partition becomes $M^{new} = \frac{V}{N+k}$. We label these new partitions P_i^{new}, $1 \leq i \leq N+k$, and assign each jth vertex, as usual, to a new partition by first hashing the vertex ID and using the resulting hash V_j to partition P_i^{new}, where $i = \lceil * \rceil \frac{V_j}{M^{new}}$.

Next, we create a bipartite graph B, which contains (i) a left set of vertices, with one vertex per new partition P_i^{new}, and (ii) a right set of vertices, with one vertex per server S_j. The left and right sets each contain $(N+k)$ vertices. The result is a complete bipartite graph, with the edge joining a partition P_i^{new} and a server S_j associated with a cost. The cost is equal to the number of vertices that must be transferred over the network if partition P_i^{new} is assigned to server S_j after scale-out. In other words, the cost is equal to $\left| P_j^{old} \cap P_i^{new} \right| = \left| P_j^{old} \right| + \left| P_i^{new} \right| - \left| P_j^{old} \cup P_i^{new} \right|$.

The problem of minimizing network transfer volume now reduces to that of finding a minimum-cost perfect matching in B. This is a well-studied problem, and an optimal solution can be obtained by using the Hungarian algorithm [128]. However, the Hungarian algorithm has $O(N^3)$ complexity [129], which may be prohibitive for large clusters.

As a result, we propose a greedy algorithm that iterates sequentially through $S_1, \ldots, S_N$, in that order.[9] For each server S_j, the algorithm considers the new partitions with which it has a nonzero overlap; because of the contiguity of partitions, there are only $O(1)$ such partitions. Among these partitions, the one with the largest number of overlapping vertices with P_j^{old} is assigned to server S_j. Because of the linear order of traversal, when S_j is considered, S_j is guaranteed to have at least one (overlapping) candidate position. This makes the greedy algorithm run efficiently in $O(N)$. For example, in Figure 6.14b, to determine the new partition for S_1, we need to consider only two partitions, $[V_1, V_{20}]$ and $[V_{21}, V_{40}]$; next, we need to consider partitions $[V_{21}, V_{40}]$ and $[V_{41}, V_{60}]$, and so on.

Ring-Based Vertex Repartitioning (RVR)

In this technique, we assume an underlying hash-based partitioning that leverages Chord-style consistent hashing [127]. To maintain load balance, servers are not hashed directly to the ring; instead (as in Cassandra [2] and Riak [115]), we assume that each server is assigned an equisized segment of the ring. Specifically, a server with ID n_i is responsible for vertices hashed in the interval $(n_{i-1}, n_i]$, where n_{i-1} is n_i's predecessor.

---

9 A similar problem was studied in Ref. [114], but its approach is not load-balanced.

Under that assumption, performing a scale-out/scale-in operation is straight-forward: a joining server splits a segment with its successor, while a leaving server gives up its segment to its successor. For instance, in a scale-out operation involving one server, the affected server receives its set of vertices from its successor in the ring, that is, a server n_i takes the set of vertices $(n_{i-1}, n_i]$ from its successor n_{i+1}. Scale-in operations occur symmetrically: a leaving server n_i migrates its vertex set $(n_{i-1}, n_i]$ to its successor n_{i+1}, which is then responsible for the set of vertices in $(n_{i-1}, n_{i+1}]$.

More generally, we can state that a scale-out/scale-in operation that involves simultaneous addition or removal of k servers affects at most k existing servers. If some of the joining or leaving servers have segments that are adjacent, the number of servers affected would be smaller than k.

While the technique is minimally invasive to existing servers and the ongoing graph computation, it may result in load imbalance. We can mitigate load imbalance for the scale-out case by choosing intelligently the point on the ring to which the new server(s) should be added. For the scale-in case, we can intelligently decide which server(s) to remove.

Consider a cluster with N servers, each with V/N vertices. If we use CVR to add $m \times N$ servers or remove $(m \times N)/(m+1)$ servers (for $m \geq 1$), then the optimal position of servers to be added or removed is same as their position with RVR.

6.5.2.3   When to Migrate?

Given the knowledge of which vertices must be migrated and to where, we must now decide *when* to migrate them in a way that minimizes interference with normal execution. Two types of data need to be migrated between servers: (i) static data, including sets of vertex IDs, neighboring vertex IDs, and edge values to neighbors, and (ii) dynamic data, including the latest values of vertices and latest values of neighbors. Static data correspond to graph partitions, while dynamic data represent computation state. Once this migration is complete, the cluster can switch to the new partition assignment.

Executing Migration

LFGraph uses a publish–subscribe mechanism. Before the iterations start, each server subscribes to in-neighbors of the vertices hosted by the server. Based on these subscriptions, each server builds a publish list for every other server in the cluster. After each iteration, servers send updated values of the vertices present in the publish lists to the respective servers. After a scale-out/scale-in operation, we perform the publish–subscribe phase again to update the publish lists of servers.

First-Cut Migration

A first-cut approach is to perform migration of both static and dynamic data during the next available barrier synchronization interval. However, when we implemented this approach, we found that it added significant overheads by prolonging that iteration. As a result, we introduced two further optimizations as follows:

Static Data Migration

This technique is based on the observation that static data can be migrated in the background while computation is going on. Recall that static data consists of vertex IDs, their neighboring vertex IDs, and edge values to neighbors. Only dynamic data (vertex values and neighboring vertex values) need to wait to be migrated during a barrier synchronization interval (i.e., after such data are last updated). This reduces the overhead on that iteration.

Dynamic Data Migration

LFGraph has two barrier synchronization intervals. One interval is between the gather and scatter phases, and the other is after the scatter phase. That gives us two options for the transfer of dynamic data. We choose to perform dynamic data transfer and cluster reconfiguration in the barrier synchronization interval *between* the gather and scatter phases. This enables us to leverage the default scatter phase to migrate neighboring vertex values. The scatter phase simply considers the new set of servers in the cluster while distributing updated vertex values. This optimization further reduces the overhead on the iteration.

A scale-out/scale-in operation that starts in iteration i ends in iteration $i + 2$. Background static data migration occurs in iterations i and $i + 1$, while vertex value migration occurs after the gather phase of iteration $i + 2$. At that point, computation continues on the new set of servers. The performance impact due to background data migration is greater in iteration i than in iteration $i + 1$, that is, iteration times are longer in iteration i. The reason is that a majority of the migration happens in iteration i. In iteration $i + 1$ servers build their new subscription lists for the publish–subscribe phase.

To explain further, we will describe the steps involved in a scale-out as follows: (i) The joining server sends a *Join* message containing its IP address and port to the barrier server at the start of iteration i. (ii) The barrier server responds with a *Cluster Info* message assigning the joining server an ID and the contact information of the servers from which it should request its vertices. (iii) In addition, the barrier server sends an *Add Host* message to all servers, informing them about the new server in the cluster. (iv) The joining server requests its vertices with a *Vertex Request* message. (v) After receiving its vertices, it informs the barrier server with a *Ready* message that it can join the cluster. Concurrently, the servers start updating their subscription lists to reflect the modifications in the cluster servers. (vi) The barrier server sends a *Reconfigure* message to the servers in the synchronization interval after the gather phase of iteration $i + 2$. (vii) Upon receiving the Reconfigure message, joining servers request the vertex values with a *Vertex Value Request* message. In addition, all servers update their vertex-to-server mapping to reflect newly added servers. (viii) The scatter phase of iteration $i + 2$ executes with this new mapping. From then on, computation proceeds on the new set of servers.

Role of Barrier Server

In our repartitioning techniques, the barrier server accepts join and leave requests and determines an optimal partition assignment. We adopted this approach, instead of a fully decentralized reassignment, for two reasons: (i) fully decentralized reassignment may lead to complex race conditions, and (ii) the barrier server, once initialized, has the capability to obtain per-server iteration run times via the barrier synchronization messages and assigns new servers to alleviate the load on the busiest servers.

6.5.2.4 Evaluation

In this section, we describe our experimental evaluation of the efficiency and overhead of our elasticity techniques. We present only selected experimental results in this chapter and refer the reader to Ref. [123] for extensive information on experiments and evaluation of our system.

Experimental Setup

We performed our experiments with both our CVR and RVR techniques on virtual instances, each with 16 GB RAM and 8 VCPUs. We used a Twitter graph [130] containing 41.65 million vertices and 1.47 billion edges. (With larger graphs, we expect similar performance improvements.) We evaluated our techniques using five graph benchmarks: PageRank, single-source shortest paths (SSSP), connected components, k-means clustering, and multiple-source shortest paths (MSSP).

Scale-Out and Scale-In

Our first set of experiments measured the overhead experienced by the computation because of a scale-out operation. Figure 6.15 illustrates two experiments in which a scale-out from X servers to $2X$ servers (for $X \in \{5, 10, 15\}$) was performed, with the scale-out starting at iteration $i = 1$ and ending at

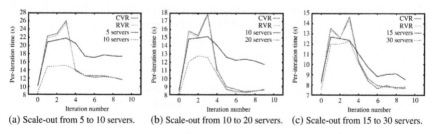

(a) Scale-out from 5 to 10 servers. (b) Scale-out from 10 to 20 servers. (c) Scale-out from 15 to 30 servers.

Figure 6.15 Per-iteration execution time with scale-out at iterations $i = 1$ to $i = 3$, for different repartitioning strategies and cluster sizes.

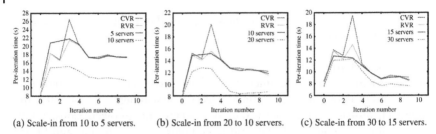

(a) Scale-in from 10 to 5 servers.　(b) Scale-in from 20 to 10 servers.　(c) Scale-in from 30 to 15 servers.

Figure 6.16 Per-iteration execution time with scale-out at iterations $i = 1$ to $i = 3$, for different repartitioning strategies and cluster sizes.

iteration 3. The vertical axis plots the *per-iteration* run time. For comparison, we plot the per-iteration times for a run with X servers throughout, and a run with $2X$ servers throughout.

In Figure 6.15a–c, we can observe that (i) both CVR and RVR appear to perform similarly, and (ii) after the scale-out operation is completed, the performance of the scaled-out system converges to that of a cluster with $2X$ servers, demonstrating that our approaches converge to the desired throughput after scale-out.

Similarly, Figure 6.16 shows the plots for scale-in from $2X$ servers to X servers (for $X \in \{5, 10, 15\}$). Once again, the cluster converges to the performance of X servers.

6.6 Priorities and Deadlines in Batch Processing Systems

6.6.1 Natjam: Supporting Priorities and Deadlines in Hadoop

This section is based on Ref. [131], and we encourage the reader to refer to that publication for further details on design, implementation, and experiments.

6.6.1.1 Motivation

Today, computation clusters running engines such as Apache Hadoop [3,132], DryadLINQ [87], DOT [133], Hive [134], and Pig Latin [135] are used to process a variety of big data sets. The batch MapReduce jobs in these clusters have priority levels or deadlines. For instance, a job with a high priority (or short deadline) may be one that processes click-through logs and differentiates ads that have reached their advertiser targets from ads that it would be good to display. For such jobs, it is critical to produce timely results, since they directly affect revenue. On the other hand, a lower priority (or long-deadline) job may, for instance, identify more lucrative ad placement patterns via a machine learning algorithm on long-term historical click data. Such jobs affect revenue indirectly and therefore need to complete soon, but they must be treated as lower priority.

The most common use case is a *dual-priority* setting, with only two priority levels: high-priority jobs and low-priority jobs. We call the high-priority jobs *production jobs* and the low-priority ones *research jobs*.[10] A popular approach among organizations is to provision two physically separate clusters: one for production jobs and one for research jobs. Administrators tightly restrict the workloads allowed on the production cluster, perform admission control manually based on deadlines, keep track of deadline violations via alert systems such as pagers, and subsequently readjust job and cluster parameters manually.

In addition to requiring intensive human involvement, the above approach suffers from (i) long job completion times, and (ii) inefficient resource utilization. For instance, jobs in an overloaded production cluster might take longer, even though the research cluster is underutilized (and vice versa). In fact, MapReduce cluster workloads are time-varying and unpredictable, for example, in the Yahoo! Hadoop traces we used in the work described here, hourly job arrival rates exhibited a max–min ratio as high as 30. Thus, there are times when the cluster is *resource-constrained*, that is, it has insufficient resources to meet incoming demand. Since physically separate clusters cannot reclaim resources from each other, the infrastructure's overall resource utilization stays suboptimal.

The goals of the work described here are (i) to run a consolidated MapReduce cluster that supports all jobs, regardless of their priority or deadline; (ii) to achieve low completion times for higher priority jobs; and (iii) to do so while still optimizing the completion times of lower priority jobs. The benefits are high cluster resource utilization, and, thus, reduced capital and operational expenses.

Natjam[11] achieves the above goals, and we have integrated it into the Hadoop YARN scheduler (Hadoop 0.23). Natjam's first challenge is to build a unified scheduler for all job priorities and deadlines in a way that fluidly manages resources among a heterogeneous mix of jobs. When a higher priority job arrives in a full cluster, today's approaches involve either killing lower priority jobs' tasks [59,84] or waiting for them to finish [83]. The former approach prolongs low-priority jobs because they repeat work, while the latter prolongs high-priority jobs. Natjam solves those problems by using an *on-demand checkpointing* technique that saves the state of a task when it is preempted, so that it can resume where it left off when resources become available. This checkpointing is fast, inexpensive, and automatic in that it requires no programmer involvement.

Natjam's second challenge is to enable quick completion of high-priority jobs, but not at the expense of extending many low-priority jobs' completion times. Natjam addresses this by leveraging smart *eviction* policies that select which

---

10 This terminology is from our use cases.
11 *Natjam* is the Korean word for nap.

low-priority jobs and their constituent tasks are affected by arriving high-priority jobs. Natjam uses a two-level eviction approach: It first selects a victim job (via a job eviction policy) and then, within that job, one or more victim tasks (via a task eviction policy). For the dual-priority setting with only two priority levels, our eviction policies take into account (i) resources utilized by a job, and (ii) time remaining in a task. We then generalize to arbitrary real-time job deadlines via eviction policies based on both a job's deadline and its resource usage.

We provide experimental results from deployments on a test cluster, both on Emulab and on a commercial cluster at Yahoo!. Our experiments used both synthetic workloads and Hadoop workloads from Yahoo! Inc. We evaluated various eviction policies and found that compared to their behavior in traditional multiprocessor environments, eviction policies have counterintuitive behavior in MapReduce environments; for example, we discovered that longest-task-first scheduling is optimal for MapReduce environments. For the dual-priority setting, Natjam incurs overheads of under 7% for all jobs. For the real-time setting with arbitrary deadlines, our generalized system, called Natjam-R, meets deadlines with only 20% extra laxity in the deadline compared to the job runtime.

In brief, at a high level, our work is placed within the body of related work as follows (see Section 6.3.4 for more details). Our focus is on batch jobs rather than streaming or interactive workloads [71–74,76,77]. Some systems have looked at preemption in MapReduce [55], with respect to fairness [86], at intelligent killing of tasks [59] (including the Hadoop Fair Scheduler [84]), and in SLOs (service level objectives) in generic cluster management [90,93,94]. In comparison, our work is the first to study the effects of eviction policies and deadline-based scheduling for resource-constrained MapReduce clusters. Our strategies can be applied orthogonally in systems such as Amoeba [55]. We are also the first to incorporate such support directly into Hadoop YARN. Finally, MapReduce deadline scheduling has been studied in infinite clusters [62–65] but not in resource-constrained clusters.

6.6.1.2 Eviction Policies for a Dual-Priority Setting

This section presents the eviction policies, and the following section describes the systems architecture. Section 6.1.4 generalizes the solution to the case where jobs have multiple priorities.

Eviction policies lie at the heart of Natjam. When a production (high-priority) MapReduce job arrives at a resource-constrained cluster and there are insufficient resources to schedule it, some tasks of research (low-priority) MapReduce jobs need to be preempted. Our goals here are to minimize job completion times both for production and for research jobs. This section addresses the twin questions of (i) how to choose a victim (research) job so that some of its tasks can be preempted, and (ii) within a given victim job, how to choose victim task (s) for preemption. We call these *job eviction* and *task eviction* policies, respectively.

The job and task eviction policies are applied in tandem, that is, for each required task of the arriving production job, a running research task is evicted through application of the job eviction policy followed by the task eviction policy. A research job chosen as victim may be evicted only partially; in other words, some of its tasks may continue running, for example, if the arriving job is relatively small, or if the eviction policy also picks other victim research jobs.

Job Eviction Policies

The choice of victim job affects the completion time of lower priority research jobs by altering resources already allocated to them. Thus, job eviction policies need to be sensitive to current resource usage of individual research jobs. We discuss three resource-aware job eviction policies:

Most Resources (MR): This policy chooses as victim the research job that is currently using the most resources inside the cluster. In Hadoop YARN, resource usage would be in terms of the number of containers used by the job, while in other versions of MapReduce, it would be determined by the number of cluster slots.[12]

The MR policy, which is loosely akin to the worst-fit policy in OS segmentation, is motivated by the need to evict as few research jobs as possible; a large research job may contain sufficient resources to accommodate one large production job or multiple small production jobs. Thus, fewer research jobs are deferred, more of them complete earlier, and average research job completion time is minimized.

The downside of the MR policy is that when there is one large research job (as might be the case with heavy tailed distributions), it is always victimized whenever a production job arrives. This may lead to starvation and thus longer completion times for large research jobs.

Least Resources (LR): In order to prevent starving of large research jobs, this policy chooses as victim the research job that is currently using the least resources inside the cluster. The rationale here is that small research jobs that are preempted can always find resources if the cluster frees up even a little in the future. However, the LR policy can cause starvation for small research jobs if the cluster stays overloaded; for example, if a new production job arrives whenever one completes, LR will pick the same smallest jobs for eviction each time.

Probabilistically Weighted on Resources (PR): In order to address the starvation issues of LR and MR, our third policy selects a victim job using a probabilistic metric based on resource usage. In PR, the probability of choosing a job as a victim is directly proportional to the resources it currently holds. In effect, PR treats all tasks identically in choosing ones to evict, that is, if the task

---

12 While our approach is amenable to more fine-grained notions of resources, resulting issues such as fragmentation are beyond the scope of this chapter. Thus, we use containers that are equi-sized.

eviction policy were random, the chances of eviction for all tasks would be identical and independent of their jobs. The downside of PR is that it spreads out evictions across multiple jobs; in PR, unlike MR, one incoming production job may slow down multiple research jobs.

Task Eviction Policies

Once a victim job has been selected, the task eviction policy is applied within that job to select one task that will be preempted (i.e., suspended).

Our approach makes three assumptions, which are based on use case studies: (i) reduces are long enough that preemption of a task takes less time than the task itself; (ii) only reduce tasks are preempted; and (iii) reduces are stateless in between keys. For instance, in Facebook workloads the median reduce task takes 231 s [58], substantially longer than the time needed to preempt a task (see Section 6.1.5). There are two reasons why we focus on preemption only of reduces. First, the challenge of checkpointing reduce tasks subsumes that of checkpointing map tasks, since a map processes individual key-value pairs, while a reduce processes batches of them. Second, several use case studies have revealed that reduces are substantially longer than maps and thus have a bigger effect on the job tail. In the same Facebook trace already mentioned, the median map task time is only 19 s. While 27.1 map containers are freed per second, only 3 (out of 3100) reduce containers are freed per second. Thus, a small production job with 30 reduces would wait on average 10 s, and a large job with 3000 reduces would wait 1000 s. Finally, the traditional stateless reduce approach is used in many MapReduce programs; however, Natjam could be extended to support stateful reducers.

A MapReduce research job's completion time is determined by its last-finishing reduce task. A long tail, or even a single task that finishes late, will extend the research job's completion time. This concern implies that tasks with a shorter remaining time (for execution) must be evicted first. However, in multiprocessors, shortest-task-first scheduling is known to be optimal [53]. In our context, this means that the task with the longest remaining time must be evicted first. That motivated two contrasting task eviction policies, SRT and LRT:

Shortest Remaining Time (SRT): In this policy, tasks that have the shortest remaining time are selected to be suspended. This policy aims to minimize the impact on the tail of a research job. Further, a task suspended by SRT will finish quickly once it has been resumed. Thus, SRT is loosely akin to the longest-task-first strategy in multiprocessor scheduling. Rather counterintuitively, SRT is provably optimal under certain conditions:

Theorem 6.1 Consider a system in which a production job arrival affects exactly one victim job and evicts several tasks from it. If all the evicted tasks are resumed simultaneously in the future, and we ignore speculative execution, then the SRT eviction policy results in an optimal (lowest) completion time for that research job.

Proof: No alternative policy can do better than SRT in two submetrics: (i) the sum of time remaining for evicted tasks, and (ii) the tail (i.e., max) of time remaining among the evicted tasks. Thus, when the evicted tasks resume simultaneously, an alternative eviction policy can do only as well as SRT in terms of the completion time of the research job.

We note that the assumption – that tasks of the victim job will be resumed simultaneously – is reasonable in real-life scenarios in which production job submission times and sizes are unpredictable. Our experiments also validated this theorem.

Longest Remaining Time (LRT): In this policy, the task with the longest remaining time is chosen to be suspended earlier. This policy is loosely akin to shortest-task-first scheduling in multiprocessors. Its main advantage over SRT is that it is less selfish and frees up more resources earlier. LRT might thus be useful in scenarios in which production job arrivals are bursty. Consider a victim job containing two tasks: one short and one with a long remaining time. SRT evicts the shorter task, freeing up resources for one production task. LRT evicts the longer task, but the shorter unevicted task will finish soon anyway, thus releasing resources for two production tasks, while incurring the overhead for only one task suspension. However, LRT can lengthen the tail of the research job, increasing its completion time.

6.6.1.3 Natjam Architecture

In order to understand the design decisions required to build eviction policies into a MapReduce cluster management system, we incorporated Natjam into the popular Hadoop YARN framework in Hadoop 0.23. We now describe Natjam's architecture, focusing on the dual-priority setting (for production and research jobs).

Preemption in Hadoop YARN

Background: Hadoop YARN Architecture: In the Hadoop YARN architecture, a single cluster-wide Resource Manager (RM) performs resource management. It is assisted by one Node Manager (NM) per node (server). The RM receives periodic heartbeats from each NM containing status updates about resource usage and availability at that node.

The RM runs the Hadoop Capacity Scheduler. The Capacity Scheduler maintains multiple queues that contain jobs. An incoming job is submitted to one of these queues. An administrator can configure two capacities per queue: a minimum (guaranteed) capacity, and a maximum capacity. The scheduler varies the queue capacity between these two queues based on the jobs that have arrived at them.

The basic unit of resource allocation for a task is called a *container*. A container is effectively a resource slot that contains sufficient resources (primarily memory) to run one task: a map, a reduce, or a master task. An example master task is the Application Master (AM), which is allocated one container. One AM is assigned to each MapReduce job and performs job management functions.

An AM requests and receives, from the RM, container allocations for its tasks. The AM assigns a task to each container it receives and sends launch requests to the container's NM. It also performs speculative execution when needed.

An AM sends heartbeats to the RM. The AM also receives periodic heartbeats from its tasks. For efficiency, YARN piggybacks control traffic (e.g., container requests and task assignments) atop heartbeat messages.

Natjam Components: Natjam entails changes to the Hadoop Capacity Scheduler (at the RM) and the AM, while the NM stays unchanged. Specifically, Natjam adds the following new components to Hadoop YARN:

1) *Preemptor:* The preemptor is a part of the RM. We configured the Capacity Scheduler to contain two queues: one for production jobs and one for research jobs. The preemptor makes preemption decisions by using job eviction policies.

2) *Releaser:* This component is a part of each AM and is responsible for running the task eviction policies.

When we modify Hadoop, instead of adding new messages that will incur overhead, Natjam leverages and piggybacks atop YARN's existing heartbeats for efficiency. The trade-off is a small scheduling delay, but our experiments show that such delays are small.

We will detail those two components in Section 6.1.3. Now we show how preemption and checkpointing work.

Natjam Preemption Mechanism Example: We illustrate how Natjam's preemption works in YARN. Figure 6.17 depicts an example in which a research job 2 is initially executing in a full cluster, when a production Job 1 requires a single container.[13] The steps in the figure are as follows:

Step 1: On AM1's heartbeat, Natjam asks the RM to allocate one container.

Steps 2 and 3: The cluster is full, so the RM applies the job eviction policies and selects Job 2 as victim.

Step 4: The Preemptor waits for AM2's next heartbeat, and in response to the heartbeat sends AM2 the number and type of containers to be released.

Step 5: The Releaser at AM2 uses the task eviction policy to select a victim task.

Step 6: When the victim task (still running) sends its next usual heartbeat to AM2, it is asked to suspend.

---

13 For simplicity, we assume that AM1 (of Job 1) already has a container.

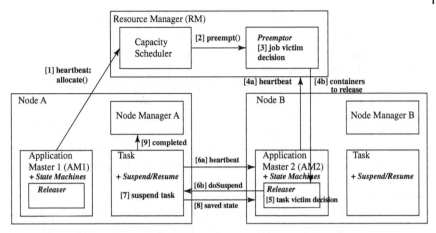

Figure 6.17 Example: Container suspend in Natjam. New components are shown in bold font; others are from YARN. AM1 is a production job, and AM2 is a research job.

Step 7:   The victim task suspends and saves a checkpoint.
Step 8:   The victim task sends the checkpoint to AM2.
Step 9:   The task indicates to NM-A that it has completed and it exits, freeing the container.

With that, the Natjam-specific steps are done. For completeness, we list below the remaining steps, which are taken by YARN by default to give AM1 the new container.

Step 10:   NM-A's heartbeat sends the container to RM.
Step 11:   AM1's next RM heartbeat gets the container.
Step 12:   AM1 sends NM-A the task request.
Step 13:   NM-A launches the task on the container.

Checkpoint Saved and Used by Natjam: When Natjam suspends a research job's reduce task, an on-demand checkpoint is saved automatically. It contains the following items: (i) an ordered list of past *suspended container IDs,* one for each attempt, that is, each time this task was suspended in the past; (ii) a *key counter,* that is, the number of keys that have been processed so far; (iii) *reduce input paths,* that is, local file paths; and (iv) the *hostname* associated with the last suspended attempt, which is useful for preferably resuming the research task on the same server. Natjam also leverages intermediate task data already available via Hadoop [136], including (v) reduce inputs, which are stored at a local host, and (vi) reduce outputs, which are stored on HDFS.

Task Suspend: We modified YARN so that the reduce task keeps track of two pieces of state: paths to files in the local file system that hold reduce input and the key counter, that is, the number of keys that have been processed by the reduce function so far. When a reduce task receives a suspend request from its AM, the task checks whether it is in the middle of processing a particular key, and finishes that key. Second, it writes the input file paths to a local log file. Third, Hadoop maintains a *partial output file* per reduce attempt, in the HDFS distributed file system. It holds the output so far generated from the current attempt. The partial output file is given a name that includes the container ID. When a task suspends, this partial output file is closed. Finally, the reduce compiles its checkpoint and sends the result to its AM, and the reduce task exits.

Task Resume: On a resume, the task's AM sends the saved checkpoint state as launch parameters to the chosen NM. The Preemptor is in charge of scheduling the resuming reduce on a node. The Preemptor prefers the old node on which the last attempt ran (available from the hostname field in the checkpoint). If the resumed task is assigned to its old node, the reduce input can be read without network overhead, that is, from local disk. If it is resumed on a different node, the reduce input is assembled from map task outputs, much like a new task.

Next, the reduce task creates a new partial output file in HDFS. It skips over the input keys that the checkpoint's key counter field indicates have already been processed. It then starts execution as a normal reduce task.

Commit after Resume: When a previously suspended reduce task finishes, it needs to assemble its partial output. It starts that by finding, in HDFS, all its past partial output files; it does so by using the ordered list of past suspended container IDs from its checkpoint. It then accumulates their data into output HDFS files that are named in that order. This order is critical so that the output is indistinguishable from that of a reduce task that was never suspended.

Implementation Issues

This section first explains how we modify the AM state machines in Hadoop YARN, and then describes the Preemptor and Releaser. As mentioned earlier, we leverage existing Hadoop mechanisms such as heartbeats.

Application Master's State Machines: For job and task management, Hadoop YARN's AM maintains separate state machines per job, per task, and per task attempt. Natjam does not change the job state machine; we enabled this state machine only to handle the checkpoint. Thus, both the suspend and the resume occur during the *Running* state in this state machine.

We modify the task state machine very little. When the AM learns that a task attempt has been suspended (from step 8 in Figure 6.17), the task state machine goes ahead and creates a new task attempt to resume the task. However, this does not mean that the task is scheduled immediately; the transitions of the task attempt state machine determine whether it does.

The task attempt state machine is used by YARN to assign the container, set up execution parameters, monitor progress, and commit output. Natjam adds two states to the task attempt state machine, as shown in Figure 6.18: *Suspend-Pending* and *Suspended*. The task attempt has a state of Suspend-Pending when it wishes to suspend a task but has not received suspension confirmation from

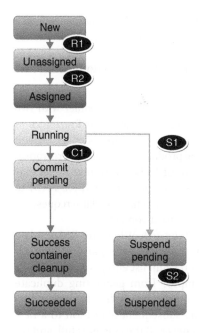

Figure 6.18 Modified task attempt state machine: At application master. Failure states are omitted.

the local task (steps 6b–7 from Figure 6.17). The state becomes Suspended when the saved checkpoint is received (step 8), and this is a terminal state for that task attempt.

The new transitions for suspension in Figure 6.18 are as follows:

- *S1:* AM asks the task to suspend and requests a checkpoint.
- *S2:* AM receives the task checkpoint and saves it in the task attempt state machine.

A resuming reduce task starts from the *New* state in the task attempt state machine. However, we modify some transitions to distinguish a resuming task from a new (nonresuming) task attempt as follows:

- *R1:* Just as for any reduce task attempt, every heartbeat from the AM to RM requests a container for the resuming reduce. If the RM cannot satisfy the request, it ignores it (since the next heartbeat will resend the request anyway). Suppose a container frees up (e.g., as production jobs complete), so that the RM can now schedule a research task. In doing so, the RM prefers to respond to a resuming reduce's request, rather than one from a non-resuming research reduce. The AM to RM requests also carry the hostname field from the task checkpoint; the RM uses it to declare a preference for allocating that container at that hostname.

- *R2:* Once the AM has received a container from the RM, it launches a task attempt on the allocated container. For resuming reduces, the AM also sends the saved checkpoint to the container.
- *C1:* On commit, the AM accumulates partial output files into the final task output in HDFS (see Section 6.1.3).

Preemptor: Recall that Natjam sets up the RM's Capacity Scheduler with two queues: one for production jobs and one for research jobs. The Preemptor is implemented as a thread within the Capacity Scheduler. In order to reclaim resources from the research queue for use by the production queue, the Preemptor periodically runs a *reclaim algorithm*, with sleeps of 1 s between runs. A run may generate reclaim requests, each of which is sent to some research job's AM to reclaim a container (which is step 4 in Figure 6.17). In a sense, a reclaim request is a statement of a production job's intention to acquire a container.

We keep track of a *per-production job reclaim list*. When the RM sends a reclaim request on behalf of a job, an entry is added to the job's reclaim list. When a container is allocated to that job, that reclaim list entry is removed. The reclaim list is needed to prevent the Preemptor from generating duplicate reclaim requests, which might occur because our reliance on heartbeats entails a delay between a container suspension and its subsequent allocation to a new task. Thus, we generate a reclaim request whenever (i) the cluster is full, and (ii) the number of pending container requests from a job is greater than the number of requests in its reclaim list.

In extreme cases, the Preemptor may need to kill a container, for example, if the AM has remained unresponsive for too long. Our threshold to kill a container is reached when a reclaim request has remained in the reclaim list for longer than a killing timeout (12 s). A kill request is sent directly to the NM to kill the container. This bypasses the AM, ensuring that the container will indeed be killed. When a kill request is sent, the reclaim request is added to an expired list. It remains there for an additional time interval (2 s), after which it is assumed the container is dead, and the reclaim request is thus removed from the expired list. With those timeouts, we have never observed killings of any tasks in any of our cluster runs.

Releaser: The Releaser runs at each job's AM and decides which tasks to suspend. Since the task eviction policies discussed in Section 6.1.2 (i.e., SRT and LRT) use the time remaining at the task, the Releaser needs to estimate it. We use Hadoop's default exponentially smoothed task runtime estimator, which relies on the task's observed progress [69]. However, calculating this estimate on demand can be expensive due to the large numbers of tasks. Thus, we have the AM only periodically estimate the progress of all tasks in the job (once a second), and use the latest complete set of estimates for task selection. While the estimates might be stale, our experiments show that this approach works well in practice.

Interaction with Speculative Execution: Our discussion so far has ignored speculative execution, which Hadoop uses to replicate straggler task attempts.

Natjam does not change speculative execution and works orthogonally, that is, speculative task attempts are candidates for eviction. When all attempts of a task are evicted, the progress rate calculation of the task is not skewed, because speculative execution tracks the progress of task attempts rather than the tasks themselves. While this interaction could be optimized further, it works well in practice. Natjam can be further optimized to support user-defined (i.e., per job) task eviction policies that would prioritize the eviction of speculative tasks, but discussion of that is beyond the scope of this chapter.

6.6.1.4 Natjam-R: Deadline-Based Eviction

We have created Natjam-R, a generalization of Natjam that targets environments in which each job has a hard and fixed real-time deadline. Unlike Natjam, which allowed only two priority levels, Natjam-R supports multiple priorities; in the real-time case, a job's priority is derived from its deadline. While Natjam supported inter-queue preemption (with two queues), Natjam-R uses only *intra-queue* preemption. Thus, all jobs can be put into one queue; there is no need for two queues. Jobs in the queue are sorted based on priority.

Eviction Policies

First, for job eviction, we explored two deadline-based policies inspired by the classical real-time literature [78,81]: *Maximum Deadline First* (MDF) and *Maximum Laxity First* (MLF). MDF chooses as victim the running job that has the longest deadline. On the other hand, MLF evicts the job with the highest laxity, where *laxity* = the *deadline* minus the *job's projected completion time*. For MLF, we extrapolate Hadoop's reported job progress rate to calculate a job's projected completion time.

While MDF is a static scheduling policy that accounts only for deadlines, MLF is a dynamic policy that also accounts for a job's resource needs. If a job has an unsatisfactory progress rate, MLF may give it more resources closer to its deadline. It may do so by evicting small jobs with long deadlines. In essence, while MLF may run some long-deadline, high-resource jobs, MDF might starve all long-deadline jobs equally. Further, MLF is fair in that it allows many jobs with similar laxities to make simultaneous progress. However, this fairness can be a shortcoming in scenarios with many short deadlines; MLF results in many deadline misses, while MDF would meet at least some deadlines. Section 6.1.6 describes our experimental evaluation of this issue.

Second, our task eviction policies remain the same as before (SRT, LRT) because the deadline is for the job, not for individual tasks.

In addition to the job and task eviction policies, we need to have a job selection policy. When resources free up, this policy selects a job from among the suspended ones and gives it containers. Possible job selection policies are earliest deadline first (EDF) and least laxity first (LLF). In fact, we implemented these but observed thrashing-like scheduling behavior if the job eviction policy was inconsistent with the job selection policy. For instance, if we used MDF job eviction and LLF job selection, a job selected for eviction by MDF would soon after be selected for resumption by LLF, and thus enter a suspend-resume loop. We concluded that the job selection policy needed to be dictated by the job eviction policy, that is, MDF job eviction implies EDF job selection, while MLF implies LLF job selection.

Implementation

The main changes that differentiate Natjam-R from Natjam are in the RM. In Natjam-R, the RM keeps one Capacity Scheduler queue sorted by decreasing priority. A priority is inversely proportional to the deadline for MDF, and to laxity for MLF. The Preemptor periodically (once a second) examines the queue and selects the first job (say J_i) that still has tasks waiting to be scheduled. Then it considers job eviction candidates from the queue, starting with the lowest priority (i.e., later deadlines or larger laxities) up to J_i's priority. If it encounters a job that still has allocated resources, that job is picked as the victim; otherwise, no further action is taken. To evict the job, the Releaser from Natjam uses the task eviction policy to free a container. Checkpointing, suspend, and resume work in Natjam-R as described earlier for Natjam (see Section 6.1.3).

6.6.1.5 Microbenchmarks

Experimental Plan

We present two sets of experiments, increasing in complexity and scale. This section presents microbenchmarking results for a small Natjam cluster. Section 6.1.6 evaluates Natjam-R. While we present only selected experimental results in this chapter, we refer the reader to Ref. [131] for extensive information on experiments and evaluation of our system.

Microbenchmark Setup

We first evaluated the core Natjam system that supports a dual-priority workload, that is, research and production jobs. We addressed the following questions: (i) How beneficial is Natjam relative to existing techniques? (ii) What is the overhead of the Natjam suspend mechanism? (iii) What are the best job eviction and task eviction policies?

We used a small-scale test bed and a representative workload because this first experimental stage involved exploration of different parameter settings and study of many fine-grained aspects of system performance. A small test bed gave us flexibility.

Our test cluster had seven servers running on a 1 GigE network. Each server had two quad-core processors and 16 GB of RAM, of which 8 GB were configured to run 1 GB-sized Hadoop containers. (Thus, 48 containers were available in the cluster.) One server acted as the Resource Manager, while the

Table 6.4 Microbenchmark settings.

Job	Number of reduces	Average time (s)
Research-XL	47	192.3
Research-L	35	193.8
Research-M	23	195.6
Research-S	11	202.6
Production-XL	47	67.2
Production-L	35	67.0
Production-M	23	67.6
Production-S	11	70.4

other six were workers. Each entity (AM, map task, and reduce task) used one container.

In our experiments, we injected a mix of research and production jobs, as shown in Table 6.4. To reflect job size variation, the job sizes ranged from XL (filling the entire cluster) to S (filling a fourth of the cluster). To mimic use case studies [58], each job had a small map execution time, and was dominated by the reduce execution time. To model variance in task running times, we selected reduce task lengths uniformly from the interval (0.5, 1.0], where 1.0 is the normalized largest reduce task. To emulate computations, we used SWIM [137] to create random keys and values, with thread sleeps called between keys. Shuffle and HDFS traffic were incurred as usual.

The primary metric was job completion time. Each of our data points shows an average and standard deviation over five runs. Unless otherwise noted, Natjam used MR job eviction and SRT task eviction policies.

Natjam versus Existing Techniques

Figure 6.19 compares Natjam to several alternatives versus an ideal setting versus two existing mechanisms in the Hadoop Capacity scheduler, and versus pessimistic killing of tasks (instead of saving the cheap checkpoint). The ideal setting measures each job's completion time when it is executed on an otherwise empty cluster; thus, it ignores resource sharing and context switch overheads. For the second setting, we chose the Hadoop Capacity Scheduler because it represents approaches that we might take with two physically separate clusters sharing the same scheduler. Finally, killing of tasks is akin to approaches such as those described in Ref. [59] and for the Hadoop Fair Scheduler [84].

In this experiment, a Research-XL job was submitted initially to occupy the entire cluster. Then, 50 s later, a Production-S job was submitted. Figure 6.19 shows that killing of tasks (the fourth pair of bars) finished production jobs fast,

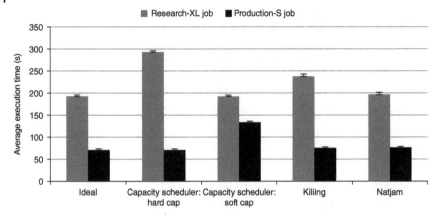

Figure 6.19 Natjam versus existing techniques. At $t = 0$ s, the Research-XL job was submitted, and at $t = 50$ s, the Production-S job was submitted.

but prolonged research jobs by 23% compared to the ideal case (the first pair of bars). Thus, saving the overhead of checkpoints is not worth the repeated work due to task restarts.

We next examined two popular Hadoop Capacity Scheduler approaches called *Hard cap* and *Soft cap*. Recall that the Capacity Scheduler allows the administrator to set a maximum cap on the capacity allocated to each of the two queues (research and production). In Hard cap, that cap is used as a hard limit for each queue. In the Soft cap approach, each queue is allowed to expand to the full cluster if there are unused resources, but it cannot scale down without waiting for its scheduled tasks to finish (e.g., if the production queue needs resources from the research queue). We configured these two approaches with the research queue set to 75% capacity (36 containers) and production queue to 25% capacity (12 containers), as these settings performed well.

Figure 6.19 shows that in Hard cap (the second pair of bars), the research job took 52% longer than ideal, while the production job was unaffected. Under Soft cap (the third pair of bars), the production job could obtain containers only when the research job freed them; this resulted in an 85% increase in production job completion time, while the research job was unaffected.

The last pair of bars shows that when Natjam was used, the production job's completion time was 7% worse (5.4 s longer) than ideal and 77% better than the result for the Hadoop Capacity Scheduler's Soft cap. The research job's completion time was only 2% worse (4.7 s longer) than ideal, 20% better than that of killing, and 49% better than that of Hadoop Hard cap. One of the reasons the research job was close to ideal is that it was able to make progress in parallel with the production job. There are other internal reasons for the performance benefit, which we explore next.

Suspend Overhead

We measured Natjam's suspend overhead on a fully loaded cluster. We observed that it took an average of 1.35 s to suspend a task and 3.88 s to resume a task. Standard deviations were low. In comparison, default Hadoop took an average of 2.63 s to schedule a task on an empty cluster. From this it might appear that Natjam incurs a higher total overhead of 5.23 s per task suspend-resume. However, in practice the effective overhead is lower; for instance, Figure 6.19 showed only a 4.7 s increase in research job completion time. The reason is that task suspends typically occur in parallel, and in some cases, task resumes do too. Thus, the time overheads are parallelized rather than aggregated.

Task Eviction Policies

We now compare the two task eviction policies (SRT and LRT) from Section 6.1.2 against each other, and against a random eviction strategy that we also implemented. We performed two sets of experiments: one with Production-S and another with Production-L. The production job was injected 50 s after a Research-XL job.

Table 6.5 tabulates the results. In all cases the production job incurred overhead similar to that for an empty cluster. Thus, we discuss only research job completion time (last column). As shown in the top half of the table, a random task eviction strategy resulted in a 45 s increase in completion time compared to the ideal; we observed that a fourth of the tasks were suspended, leading to a long job tail. Evicting the LRT incurred a higher increase of 55 s because LRT prolongs the tail. Evicting the SRT emerged as the best policy and was only 4.7 s worse than the ideal because it respects the job tail.

In the lower half of Table 6.5, it can be seen that a larger production job caused more suspensions. The research job completion times for the random and LRT eviction policies are similar to those in the top half because the job's tail was already long for the small production job, and was not much longer for the larger job. SRT is worse than it was for a small production job, yet it outperformed the other two eviction strategies.

Table 6.5 Task eviction policies.

Task eviction policy	Production job	Mean (S.D.) runtime (s)	Research job	Mean (S.D.) runtime (s)
Random	Production-S	76.6 (3.0)	Research-XL	237.6 (7.8)
LRT	Production-S	78.8 (1.8)	Research-XL	247.2 (6.3)
SRT	Production-S	75.6 (1.5)	Research-XL	197.0 (5.1)
Random	Production-L	75.0 (1.9)	Research-XL	244.2 (5.6)
LRT	Production-L	75.8 (0.4)	Research-XL	246.6 (6.8)
SRT	Production-L	74.2 (1.9)	Research-XL	234.6 (3.4)

At $t = 0$ s, a Research-XL job was submitted; at $t = 50$ s, the production job was submitted. Job completion times are shown. The ideal job completion times are shown in Table 6.4.

We conclude that SRT is the best task eviction policy, especially when production jobs are smaller than research jobs. We believe this is a significant use case since research jobs run longer and process more data, while production jobs are typically small due to the need for faster results.

Job Eviction Policies

We next compare the three job eviction policies discussed in Section 6.1.2. Based on the previous results, we always used SRT task eviction.[14] We initially submitted two research jobs and followed 50 s later with a small production job. We examined two settings: one in which the initial research jobs were comparable in size and another in which they were different. We observed that the production job completion time was close to ideal; hence Table 6.6 shows only research job completion times.

The top half of Table 6.6 shows that when research job sizes were comparable, probabilistic weighing of job evictions by resources (PR) and eviction of the job with the MR performed comparably: research job completion times for the two policies were within 2 s (0.5%) of each other. This is desirable due to the matching job sizes. On the other hand, eviction of the job with the LR performed the worst, because it caused starvation in one of the jobs. Once tasks start getting evicted from a research job (which may at first have been picked randomly by LR if all jobs had the same resource usage), subsequently LR will always pick the same job (until it is fully suspended).

That behavior of LR is even more pronounced on small research jobs in a heterogeneous mix, as can be seen in the bottom half of Table 6.6. The Research-S

Table 6.6 Job eviction policies.

Job eviction policy	Research job	Mean (S.D.) runtime (s)	Research job	Mean (S.D.) runtime (s)
PR	Research-M	195.8 (1.3)	Research-M	201.2 (0.8)
MR	Research-M	196.2 (1.3)	Research-M	200.6 (2.1)
LR	Research-M	200.6 (1.3)	Research-M	228.8 (12.7)
PR	Research-L	201.6 (8.3)	Research-S	213.8 (18.8)
MR	Research-L	195.8 (1.1)	Research-S	204.8 (2.2)
LR	Research-L	195.8 (0.4)	Research-S	252.4 (9.3)

At $t = 0$ s, two research jobs were submitted (either two Research-M's or a Research-S and a Research-L); at $t = 50$ s, a Production-S job was submitted. Only the research job completion times are shown. The ideal job completion times are shown in Table 6.5.

---

14 Using LRT always turned out to be worse.

job is picked as a victim by PR less often than by LR, and thus PR outperforms LR. PR penalizes the Research-L job slightly more than LR does, since PR evicts more tasks from a larger job. Even so, PR and MR are within 10 s (5%) of each other; any differences are due to the variable task lengths, and the effectiveness of the SRT task eviction policy. We observed that MR evicted no tasks at all from the Research-S job.

We conclude that when the best task eviction policy (SRT) is used, the PR and MR job eviction policies are preferable to LR, and MR is especially good under heterogeneous mixes of research job sizes.

6.6.1.6 Natjam-R Evaluation
We evaluated the real-time support of our Natjam-R system (described in Section 6.1.4). Our experiments addressed the following questions: (i) How do MDF and MLF job eviction strategies compare? (ii) How good is Natjam-R at meeting deadlines? And (iii) Do Natjam-R's benefits hold under realistic workloads?

We used eight Emulab servers [111,138], each with eight-core Xeon processors and 250 GB disk space. One server was the Resource Manager, and each of the other seven servers ran three containers of 1 GB each. (Thus, there were 21 containers in total.)

MDF versus MLF
We injected three identical jobs, Job 1 to Job 3, each with 8 maps and 50 reduces. (Each job took 87 s on an empty cluster.) They were submitted in numerical order starting at $t = 0$ s and 5 s apart, thus overloading the cluster. Since MDF and MLF will both meet long deadlines, we chose shorter deadlines. To force preemption, the deadlines of job 1, job 2, and job 3 were set 10 s apart: 200 s, 190 s, and 180 s, respectively.

Figure 6.20 depicts the progress rate for the MDF cluster and the MLF cluster. Our first observation is that while MDF allowed the short-deadline jobs to run earlier and thus satisfy all deadlines, MLF missed all deadlines (see Figure 6.20b). In the reduce phase for MLF, after a while jobs proceeded in lockstep in the

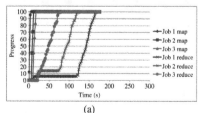
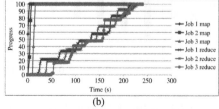

(a)   (b)

Figure 6.20 Natjam-R: (a) MDF versus (b) MLF. Lower index jobs have shorter deadlines but arrive later.

reduce phase, because when a lower laxity job (e.g., job 3) has run for a while in lieu of a higher laxity job (e.g., job 1), their laxities become comparable. Thereafter, the two jobs take turns preempting each other. Breaking ties, for example, by using a deadline, does not eliminate this behavior. In a sense, MLF tries to be fair to all jobs by allowing them all to make progress simultaneously, but this fairness is in fact a drawback.

MLF also takes longer to finish all jobs, that is, 239 s compared to MDF's 175 s. MLF's lockstep behavior incurs a high context switch overhead. We conclude that MDF is preferable to MLF, especially under short deadlines.

Varying the Deadline

We submitted a job (job 1) just as described previously, and 5 s later submitted an identical job (job 2) whose deadline was 1 s earlier than job 1's. We measured job 1's clean compute time as the time to run the job in an empty cluster. Then, we set its $deadline = submission\ time + (clean\ compute\ time \times (1 + \varepsilon))$. Figure 6.21 shows the effect of ε on a metric called *margin*. We define a job's *margin* = (*deadline*) minus (*job completion time*). A negative margin implies a deadline miss. We observe that an ε as low as 0.8 still meets both deadlines, while an ε as low as 0.2 meets at least the shorter deadline. This means that given one critical job with a very short deadline, Natjam-R can satisfy it if it has at least 20% more time than the job's clean compute time. This percentage is thus an estimate of Natjam-R's overhead. We also performed experiments that varied the second job's size as a fraction of the first job from 0.4 to 2.0, but we saw little effect on margin.

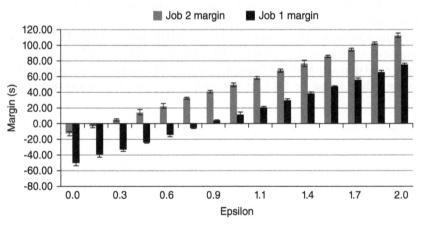

Figure 6.21 Natjam-R: Effect of deadlines: Margin = Deadline – Job completion time; thus, a negative margin implies a deadline miss. Job 2 has a deadline 1 s earlier but is submitted 5 s after Job 1.

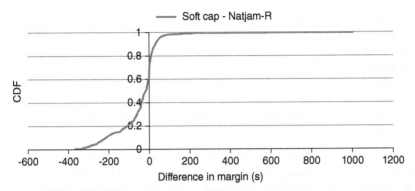

Figure 6.22 Natjam-R: Effect of real Yahoo! Hadoop trace: Margin = Deadline – Job completion time. Negative values imply that Natjam-R is better.

Trace-Driven Experiments

We used the Yahoo! Hadoop traces to evaluate Natjam-R's deadline satisfaction. We used only the production cluster trace, scaled so as to overload the target cluster. Since the original system did not support deadline scheduling, no deadlines were available from the traces. Thus, we chose ϵ randomly for each job from the interval [0, 2.0], and used it to set the job's deadline forward from its submission time (as described earlier). A given job's deadline was selected to be the same in all runs.

Figure 6.22 compares Natjam-R against Hadoop Soft cap. It shows the CDF of the difference between the margins of the two approaches; a negative difference implies that Natjam-R is better. Natjam-R's margin is better than Soft cap's for 69% of jobs. The largest improvement in margin was 366 s. The plot is biased by one outlier job that took 1000 s longer in Natjam-R; the next greatest outlier is only −287 s. The first outlier job suffered in Natjam-R because the four jobs submitted just before it and one job right after had much shorter deadlines. Yet the conclusion is positive: among the 400 jobs with variable deadlines, there was only one such outlier. We conclude that Natjam-R satisfies deadlines well under a realistic workload.

6.7 Summary

In this chapter, we have given an overview of five systems we created that are oriented toward offering performance assuredness in cloud computing frameworks, even while the system is under change.

1) Morphus, which supports reconfigurations in sharded distributed NoSQL databases/storage systems.

2) Parqua, which supports reconfigurations in distributed ring-based key-value stores.
3) Stela, which supports scale-out/scale-in in distributed stream processing systems.
4) A system to support scale-out/scale-in in distributed graph processing systems.
5) Natjam, which supports priorities and deadlines for jobs in batch processing systems.

For each system, we described its motivations, design, and implementation, and presented experimental results. Our systems are implemented in popular open-source cloud computing frameworks, including MongoDB (Morphus), Cassandra (Parqua), Storm (Stela), LFGraph, and Hadoop (Natjam). Readers who are interested in more detailed design and implementation and more extensive experimental findings are advised to see our original papers that introduced these systems [102,114,120,123,131].

6.8 The Future

Overall, building systems that perform *predictably* in the cloud remains one of the biggest challenges today, both in mission-critical scenarios and in non-real-time scenarios. The work outlined in this chapter has made deep inroads toward solving key issues in this area.

More specifically, the work described in this chapter constitutes the starting steps toward realization of a *truly autonomous and self-aware cloud system* for which the mission team merely needs to specify SLAs/SLOs (service level agreements and objectives), and the system will reconfigure itself automatically and continuously over the lifetime of the mission to ensure that these requirements are always met. For instance, as of this writing, we are currently building on our scale-out/scale-in work in the areas of distributed stream processing and distributed graph processing, by adding in an extra layer of *adaptive scale-out/scale-in* that seeks to meet SLA/SLO requirements such as latency or throughput (for stream processing), and completion time deadlines or throughput (for graph processing). These adaptive techniques will automatically give resources to a job that is facing a higher workload or more stringent deadlines and take away resources from a job that has more relaxed needs. Adaptivity implies that there is no human involvement in making decisions on, for example, the number of machines to give or take away from a job, or when to do so; such decisions will be made automatically by the system. Such adaptive logic may be able to leverage machine learning techniques that will learn the system's performance characteristics and adjust the resource allocation changes over time to ensure that the best possible performance is gained, given the cloud resources at hand. It is also potentially possible to add an adaptive layer atop our database reconfiguration systems (Morphus and Parqua); however, that would need to be done wisely and relatively rarely because of the enormous cost of each reconfiguration operation for large databases.

References

1 MongoDB. Available at https://www.mongodb.com/ (accessed Jan. 1, 2015).
2 Lakshman, A. and Malik, P. (2010) Cassandra: a decentralized structured storage system. *ACM SIGOPS Operating Systems Review*, **44** (2), 35–40.
3 Apache Hadoop. Available at http://hadoop.apache.org/.
4 Apache Storm. The Apache Software Foundation. Available at http://storm. apache.org/ (accessed 2016).
5 Kulkarni, S., Bhagat, N., Fu, M., Kedigehalli, V., Kellogg, C., Mittal, S., Patel, J.M., Ramasamy, K., and Taneja, S. (2015) Twitter Heron: stream processing at scale, in Proceedings of the 2015 ACM SIGMOD International Conference on Management of Data, pp. 239–250.
6 Samza. Available at http://samza.apache.org/ (accessed Nov. 14, 2016).
7 Carey, M.J. and Lu, H., Load balancing in a locally distributed DB system, in Proceedings of the 1986 ACM SIGMOD International Conference on Management of Data, pp. 108–119.
8 Kemme, B., Bartoli, A., and Babaoglu, O., Online reconfiguration in replicated databases based on group communication, in Proceedings of the 2001 International Conference on Dependable Systems and Networks, pp. 117–126.
9 Rae, I., Rollins, E., Shute, J., Sodhi, S., and Vingralek, R. (2013) Online, asynchronous schema change in F1. *Proceedings of the VLDB Endowment*, **6** (11), 1045–1056.
10 Elmore, A.J., Arora, V., Taft, R., Pavlo, A., Agrawal, D., and El Abbadi, A. (2015) Squall: fine-grained live reconfiguration for partitioned main memory databases, in Proceedings of the ACM SIGMOD International Conference on Management of Data, pp. 299–313.
11 Copeland, G., Alexander, W., Boughter, E., and Keller, T. (1988) Data placement in Bubba, in Proceedings of the ACM SIGMOD International Conference on Management of Data, pp. 99–108.
12 Mehta, M. and DeWitt, D.J. (1997) Data placement in shared-nothing parallel database systems. *Proceedings of the VLDB Endowment*, **6** (1), 53–72.
13 Das, S., Nishimura, S., Agrawal, D., and El Abbadi, A. (2011) Albatross: lightweight elasticity in shared storage databases for the cloud using live data migration. *Proceedings of the VLDB Endowment*, **4** (8), 494–505.

14 Elmore, A.J., Das, S., Agrawal, D., and El Abbadi, A. (2011) Zephyr: live migration in shared nothing databases for elastic cloud platforms, in Proceedings of the 2011 ACM SIGMOD International Conference on Management of Data, pp. 301–312.

15 Barker, S., Chi, Y., Hacıgümüş, H., Shenoy, P., and Cecchet, E. (2014) ShuttleDB: database-aware elasticity in the cloud, in Proceedings of the 11th International Conference on Autonomic Computing, pp. 33–43. Available at https://www.usenix.org/node/183068.

16 Curino, C., Jones, E.P.C., Popa, R.A., Malviya, N., Wu, E., Madden, S., Balakrishnan, H., and Zeldovich, N. (2011) Relational cloud: a database-as-a-service for the cloud, in Proceedings of the 5th Biennial Conference on Innovative Data Systems Research, pp. 235–240. Available at http://cidrdb.org/cidr2011/program.html.

17 Ardekani, M.S. and Terry, D.B. (2014) A self-configurable geo-replicated cloud storage system, in Proceedings of the 11th USENIX Conference on Operating Systems Design and Implementation, pp. 367–381. Available at https://www.usenix.org/node/186187.

18 Clark, C., Fraser, K., Hand, S., Hansen, J.G., Jul, E., Limpach, C., Pratt, I., and Warfield, A. (2005) Live migration of virtual machines, in Proceedings of the 2nd Symposium on Networked Systems Design and Implementation, vol. 2, pp. 273–286.

19 Bradford, R., Kotsovinos, E., Feldmann, A., and Schiöberg, H. (2007) Live wide-area migration of virtual machines including local persistent state, in Proceedings of the 3rd International Conference on Virtual Execution Environments, pp. 169–179.

20 Barker, S., Chi, Y., Moon, H.J., Hacıgümüş, H., and Shenoy, P. (2012) 'Cut me some slack': latency-aware live migration for databases, in Proceedings of the 15th International Conference on Extending Database Technology, pp. 432–443.

21 Chowdhury, M., Zaharia, M., Ma, J., Jordan, M.I., and Stoica, I., Managing data transfers in computer clusters with Orchestra, in Proceedings of the ACM SIGCOMM 2011 Conference, pp. 98–109.

22 Al-Fares, M., Radhakrishnan, S., Raghavan, B., Huang, N., and Vahdat, A. (2010) Hedera: dynamic flow scheduling for data center networks, in Proceedings of the 7th USENIX Conference on Networked Systems Design and Implementation, p. 19.

23 Abadi, D.J., Carney, D., Çetintemel, U., Cherniack, M., Convey, C., Lee, S., Stonebraker, M., Tatbul, N., and Zdonik, S. (2003) Aurora: a new model and architecture for data stream management. *VLDB J.*, **12** (2), 120–139.

24 Tatbul, N., Ahmad, Y., Çetintemel, U., Hwang, J.-H., Xing, Y., and Zdonik, S. (2008) Load management and high availability in the Borealis distributed stream processing engine, in *GeoSensor Networks: 2nd International Conference, GSN 2006, Boston, MA, USA, October 1–3, 2006: Revised Selected* *and Invited Papers*, Lecture Notes in Computer Science, vol. 4540 (eds. S. Nittel, A. Labrinidis, and A. Stefanidis), Springer, Berlin, Germany, pp. 66–85.

25 Loesing, S., Hentschel, M., Kraska, T., and Kossmann, D., Stormy: an elastic and highly available streaming service in the cloud, in Proceedings of the 2012 Joint EDBT/ICDT Workshops, pp. 55–60.

26 Gulisano, V., Jimenez-Peris, R., Patino-Martinez, M., Soriente, C., and Valduriez, P. (2012) StreamCloud: an elastic and scalable data streaming system. *IEEE Transactions on Parallel and Distributed Systems*, **23** (12), 2351–2365.

27 Abadi, D.J., Ahmad, Y., Balazinska, M., Çetintemel, U., Cherniack, M., Hwang, J.-H., Lindner, W., Maskey, A.S., Rasin, A., Ryvkina, E., Tatbul, N., Xing, Y., and Zdonik, S. (2005) The design of the Borealis stream processing engine, in Proceedings of the 2nd Biennial Conference on Innovative Data Systems Research, pp. 277–289.

28 Castro Fernandez, R., Migliavacca, M., Kalyvianaki, E., and Pietzuch, P., Integrating scale out and fault tolerance in stream processing using operator state management, in Proceedings of the 2013 ACM SIGMOD International Conference on Management of Data, pp. 725–736.

29 Gedik, B., Schneider, S., Hirzel, M., and Wu, K.-L. (2014) Elastic scaling for data stream processing. *IEEE Transactions on Parallel and Distributed Systems*, **25** (6), 1447–1463.

30 Schneider, S., Andrade, H., Gedik, B., Biem, A., and Wu, K.-L. (2009) Elastic scaling of data parallel operators in stream processing, in Proceedings of the IEEE International Symposium on Parallel and Distributed Processing, pp. 1–12.

31 Amini, L., Andrade, H., Bhagwan, R., Eskesen, F., King, R., Selo, P., Park, Y., and Venkatramani, C. (2006) SPC: a distributed, scalable platform for data mining, in Proceedings of the 4th International Workshop on Data Mining Standards, Services and Platforms, pp. 27–37.

32 Jain, N., Amini, L., Andrade, H., King, R., Park, Y., Selo, P., and Venkatramani, C., Design, implementation, and evaluation of the linear road benchmark on the stream processing core, in Proceedings of the 2006 ACM SIGMOD International Conference on Management of Data, pp. 431–442.

33 Wu, K.-L., Yu, P.S., Gedik, B., Hildrum, K.W., Aggarwal, C.C., Bouillet, E., Fan, W., George, D.A., Gu, X., Luo, G., and Wang, H. (2007) Challenges and experience in prototyping a multi-modal stream analytic and monitoring application on System S, in Proceedings of the 33rd International Conference on Very Large Data Bases, pp. 1185–1196.

34 Gedik, B., Andrade, H., Wu, K.-L., Yu, P.S., and Doo, M. (2008) SPADE: the system S declarative stream processing engine, in Proceedings of the ACM SIGMOD International Conference on Management of Data, pp. 1123–1134.

35 Lohrmann, B., Janacik, P., and Kao, O. (2015) Elastic stream processing with latency guarantees, in Proceedings of the IEEE 35th International Conference on Distributed Computing Systems, pp. 399–410.

36 Heinze, T., Roediger, L., Meister, A., Ji, Y., Jerzak, Z., and Fetzer, C. (2015) Online parameter optimization for elastic data stream processing, in Proceedings of the 6th ACM Symposium on Cloud Computing, pp. 276–287.

37 Aniello, L., Baldoni, R., and Querzoni, L. (2013) Adaptive online scheduling in Storm, in Proceedings of the 7th ACM International Conference on Distributed Event-Based Systems, pp. 207–218.

38 Gandhi, A., Harchol-Balter, M., Raghunathan, R., and Kozuch, M.A. (2012) AutoScale: dynamic, robust capacity management for multi-tier data centers. *ACM Transactions on Computer Systems*, **30** (4), 14:1–14:26.

39 Zhang, Q., Zhani, M.F., Zhang, S., Zhu, Q., Boutaba, R., and Hellerstein, J.L. (2012) Dynamic energy-aware capacity provisioning for cloud computing environments, in Proceedings of the 9th International Conference on Autonomic Computing, pp. 145–154.

40 Shen, Z., Subbiah, S., Gu, X., and Wilkes, J. (2011) CloudScale: elastic resource scaling for multi-tenant cloud systems, in Proceedings of the 2nd ACM Symposium on Cloud Computing, pp. 5:1–5:14.

41 Jiang, J., Lu, J., Zhang, G., and Long, G. (2013) Optimal cloud resource auto-scaling for web applications, in Proceedings of the 13th IEEE/ACM International Symposium on Cluster, Cloud, and Grid Computing, pp. 58–65.

42 Nguyen, H., Shen, Z., Gu, X., Subbiah, S., and Wilkes, J. (2013) AGILE: elastic distributed resource scaling for infrastructure-as-a-service, in Proceedings of the 10th International Conference on Autonomic Computing, pp. 69–82.

43 Pujol, J.M., Erramilli, V., Siganos, G., Yang, X., Laoutaris, N., Chhabra, P., and Rodriguez, P. (2010) The little engine(s) that could: scaling online social networks, in Proceedings of the ACM SIGCOMM Conference, pp. 375–386.

44 Tsoumakos, D., Konstantinou, I., Boumpouka, C., Sioutas, S., and Koziris, N. (2013) Automated, elastic resource provisioning for NoSQL clusters using TIRAMOLA, in Proceedings of the 13th IEEE/ACM International Symposium on Cluster, Cloud, and Grid Computing. pp. 34–41.

45 Didona, D., Romano, P., Peluso, S., and Quaglia, F. (2012) Transactional Auto Scaler: elastic scaling of in-memory transactional data grids, in Proceedings of the 9th International Conference on Autonomic Computing, pp. 125–134.

46 Herodotou, H., Lim, H., Luo, G., Borisov, N., Dong, L., Cetin, F.B., and Babu, S. (2011) Starfish: a self-tuning system for big data analytics, in Proceedings of the 5th Biennial Conference on Innovative Data Systems Research, pp. 261–272. Available at http://cidrdb.org/cidr2011/program.html.

47 Herodotou, H., Dong, F., and Babu, S. (2011) No one (cluster) size fits all: automatic cluster sizing for data-intensive analytics, in Proceedings of the 2nd ACM Symposium on Cloud Computing, pp. 18:1–18:14.

48 Gonzalez, J.E., Low, Y., Gu, H., Bickson, D., and Guestrin, C. (2012) PowerGraph: Distributed graph-parallel computation on natural graphs, in Proceedings of the 10th USENIX Symposium on Operating Systems Design and Implementation, pp. 17–30. Available at https://www.usenix.org/node/170825.

49 Low, Y., Bickson, D., Gonzalez, J., Guestrin, C., Kyrola, A., and Hellerstein, J.M. (2012) Distributed GraphLab: a framework for machine learning and data mining in the cloud. *Proceedings of the VLDB Endowment*, **5** (8), 716–727.

50 Stanton, I. and Kliot, G. (2012) Streaming graph partitioning for large distributed graphs, in Proceedings of the 18th ACM SIGKDD International Conference on Knowledge Discovery and Data Mining, pp. 1222–1230.

51 Vaquero, L., Cuadrado, F., Logothetis, D., and Martella, C. (2013) Adaptive partitioning for large-scale dynamic graphs, in Proceedings of the 4th Annual Symposium on Cloud Computing, pp. 35:1–35:2.

52 Salihoglu, S. and Widom, J. (2013) GPS: a graph processing system, in Proceedings of the 25th International Conference on Scientific and Statistical Database Management, pp. 22:1–22:12.

53 Tanenbaum, A.S. (2008) The operating system as a resource manager, in *Modern Operating Systems*, 3rd edn, Pearson Prentice Hall, Upper Saddle River, NJ, Chapter 1, Section 1.1.2, pp. 6–7.

54 Ananthanarayanan, G., Ghodsi, A., Wang, A., Borthakur, D., Kandula, S., Shenker, S., and Stoica, I. (2012) PACMan: coordinated memory caching for parallel jobs, Proceedings of the 9th USENIX Symposium on Networked Systems Design and Implementation. Available at https://www.usenix.org/conference/nsdi12/technical-sessions/presentation/ananthanarayanan.

55 Ananthanarayanan, G., Douglas, C., Ramakrishnan, R., Rao, S., and Stoica, I. (2012) True elasticity in multi-tenant data-intensive compute clusters, in Proceedings of the 3rd ACM Symposium on Cloud Computing, Article No. 24.

56 Rao, S., Ramakrishnan, R., Silberstein, A., Ovsiannikov, M., and Reeves, D. (2012) Sailfish: a framework for large scale data processing, in Proceedings of the 3rd ACM Symposium on Cloud Computing, Article No. 4.

57 Netty (2013) Available at http://netty.io/.

58 Zaharia, M., Borthakur, D., Sarma, J.S., Elmeleegy, K., Shenker, S., and Stoica, I. (2010) Delay scheduling: a simple technique for achieving locality and fairness in cluster scheduling, in Proceedings of the 5th European Conference on Computer Systems, pp. 265–278.

59 Cheng, L., Zhang, Q., and Boutaba, R. (2011) Mitigating the negative impact of preemption on heterogeneous MapReduce workloads, in Proceedings of the 7th International Conference on Network and Service Management.

60 Preemption of Reducer (and Shuffle) via checkpointing. Hadoop MapReduce, MAPREDUCE-5269, the Apache Software Foundation, 2013. Available at https://issues.apache.org/jira/browse/MAPREDUCE-5269.

61 Power, R. and Li, J. (2010) Piccolo: building fast, distributed programs with partitioned tables, in Proceedings of the 9th USENIX Symposium on Operating Systems Design and Implementation. Available at https://www.usenix.org/legacy/event/osdi10/tech/.

62 Verma, A., Cherkasova, L., and Campbell, R.H. (2011) ARIA: automatic resource inference and allocation for MapReduce environments, in Proceedings of the 8th ACM International Conference on Autonomic Computing, pp. 235–244.

63 Wieder, A., Bhatotia, P., Post, A., and Rodrigues, R. (2012) Orchestrating the deployment of computations in the cloud with conductor, in Proceedings of the 9th USENIX Symposium on Networked Systems Design and Implementation. Available at https://www.usenix.org/conference/nsdi12/technical-sessions/presentation/wieder.

64 Phan, L.T.X., Zhang, Z., Loo, B.T., and Lee, I. (2010) Real-time MapReduce scheduling, University of Pennsylvania Department of Computer and Information Science Technical Report MS-CIS-10-32, Jan. 1. Available at http://repository.upenn.edu/cgi/viewcontent.cgi?article=1988&context=cis_reports.

65 Ferguson, A.D., Bodik, P., Kandula, S., Boutin, E., and Fonseca, R. (2012) Jockey: guaranteed job latency in data parallel clusters, in Proceedings of the 7th ACM European Conference on Computer Systems, pp. 99–112.

66 Ganapathi, A., Chen, Y., Fox, A., Katz, R., and Patterson, D. (2010) Statistics-driven workload modeling for the cloud, in Proceedings of the IEEE 26th International Conference on Data Engineering Workshops, pp. 87–92.

67 Kambatla, K., Pathak, A., and Pucha, H. (2009) Towards optimizing Hadoop provisioning in the cloud, in Proceedings of the USENIX Workshop Hot Topics in Cloud Computing. Available at https://www.usenix.org/legacy/event/hotcloud09/tech/.

68 Ananthanarayanan, G., Kandula, S., Greenberg, A., Stoica, I., Lu, Y., Saha, B., and Harris, E. (2010) Reining in the outliers in Map-Reduce clusters using Mantri, in Proceedings of the 9th USENIX Symposium on Operating Systems Design and Implementation. Available at https://www.usenix.org/conference/osdi10/reining-outliers-map-reduce-clusters-using-mantri.

69 Zaharia, M., Konwinski, A., Joseph, A.D., Katz, R., and Stoica, I. (2008) Improving MapReduce performance in heterogeneous environments, in Proceedings of the 8th USENIX Symposium on Operating Systems Design and Implementation, pp. 29–42. Available at https://www.usenix.org/legacy/event/osdi08/tech/.

70 Sandholm, T. and Lai, K. (2010) Dynamic proportional share scheduling in Hadoop, in Proceedings of the 15th International Conference on Job Scheduling Strategies for Parallel Processing, pp. 110–131.

71 Condie, T., Conway, N., Alvaro, P., Hellerstein, J.M., Elmeleegy, K., and Sears, R. (2010) MapReduce online, in Proceedings of the 7th USENIX Conference on Networked Systems Design and Implementation.

72 Zaharia, M., Chowdhury, M., Franklin, M.J., Shenker, S., and Stoica, I. (2010) Spark: cluster computing with working sets, in Proceedings of the 2nd USENIX Workshop Hot Topics in Cloud Computing. Available at https://www.usenix.org/legacy/event/hotcloud10/tech/.

73 Storm. Available at http://www.storm-project.net/.

74 Qian, Z., He, Y., Su, C., Wu, Z., Zhu, H., Zhang, T., Zhou, L., Yu, Y., and Zhang, Z. (2013) TimeStream: reliable stream computation in the cloud, in Proceedings of the 8th ACM European Conference on Computer Systems.

75 IBM InfoSphere Platform. Available at https://www-01.ibm.com/software/au/data/infosphere/.

76 Agarwal, S., Mozafari, B., Panda, A., Milner, H., Madden, S., and Stoica, I. (2013) BlinkDB: queries with bounded errors and bounded response times on very large data, in Proceedings of the 8th ACM European Conference on Computer Systems, pp. 29–42.

77 Cruz, F., Maia, F., Matos, M., Oliveira, R., Paulo, J., Pereira, J., and Vilaça, R. (2013) MeT: workload aware elasticity for NoSQL, in Proceedings of the 8th ACM European Conference on Computer Systems, pp. 183–196.

78 Liu, C.L. and Layland, J.W. (1973) Scheduling algorithms for multiprogramming in a hard-real-time environment. *Journal of ACM*, **20** (1), 46–61.

79 Liu, J.W.S. (2000) *Real-Time Systems*, Prentice Hall.

80 Goossens, J., Funk, S., and Baruah, S. (2003) Priority-driven scheduling of periodic task systems on multiprocessors. *Real-Time Systems*, **25** (2), 187–205.

81 Dertouzos, M.L. and Mok, A.K.-L. (1989) Multiprocessor on-line scheduling of hard-real-time tasks. *IEEE Transactions on Software Engineering*, **15** (12), 1497–1506.

82 Santhoshkumar, I., Manimaran, G., and Murthy, C.S.R. (1999) A pre-run-time scheduling algorithm for object-based distributed real-time systems. *Journal of Systems Architecture*, **45** (14), 1169–1188.

83 Hadoop Capacity Scheduler. Available at http://hadoop.apache.org/docs/r1.2.1/capacity_scheduler.html.

84 Fair Scheduler, Hadoop, Apache Software Foundation. Available at https://hadoop.apache.org/docs/r1.2.1/fair_scheduler.html (accessed Aug. 4, 2013).

85 Remove pre-emption from the capacity scheduler code base. Hadoop Common, HADOOP-5726, the Apache Software Foundation. Available at https://issues.apache.org/jira/browse/HADOOP-5726 (updated June 22, 2012).

86 Isard, M., Prabhakaran, V., Currey, J., Wieder, U., Talwar, K., and Goldberg, A. (2009) Quincy: fair scheduling for distributed computing clusters, in Proceedings of the ACM SIGOPS 22nd Symposium on Operating Systems Principles, pp. 261–276.

87 Yu, Y., Isard, M., Fetterly, D., Budiu, M., Erlingsson, Ú., Gunda, P.K., and Currey, J. (2008) DryadLINQ: a system for general-purpose distributed data-parallel computing using a high-level language, in Proceedings of the 8th USENIX Symposium on Operating Systems Design and Implementation, pp. 1–14. Available at https://www.usenix.org/legacy/event/osdi08/tech/.

88 Andrzejak, A., Kondo, D., and Yi, S. (2010) Decision model for cloud computing under SLA constraints, in Proceedings of the 2010 IEEE International Symposium on Modeling, Analysis and Simulation of Computer and Telecommunication Systems, pp. 257–266.

89 Amirijoo, M., Hansson, J., and Son, S.H. (2004) Algorithms for managing QoS for real-time data services using imprecise computation, in *Real-Time and Embedded Computing Systems and Applications: 9th International Conference on, RTCSA 2003, Tainan, Taiwan, February 18–20, 2003: Revised Papers*, Lecture Notes in Computer Science, vol. 2968 (eds. J. Chen and S. Hong), Springer, Berlin, Germany, pp. 136–157.

90 Schwarzkopf, M., Konwinski, A., Abd-El-Malek, M., and Wilkes, J. (2013) Omega: flexible, scalable schedulers for large compute clusters, in Proceedings of the 8th ACM European Conference on Computer Systems, pp. 351–364.

91 Wang, A., Venkataraman, S., Alspaugh, S., Katz, R., and Stoica, I. (2012) Cake: enabling high-level SLOs on shared storage systems, in Proceedings of the 3rd ACM Symposium on Cloud Computing, Article No. 14.

92 Calder, B., Wang, J., Ogus, A., Nilakantan, N., Skjolsvold, A., McKelvie, S., Xu, Y., Srivastav, S., Wu, J., Simitci, H., Haridas, J., Uddaraju, C., Khatri, H., Edwards, A., Bedekar, V., Mainali, S., Abbasi, R., Agarwal, A., ul Haq, M.F., ul Haq, M.I., Bhardwaj, D., Dayanand, S., Adusumilli, A., McNett, M., Sankaran, S., Manivannan, K., and Rigas, L. (2011) Windows Azure Storage: a highly available cloud storage service with strong consistency, in Proceedings of the 23rd ACM Symposium on Operating Systems Principles, pp. 143–157.

93 Adya, A., Dunagan, J., and Wolman, A. (2010) Centrifuge: integrated lease management and partitioning for cloud services, in Proceedings of the 7th USENIX Conference on Networked Systems Design and Implementation.

94 Hindman, B., Konwinski, A., Zaharia, M., Ghodsi, A., Joseph, A.D., Katz, R., Shenker, S., and Stoica, I. (2011) Mesos: a platform for fine-grained resource sharing in the data center, in Proceedings of the 8th USENIX Symposium on Networked Systems Design and Implementation. Available at https://www.usenix.org/legacy/events/nsdi11/tech/.

95 Shue, D., Freedman, M.J., and Shaikh, A. (2012) Performance isolation and fairness for multi-tenant cloud storage, in Proceedings of the 10th USENIX

Symposium on Operating Systems Design and Implementation, pp. 349–362. Available at https://www.usenix.org/node/170865.

96 NoSQL market forecast 2015–2020. Market Research Media, 2012. Available at http://www.marketresearchmedia.com/?p=568 (accessed Jan. 1, 2015).

97 Apache HBase. The Apache Software Foundation. Available at https://hbase.apache.org (accessed Jan. 5, 2015).

98 Chang, F., Dean, J., Ghemawat, S., Hsieh, W.C., Wallach, D.A., Burrows, M., Chandra, T., Fikes, A., and Gruber, R.E. (2006) Bigtable: a distributed storage system for structured data, in Proceedings of the 7th USENIX Symposium on Operating Systems Design and Implementation, pp. 205–218. Available at https://www.usenix.org/conference/osdi-06/bigtable-distributed-storage-system-structured-data.

99 Can I change the shard key after sharding a collection? in FAQ: Sharding with MongoDB FAQ. Available at https://docs.mongodb.com/manual/faq/sharding/#can-i-change-the-shard-key-after-sharding-a-collection (accessed Jan. 5, 2015).

100 Alter Cassandra column family primary key using cassandra-cli or CQL, Stack Overflow. Available at http://stackoverflow.com/questions/18421668/alter-cassandra-column-family-primary-key-using-cassandra-cli-or-cql (accessed Jan. 5, 2015).

101 The great primary-key debate. *TechRepublic*, Mar. 22, 2012. Available at https://www.techrepublic.com/article/the-great-primary-key-debate/ (accessed Jan. 5, 2015).

102 Ghosh, M., Wang, W., Holla, G., and Gupta, I. (2015) Morphus: supporting online reconfigurations in sharded NoSQL systems, in Proceedings of the IEEE International Conference on Autonomic Computing, pp. 1–10.

103 RethinkDB. Available at https://rethinkdb.com/ (accessed Jan. 1, 2015).

104 CouchDB. The Apache Software Foundation. Available at http://couchdb.apache.org/ (accessed Jan. 5, 2015).

105 Ghosh, M., Wang, W., Holla, G., and Gupta, I. (2015) Morphus: supporting online reconfigurations in sharded NoSQL systems, in IEEE Transactions on Emerging Topics in Computing.

106 *Hungarian algorithm*. Wikipedia. Available at https://en.wikipedia.org/wiki/Hungarian_algorithm (accessed Jan. 1, 2015).

107 Al-Fares, M., Loukissas, A., and Vahdat, A., A scalable, commodity data center network architecture, in Proceedings of the ACM SIGCOMM 2008 Conference on Data Communication, pp. 63–74.

108 Kim, J., Dally, W.J., and Abts, D. (2010) Efficient topologies for large-scale cluster networks, in Proceedings of the 2010 Conference on Optical Fiber Communication collocated Natural Fiber Optic Engineers Conference, pp. 1–3.

109 Kim, J., Dally, W.J., and Abts, D. (2007) Flattened butterfly: a cost-efficient topology for high-radix networks, in Proceedings of the 34th Annual International Symposium on Computer Architecture, pp. 126–137.

110 McAuley, J. and Leskovec, J. (2013) Hidden factors and hidden topics: understanding rating dimensions with review text, in Proceedings of the 7th ACM Conference on Recommender Systems, pp. 165–172.

111 Emulab. Available at http://emulab.net/ (accessed 2016).

112 Cooper, B.F., Silberstein, A., Tam, E., Ramakrishnan, R., and Sears, R. (2010) Benchmarking cloud serving systems with YCSB, in Proceedings of the 1st ACM Symposium on Cloud Computing, pp. 143–154.

113 Google Cloud Platform. Available at https://cloud.google.com/ (accessed Jan. 5, 2015).

114 Shin, Y., Ghosh, M., and Gupta, I., Parqua: online reconfigurations in virtual ring-based NoSQL systems, in Proceedings of the 2015 International Conference on Cloud and Autonomic Computing, pp. 220–223.

115 Riak. Available at http://basho.com/products/ (accessed Jan. 1, 2015).

116 Amazon DynamoDB. Amazon Web Services, Inc. Available at https://aws .amazon.com/dynamodb/ (accessed May 5, 2015).

117 Voldemort. Available at http://www.project-voldemort.com/voldemort/ (accessed May 12, 2014).

118 C An introduction to using custom timestamps in CQL3. Available at http:// planetcassandra.org/blog/an-introduction-to-using-custom-timestamps-in-cql3/ (accessed Apr. 25, 2015).

119 Shin, Y., Ghosh, M., and Gupta, I. (2015) Parqua: Online Reconfigurations in Virtual Ring-Based NoSQL Systems, Technical Report, University of Illinois at Urbana-Champaign. Available at http://hdl.handle.net/2142/78185.

120 Xu, L., Peng, B., and Gupta, I. (2016) Stela: enabling stream processing systems to scale-in and scale-out on-demand, in Proceedings of the IEEE International Conference on Cloud Engineering, pp. 22–31.

121 Zaharia, M., Das, T., Li, H., Hunter, T., Shenker, S., and Stoica, I. (2013) Discretized streams: fault-tolerant streaming computation at scale, in Proceedings of the 24th ACM Symposium on Operating Systems Principles, pp. 423–438.

122 Armbrust, M., Fox, A., Griffith, R., Joseph, A.D., Katz, R., Konwinski, A., Lee, G., Patterson, D., Rabkin, A., Stoica, I., and Zaharia, M. (2010) A view of cloud computing. *Communications of the ACM*, **53** (4), 50–58.

123 Pundir, M., Kumar, M., Leslie, L.M., Gupta, I., and Campbell, R.H. (2016) Supporting on-demand elasticity in distributed graph processing, in Proceedings of the IEEE International Conference on Cloud Engineering, pp. 12–21.

124 Malewicz, G., Austern, M.H., Bik, A.J.C., Dehnert, J.C., Horn, I., Leiser, N., and Czajkowski, G., Pregel: a system for large-scale graph processing, in

Proceedings of the 2010 ACM SIGMOD International Conference on Management of Data, pp. 135–146.

125 Low, Y., Gonzalez, J., Kyrola, A., Bickson, D., Guestrin, C., and Hellerstein, J. (2010) GraphLab: a new framework for parallel machine learning, in Proceedings of the 26th Conference on Uncertainty in Artificial Intelligence. Available at https://dslpitt.org/uai/displayArticles.jsp? mmnu=1&smnu=1&proceeding_id=26.

126 Hoque, I. and Gupta, I. (2013) LFGraph: simple and fast distributed graph analytics, in Proceedings of the 1st ACM SIGOPS Conference on Timely Results in Operating Systems, pp. 9:1–9:17.

127 Stoica, I., Morris, R., Karger, D., Kaashoek, M.F., and Balakrishnan, H. (2001) Chord: a scalable peer-to-peer lookup service for Internet applications, in Proceedings of the Conference on Applications, Technologies, Architectures, and Protocols for Computer Communications, pp. 149–160.

128 Kuhn, H.W. (1955) The Hungarian method for the assignment problem. *Naval Research Logistics Quarterly*, **2** (1–2), 83–97.

129 Jonker, R. and Volgenant, T. (1986) Improving the Hungarian assignment algorithm. *Operations Research Letters*, **5** (4), 171–175.

130 Kwak, H., Lee, C., Park, H., and Moon, S. (2010) What is Twitter, a social network or a news media? in Proceedings of the 19th International Conference on World Wide Web, pp. 591–600.

131 Cho, B., Rahman, M., Chajed, T., Gupta, I., Abad, C., Roberts, N., and Lin, P. (2013) Natjam: design and evaluation of eviction policies for supporting priorities and deadlines in Mapreduce clusters, in Proceedings of the 4th Annual Symposium on Cloud Computing, pp. 6:1–6:17.

132 Dean, J. and Ghemawat, S. (2008) MapReduce: simplified data processing on large clusters. *Communications of the ACM*, **51** (1), 107–113.

133 Huai, Y., Lee, R., Zhang, S., Xia, C.H., and Zhang, X. (2011) DOT: a matrix model for analyzing, optimizing and deploying software for big data analytics in distributed systems, in Proceedings of the 2nd ACM Symposium on Cloud Computing, Article no. 4.

134 Thusoo A., Sarma, J.S., Jain, N., Shao, Z., Chakka, P., Anthony, S., Liu, H., Wyckoff, P., and Murthy, R. (2009) Hive: a warehousing solution over a map-reduce framework, in Proceedings of the VLDB Endowment, 2 (2), 1626–1629.

135 Olston, C., Reed, B., Srivastava, U., Kumar, R., and Tomkins, A. (2008) Pig Latin: a not-so-foreign language for data processing, in Proceedings of the ACM SIGMOD International Conference on Management of Data, pp. 1099–1110.

136 Ko, S.Y., Hoque, I., Cho, B., and Gupta, I. (2010) Making cloud intermediate data fault-tolerant, in Proceedings of the 1st ACM Symposium on Cloud Computing, pp. 181–192.

137 Chen, Y., Ganapathi, A., Griffith, R., and Katz, R. (2011) The case for evaluating MapReduce performance using workload suites, in Proceedings of the IEEE 19th Annual International Symposium on Modeling, Analysis and Simulation of Computer and Telecommunication Systems, pp. 390–399.

138 White, B., Lepreau, J., Stoller, L., Ricci, R., Guruprasad, S., Newbold, M., Hibler, M., Barb, C., and Joglekar, A. (2002) An integrated experimental environment for distributed systems and networks, in *ACM SIGOPS Operating Systems Review – OSDI '02: Proceedings of the 5th Symposium on Operating Systems Design and Implementation*, vol. 36, No. SI, pp. 255–270.

7

Theoretical Considerations: Inferring and Enforcing Use Patterns for Mobile Cloud Assurance

Gul Agha, Minas Charalambides, Kirill Mechitov, Karl Palmskog, Atul Sandur, and Reza Shiftehfar

Department of Computer Science, University of Illinois at Urbana-Champaign, Urbana, IL, USA

The mobile cloud is the integration of smart sensors, mobile devices, and cloud computers in a well-connected ecosystem. Such integration can improve the efficiency of services. However, such integration also leads to security and trust issues. For example, the security of cloud spaces has sometimes been breached through accessing of peripheral devices, such as HVAC systems. This chapter will show how mobile cloud security and trust can be improved while maintaining the benefits of efficiency by supporting fine-grained mobility. Specifically, we discuss an actor-based programming framework that can facilitate the development of mobile cloud systems in a way that improves efficiency while enforcing security and privacy. There are two key ideas here. First, by supporting fine-grained units of computation (actors), a mobile cloud can be agile in migrating components. It does so in response to a system context (including dynamic variables such as available bandwidth, processing power, and energy) while respecting constraints on information containment boundaries. Second, by specifying constraints on interaction patterns, information flow between actors can be observed and suspicious activity flagged or prevented. We introduce the concepts and discuss their realization in notations and prototypes. Finally, we will discuss open research issues such as inference of interaction patterns.

7.1 Introduction

Mobile devices and smart sensors have become ubiquitous. Such devices are relatively inexpensive but have limited computational resources (memory, processing capability, and energy). These limitations currently preclude the use of connected devices in many complex applications [1]. At the same time, cloud computing provides elastic on-demand access to virtually unlimited resources at an affordable price. To provide the functionality and quality of service that users demand, mobile devices and smart sensors need to be

Assured Cloud Computing, First Edition. Edited by Roy H. Campbell, Charles A. Kamhoua, and Kevin A. Kwiat.

integrated into a broader context of cloud computing. We call such an integrated system a *mobile cloud.*

We propose to consider computation in the mobile cloud as an interaction between mobile agents or actors [2]. Each actor isolates state and can be accessed only through a message-passing interface. Moreover, actors are the units of migration. This makes migration efficient, as an actor's behavior is not affected by another's internal state. Thus, a mobile cloud based on actors provides many opportunities for improving energy efficiency and timeliness of services.

However, while actor (code and data) offloading can improve application user experience, it must be performed while respecting user security and privacy requirements. In an environment with both trusted (private) and untrusted (public) cloud resources, the origin and destination of data and code sent from devices, for example, during code offloading, must be taken into account. To support such *hybrid cloud* environments, actor frameworks for the mobile cloud must allow specification of fine-grained *security policies* that are monitored and enforced by the application runtime.

With actors as the unit of migration, data and code that manipulates the data are treated as a single entity. This facilitates security, for example, through prevention of actor migration when such migration violates security and privacy requirements. Actor migration can be favorably contrasted with the common practice of computation migration at the level of virtual machines (VM); virtual machines are relatively coarse-grained, incorporating a substantial amount of code and data. This not only complicates migration but also makes it more difficult to enforce boundaries between different pieces of code and data that happen to reside on the same VM.

Runtime monitoring may be used to collect observations on application intent, which in turn can be used to infer and adapt (constrain) behavior of application components. For example, actor migration may be triggered when communication patterns suggest offloading to specific cloud resources for better application latency, or to prevent violations of interaction protocols between actors. Monitoring of interaction between actors can also be used to prevent violations of interaction protocols or to flag them. Coordination constraints can represent the permitted order of operations. Such constraints may be explicitly specified, or they may be inferred. Coordination constraints can be encoded using actor synchronization constraints [3]. They can also be translated to actor session types [4,5]. Session types allow us to check whether the behavior of participant actors in an interaction conforms to specific interaction patterns that implement a protocol.

In this chapter, we discuss an actor-based approach to programming mobile hybrid clouds. Our approach facilitates holistic Assured Cloud Computing (ACC). It is realized in the Illinois Mobile Cloud Manager (IMCM), a prototype mobile hybrid cloud platform. We describe this platform, and then discuss how suitable formalisms can be used to capture interaction patterns to improve the safety and security of computation in the mobile cloud.

The chapter is organized as follows. Section 7.2 outlines the vision of the IMCM framework, and how its use can mitigate current problems in mobile cloud computing (MCC). Section 7.3 gives an overview of the relevant state of the art. Section 7.4 describes the framework's offloading model, architecture, and security policy language in more detail. Section 7.5 covers actor coordination using synchronization constraints. Section 7.6 describes actor session types that can abstractly encode many application constraints. Finally, Section 7.7 discusses future work and concludes.

7.2 Vision

Consider a mobile application that should perform *facial recognition* of a given image using a database of known faces, of which some must remain confidential. Since this kind of image processing is computationally expensive, tasks should be offloaded to the cloud whenever possible. We assume that application developers would want to deploy this application in a *hybrid* cloud environment, spanning both a public and a private cloud. Using existing frameworks, engineers would face a number of difficult issues in the development, deployment, and maintenance of such an application:

Productivity. The application may have to be decomposed in specific ways to enable fine-grained code offloading, and the decomposition may be different depending on the deployment scenario. Developers may have to translate high-level application requirements into executable imperative code. To programmatically access sensor data, knowledge of low-level interfaces may be required.

Security and Privacy. To achieve requirements on security and privacy, developers may have to use specific knowledge about the deployment environment, for example, whether a specific offloading task is sent to a certain public cloud. Developers may also need to add security checks at specific places in the application code, for example, where a photo that should remain confidential is accessed.

Maintainability. The application may have to be rearchitected and redeployed because of small changes in the environment, for example, if the cloud provider changes or the average network latency increases. When application requirements on energy consumption and availability change, developers may have to manually adjust parameters inside imperative code.

We address these issues by programming applications using the actor model. The approach does not commit developers to a specific computation offloading policy; the granularity of actor components can be adapted for offloading efficiency. When requirements are encoded as declarative constraints enforced by the framework, application evolution becomes less involved and prone to

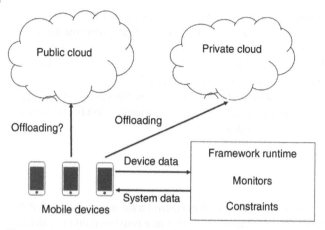

Figure 7.1 Application scenario using the IMCM framework.

failures; developers no longer carry the burden of inserting code for checking security policy conformance. The framework also hides low-level sensor interfaces. In addition, programmers need not write any logic for deciding when it is beneficial (with respect to energy consumption, latency, etc.) to offload actors into the cloud. Instead, using information about runtime variables such as energy requirements for a service, communication bandwidth and latency, and available energy on a device, as well as applicable policies, a middleware framework can make offloading decisions on-the-fly.

We realize this approach in the IMCM framework. Figure 7.1 illustrates the application scenario at a high level. The image application runs on one or more mobile devices that may offload certain actors to either a private or a public cloud. Meanwhile, the framework runtime performs monitoring of devices and can provide the data to determine when it is appropriate to perform offloading.

Actors also isolate state, allowing information flow boundaries to be mediated through message-passing interfaces. These interfaces allow us to observe and enforce interaction patterns, thus forcing actors to conform to protocols. We express the requirements of a protocol as constraints in the form of coordination constraints between actors, or as multiparty session types. We describe in the following section how such constraints may be used to improve assurance in the mobile cloud.

7.3 State of the Art

Before delving into the details of the IMCM framework, coordination constraints, and session types, we first review state-of-the-art work in these three domains so as to set the context for understanding the contributions of our work.

7.3.1 Code Offloading

In recent years, offloading of computations between mobile devices and the cloud has been proposed to address the limitations of mobile platforms. For example, Kumar *et al.* [1] discuss program partitioning to support offloading, while Rahman *et al.* [6] suggest using virtual machine migration. In mobile cloud computing systems using the former approach, the application developer manually partitions a program and then specifies how to offload parts of an application to remote servers. In MCC systems using the latter approach, the entire process or entire OS is moved to a cloud space [7]. However, program partitioning requires substantial developer effort, while VM migration can be prohibitively inefficient [8].

An alternative is to provide automatic data-partitioning solutions for fine-grained, energy-aware offloading of computations. Mobile cloud systems such as MAUI [9] use a combination of virtual machine migration and automatic code partitioning to decide at runtime which methods of an object's code should be executed remotely. The goal of MAUI is to minimize energy consumption on mobile devices by offloading computations. However, MAUI has several limitations. First, it only supports sequential execution: The object in a mobile device is paused while it awaits the results from offloaded code. Second, MAUI only supports a single remote server and requires manual annotation of methods by the programmer and offline static analysis of the source code before execution. Other systems for offloading computations include CloneCloud [10], ThinkAir [11], and COS [8]. Each of these systems has its own limitations. For a good review, we refer interested readers to Ref. [12].

IMCM supports a concurrent application model, enabling simultaneous execution on both mobile device and multiple remote cloud resources.

7.3.2 Coordination Constraints

Actors are autonomous and decentralized, so they must explicitly synchronize when they need to coordinate their actions. A number of frameworks have been developed to simplify the expression of coordination. We review some of the important frameworks. A hierarchical model of coordination was defined in Ref. [13]. In the *Directors* coordination model, actors are organized into trees. Messages sent between two actors (say A and B) that need to be coordinated are delivered through the closest common ancestor: They are forwarded to a parent node (representing a director), which forward it to its parent, and so on, until a common ancestor of A and B is reached. Then the message is sent down to subdirectors until it reaches the target. This enables enforcement of hierarchical policies. Directors do not support arbitrarily overlapping constraints. Furthermore, the model does not provide semantics for dynamic reconfiguration; actors are inserted into the tree when they are constructed.

Another approach is to use computational reflection and meta-programming. An actor's mail queue buffers messages when they arrive and dispatches them to the actor. The mail queue can be programmed to reject or reorder messages, thus enforcing constraints on acceptable ordering of messages. Similarly, messages are sent to other actors through a dispatcher. The dispatcher can also be customized to reorder messages, or to communicate with a mail queue before sending a message in order to synchronize the state. In this framework, an actor's mail queue and dispatcher are considered separate actors. Astley and Agha [14] and Sturman [15] propose using actor systems with meta-actors, which require a two-level semantics. Such a semantic formalization has been developed in Refs [16,17].

In the *Actor-Role-Coordinator* (ARC) model [18], coordination is transparent to base-level actors; coordination tasks are divided into intrarole and interrole communication. While this hierarchical design provides load balancing for highly dynamic systems, the coordination structure itself is static. ARC systems therefore avoid security issues through reconfiguration, but require a restart to adapt to changing specifications.

A scalable coordination model must be able to cope with uncooperative actors. We discuss scoping of synchronization constraints to handle scenarios such as permanent locking of a target actor because of a malicious or buggy actor.

7.3.3 Session Types

Interaction types, used to capture communication protocols among parallel processes, were first introduced in the context of the π-calculus [19] by Honda [20]. Targeting two-party interactions, Honda's system could statically ensure that the participants have compatible behavior, by requiring *dual* types; that is, behaviors in which each participant expects precisely the message sequence that the other participant sends and vice versa.

Many real-world protocols involve more than two participants, which makes their description in terms of multiple two-party sessions unnatural. The Web Services Choreography Description Language [21] is an XML-based language that allows the description – from a global perspective – of protocols that involve multiple concurrent entities (Web services and clients). The notion of an end-point *projection* of such a global specification was first studied by Carbone *et al.* [22], along with the associated correctness requirements. The idea was further studied by Honda *et al.* [23], whose session types support multiple participants: A *global type* specifies the interactions among all participants from a global perspective, and a projection algorithm then mechanically derives the behavior specification of each individual participant, that is, its *local type*. A subtyping relation for Honda's system was first studied by Gay and Hole [24], who dealt with recursive types through an *unfold* function. Their system treats the substitutability concept in a syntactic manner, making subtyping statically decidable.

Asynchronous session types were first studied by Gay and Vasconcelos [25], who proposed session types for an asynchronous flavor of the π-calculus. Parameters were introduced to session types by Yoshida *et al.* [26], enabling type-safe interactions among a fixed, but not statically known, number of participants.

Castagna *et al.* [27] proposed a specification language for global types that is closer to regular expressions; they investigated the requirements for correctness after projection, and guarantee certain liveness properties. Their work was the starting point for the full version of the material in Section 7.6, by Charalambides *et al.* [4,28]. The theoretical foundations of runtime monitoring of protocols with assertions were laid by Bocchi *et al.* [29], who studied the conditions under which localized, end-point runtime checks suffice to enforce globally specified assertions. In a similar manner, Neykova *et al.* [30] extended Scribble [31] with timing constraints, and studied the conditions that enable purely local, end-point runtime monitoring of distributed systems with real-time constraints.

Conventional session types are not suitable for typing interactions in actors, so we discuss a programming language and session type system that capture the inherent asynchrony of actor programs.

7.4 Code Offloading and the IMCM Framework

The first widespread use of computation offloading was in the context of cluster and grid computing. By migrating processes between different nodes in a computing network, it was possible to improve performance through load balancing [32]. Another impetus for cloud computing is the offloading of computations – both data and code – in order to support elasticity of resources. Moving computations onto the cloud was facilitated by virtualization technology; this allowed cloud vendors to run different applications from many customers at the same time.

IMCM, the framework we have developed, improves application performance by using information from the dynamic runtime environment, the end-user context, and the application's behavior. Overcoming the limitations of earlier systems, IMCM supports a concurrent application model that facilitates offloading of computations to multiple remote locations and simultaneous execution on both mobile devices and remote cloud resources.

Mobile cloud systems are constrained because of a number of factors, such as limited energy, communication bandwidth, latency, processor speed, and the memory available on mobile devices. However, energy is particularly interesting, since estimation of energy use by an application and its components is complicated; many applications may be running on a device. Research in our group has shown that the energy consumption of the individual components of an application can be estimated using application-related logs and regression [33]. An interesting alternative approach is to use crowd-sourced measurements of system state parameters such as CPU usage or battery voltage to build a model of energy use – an approach taken by CARAT [34].

7.4.1 IMCM Framework: Overview

Many organizations, developers, and users benefiting from cloud resources have privacy requirements, expectations, and policies in terms of how different private or public cloud resources can be used by a mobile application. Without having enough flexibility in the offloading framework to address these requirements, many users will not be able to benefit from the cloud resources. In order to accommodate these requirements, we describe a language to define policies, and explain how the IMCM framework can be customized to address them.

While addressing these policies is critical, other quantitative properties, such as performance and energy characteristics on the mobile device, greatly affect the quality of an application in meeting overall user requirements. The framework allows for configuration of policies that need to be enforced, but such policies may affect the performance and energy usage of the application. On the other hand, it may be possible to leverage optimized performance and energy to provide stronger privacy guarantees. Through the framework, we discuss mechanisms that allow developers or users to control all privacy, performance, and energy aspects of their applications via code offloading.

7.4.2 Cloud Application and Infrastructure Models

In order to formulate the application component offloading problem, a comprehensive mobile-hybrid-cloud application model is needed. This section summarizes our views on clouds, cloud applications, and mobile-cloud applications.

Over time, cloud services have moved from the model of public cloud spaces to private clouds and, recently, to a hybrid model combining both [35]. Cloud infrastructure has traditionally been provided by large third-party organizations, and thus has been referred to as the *public* cloud. However, storing data on third-party machines suffers from a potential lack of control and transparency in addition to having legal implications [36]. In order to address this problem, cryptographic methods are usually used to encrypt the data stored in public clouds, while decryption keys are disclosed only to authorized users. However, those solutions inevitably introduce heavy computational overhead for continuous encryption and decryption of data, distribution of decryption keys to authorized users, and management of data when fine-grained data access control is desired. Thus, cryptographic methods do not scale well, have significant overhead, are expensive to maintain, and are often slow, especially in widely geographically distributed environments such as clouds.

To address some of those limitations, several techniques, such as attribute-based encryption (ABE), proxy re-encryption and lazy re-encryption, and homomorphic encryption techniques, have been developed to provide attribute-based encryption while delegating part of the required decryption to the cloud or providing limited in-cloud computations on the ciphered text (encrypted data) without revealing data content. However, these efforts have had the traditional data-centric view of cloud computing, focused on storing data and providing services to access the stored data. If data storage is the primary use for the cloud, the required data access control is already mature enough that a fine-grained access control system can be implemented effectively.

However, in modern cloud applications, the resources stored in the cloud contain more than just data. These resources contain part of the application code that results in access operations for execution of the code inside the cloud. It is obvious that the certificate-based authorization systems fail to address these types of applications, as an encrypted piece of code within the cloud cannot be executed without decryption, thus revealing the content to the cloud provider. As a result, companies have gradually moved toward building their own private clouds [36]. Storing sensitive data within private clouds aligns with the traditional on-premises application deployment model whereby sensitive data reside within the enterprise boundary and are subject to its physical, logical, and personnel security and access control policies [35]. However, owning and maintaining a private cloud is usually not as efficient, scalable, reliable, or elastic as using one of the public ones that are offered, supported, and maintained by large third-party companies.

In recent years, a combination of both private and public cloud spaces has been used that allows users to benefit from all the advantages of the public cloud while keeping their confidential or sensitive data and algorithms in-house [37]. In order to cover different applications, our framework simply views the cloud as the most general form that combines one or multiple private and public cloud spaces. This allows the creation of a general, flexible, elastic hybrid cloud space to address the different needs of a specific organization while different users (including both internal staff and external clients) can access cloud resources with different access levels and limitations.

7.4.3 Cloud Application Model

Although the benefits of cloud computing for enterprise consumers and service providers have been thoroughly explored over the past few years, its effect on end users, applications, and application developers is still not clear. As mentioned in the previous section, the traditional data-centric view of cloud services needs to be replaced with a more general data/computation-centric view. To achieve that transition, the current common client-server-based, service-oriented architecture [38], which provides services on data stored in the cloud to external users, needs to be replaced with a more general, elastic architecture that dynamically and transparently leverages cloud resources to provide services and support resource limitations on the end user. In such an elastic application development environment, components that store data or perform computations are transparently scattered among the private clouds, public clouds, and end-user devices. When such an application is launched, an elasticity manager monitors the execution environment and resource requirements of the application to make decisions about where components should be launched and when they should migrate from device to cloud, or from cloud to device, according to changes in user preferences. Rather than a simple and rigid solution wherein nearly all processing and storage are either on the device or in the cloud, an elastic mechanism should have the ability to migrate functionality between the device and the cloud. This ability would allow the device to adapt to different workloads, performance goals, and network latencies.

An important design objective of this modern application development approach is to build an infrastructure with enabling functions such as network protocols, secure communication, and resource management, so that the new elastic computing model will introduce minimal extra considerations for application developers. Unnecessary details of distribution and move-around of application components should be masked from the programmers, while access to different components and resources is still restricted for different users. Elastic components need to access sensitive or restricted resources, and thus an authorization system is needed to give the required component access privileges. For access control, elastic application components on the cloud should adhere to the property of least privileges. Which permissions a component might have may depend on its execution location. Implicit access to device resources may require additional scrutiny when the component is no longer running local to the device.

End users interact with that elastic architecture system by using their mobile devices. That inevitably requires us to address the problem of minimizing energy usage on the mobile device (as detailed in Section 7.4.8) to support complex applications. In order to do so, we need to be able to attribute energy consumption to components of an application. The actor model lends itself naturally to defining the granularity for energy monitoring at the level of individual or groups of actors. We leverage it as the application programming model for the IMCM framework. Actors can be the primitive units for targeting energy measurements, while groupings of actors of a particular type can be considered for aggregations/higher level metrics like average energy consumption. Schedulers for actor-based languages also view actors as basic computational entities for scheduling decisions, so the underlying runtime can be instrumented to track actors running at different time intervals.

As a result, our view of a mobile-cloud application consists of actors distributed across local mobile devices and different cloud spaces.

7.4.4 Defining Privacy for Mobile Hybrid Cloud Applications

We want to enable developers and users to restrict access to different resources, and enable the mobility of a sensitive or confidential application component's resources, based on required policies. This requires the framework to follow authorization rules defined by the organizations, developers, or users. Elastic application components on the cloud should adhere to the property of least privileges. Which permissions a component should have may depend on its execution location, application requirements, or user concerns. Implicit access to device resources may require additional scrutiny when the component is no longer running local to the device [39]. A comprehensive security solution requires authentication, access control, and auditing. There has been a significant amount of work on authentication and auditing for cloud applications in the past, and the existing solutions are mature enough to address most applications [39–43]. Therefore, we focus on an approach for adding policy-based privacy to our framework that restricts the accesses, actions, and mobility of components.

Before we delve into the details of the authorization system and its language, policy description, and evaluation rules, we describe in detail the image recognition application that was first introduced in Section 7.2 as an example to demonstrate the usage of the privacy-related capabilities of our system.

7.4.5 A Face Recognition Application

An image-processing application, called *Image*, recognizes faces in a given image by using a database of known faces. Figure 7.2 shows a summary of different modules involved in the Image application.

In order to respect the privacy of the clients and prevent potential vulnerabilities in storing private data on third-party machines, the owner organization breaks down the database of all known faces into two main categories: (i) faces that are publicly known, such that the data contain no sensitive or confidential information and can be stored on public cloud space; and (ii) facial data that contain sensitive or confidential information and should never be placed on third-party machines, even if encrypted. In order to make this more interesting, different subcategories are also considered for each main category. Let us assume that the goal is to recognize people entering the country at an airport through image-processing of photos taken by security officers at the airport. Based on this scenario, the two main subcategories for the database of known faces would include (i) known criminals whose faces have been publicly announced, and (ii) normal people/citizens whose faces should not be stored

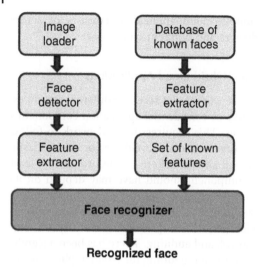

Figure 7.2 Modules involved in the Image image-processing application.

on any third-party machines. Furthermore, faces of known criminals might be divided into international criminals and national criminals. Similarly, faces of "normal" people might be divided into citizens, permanent residents, and visa holders. Figure 7.3 shows how these databases would be stored in different public and private cloud spaces.

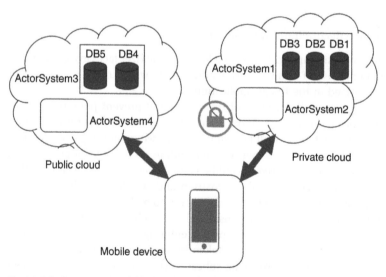

Figure 7.3 Organization of databases within private and public cloud spaces for Image application.

In the figure, DB1 stores faces of citizens, DB2 stores faces of permanent residents, DB3 stores faces of visa holders, DB4 stores faces of international criminals, and DB5 stores faces of national criminals. The image-processing application makes it possible to compare a photo taken with a phone to these databases and thereby identify the person in the photo. The main design goal of developing an authorization policy for this application is to keep confidential data away from unauthorized users and provide different levels of restrictions for different types of users, ranging from security officers (who have access to all databases) to normal airport staff (who should only be able to recognize criminals). In the next section, we use this privacy-sensitive application to demonstrate the policies that can be defined using our framework.

7.4.6 The Design of an Authorization System

Since we want to provide fine-grained authorization systems for application components, we have adopted a hierarchical approach whereby organizations can enforce an organization-wide policy, while developers and end users can fine-tune it. An organization is the primary owner of the data and resources and must be able to keep private and public cloud components separate from each other and define an overall policy in terms of resource usage for different users or different applications. Specific applications may also need to further tighten those organization-wide policy rules. End users or programmers must also be able to further restrict resource usage and component distributions for specific applications. As a result, our framework supports two types of policies: *hard* policies and *soft* policies.

Hard policies are organization-wide authorization rules defined per user or application by the organization. Users include different developers inside the organization in addition to external clients. On the other hand, *soft policies* are application-specific authorization rules defined in addition to the organization-wide hard policy. Despite the fact that these two types of policies have complementary roles in increasing system flexibility, a soft policy can only tighten the organization-wide policy and not vice versa. In other words, if the organization-wide hard policy allows a specific user or a specific application to access resources A and B, a soft policy can only further restrict the access to one of the resources A or B; it can never loosen the restrictions by allowing access to a new resource such as C. Separating the restriction policy definition from the application logic in this way allows organizations to define their hard policies without programmers' having to worry about compromising the predefined organization-wide policy.

Each application instance initially authenticates itself with a Policy Manager Machine (PMM) and receives a locked unchangeable hard policy that contains the organization-wide authorization rules defined by the organization. Each organization can define its authorization policy as one policy for all users, one policy for all applications, one policy per application, one policy per user, or one policy per application instance. In the end, each application instance can acquire one locked hard policy from the policy manager machine. In addition, each application instance can have one soft policy. Developers can define the initial soft policy per application or per application instance. They can also allow end users to change all or part of this soft policy through the application. To implement these rules, we follow the XACML usage model [44] and assume a Policy Enforcement Point (PEP) as part of our elasticity manager. The PEP is responsible for protecting authorization rules by sending a request that contains a description of the attempted action to a Policy Decision Point (PDP) for evaluation against available hard and soft policies. The PDP evaluates the requested action and returns an authorization decision for the PEP to enforce.

Our authorization framework needs to be able to apply the restriction rules at the granularity of actors. It still allows those rules to be defined at higher level entities, such as groups or sets of actors, but it recursively propagates all those specified authorizations (permissions or denials) to all actors contained within a set at runtime. This makes it easy to specify authorizations that hold for a larger set of actors (on the whole system, if "*all*" is used) and have them propagated to all the actors within that set until stopped by an explicit conflicting restriction rule [45]. Actor frameworks allow multiple actors to be placed together in a container, called the *actor system* or *theater*, to share common attributes. We respect this structuring in our language and allow authorization rules to be defined on actors, actor systems, sets of actors (each called a *Group*), a set of actor systems (called a *Location*), or a subset of multiple actors and actor systems (called a *Selection*).

While access control models restrict access to different components or resources, our mobile hybrid cloud framework provides more than access restriction. The actor programming paradigm allows an actor to send and receive messages, create new actors, or migrate to new locations. As a result, our authorization grammar must allow for definition of rules regulating all those actions. Note that these actions are usually bidirectional, meaning that if Actor 1 is allowed to send to Actor 2, then Actor 2 must also be allowed to receive from Actor 1 in order for the policy to be consistent. If either of those two actions is not explicitly allowed as part of the policy, the framework automatically rejects both actions, as they will always happen together.

7.4.7 Mobile Hybrid Cloud Authorization Language

Authorization decisions are made based on the attributes of the requester, the resource, and the requested action, using policy-defined rules. As a result, defining an authorization policy means defining the authorization entities and their required attributes in addition to defining rules and desired rule orderings.

In the cloud application model in which actors are the smallest entities in an application, actors are the finest granularity on which we can define access restriction. In order to provide location transparency, multiple actors running on one runtime instance on one machine are placed inside a container, called an *actor system* or *theater* as in SALSA [46]. Our language supports definition of both actors and actor systems. Every actor is defined by its related reference and a logical path to reach the actor in the runtime environment, in addition to its containing actor system. The authorization framework uses these attributes to bind the actors defined in the policy to their real-world application components.

In order for our language to be able to account for the existence and activities of to-be-developed application-specific components (while enabling the writing of organization-wide policies), anonymous types of entities are defined as part of the proposed language grammar. A rule called *anonymous-actor* allows restriction of the creation and number of unknown actors in a reference-actor system. Similarly, a rule called *anonymous-actor system* allows control of the creation and the number of unknown actor systems.

The following shows how the above supported entities can be used to define the components of the face recognition example from Section 7.4.5.

```
{Name: ActorSysUser, Static (URL:98.123.123.456, Port:1979)}
```

The above clause defines an actor system called *ActorSysUser* that is statically bound to the given IP address and port number.

```
{Name: UserImageLoader, Static(Reference:actor.tcp://
app@98.123.123.456/Image-Loader, ActorSystem:ActorSysUser)}
```

The above clause defines an actor called *UserImageLoader* for the component that loads an image from disk, with the reference providing the logical path to reach the element in the runtime environment. The clause also provides the actor system to which the actor belongs. We can have similar definitions for Face-Detector, Feature-Extractor, and Face-Recognizer.

```
{Name:AnonymousUser, Ref-ActorSystem:ActorSysUser,
Existence:FORBIDDEN}
```

The above clause defines an anonymous actor within the ActorSysUser actor system, to prevent the creation of additional actors within the actor system. The grammar also supports limiting of the maximum number of anonymous actors that can be created within an actor system.

```
{Name:Other-ActorSys-User, URL:98.123.123.456,
Creation:FORBIDDEN}
```

Finally, the above clause defines an anonymous actor system to restrict the creation of unknown actor systems at the given URL. Note that anonymous URLs for dynamic binding can also be created to support arbitrary architecture of to-be-developed applications.

7.4.7.1 Grouping, Selection, and Binding

Although definitions like those in the previous section can be used to define individual actors and actor systems, in many cases it is easier to group several entities and treat them as one. A *Group* definition puts several actors together into one virtual container and allows placement of both known actors and unknown, anonymous actors together into one group. Similarly, we can have a *Location* definition to provide the same grouping functionality but for actor systems. One or several previously defined actor systems, locations, or even unknown anonymous actor systems can be placed into one container location entity.

Instead of specifying individual entities to form a container, a *Selection* definition can be used to pick entities based on a condition. In order to bind previously defined dynamic actors and actor systems to specific runtime components, an *Assignment* definition can be used. Any remaining unbound dynamic actor or actor system is in a passive state and will be ignored while the policy is being enforced. An *Assignment* definition can then be used to bind the runtime components to specific actors or actor systems and change their passive state to active at any time.

The following clause utilizes some of the above definitions to define groupings for components of an application:

```
{Name:FaceActors,Actors:UserFaceDetector,UserFeatureExtractor,
UserFaceRecognizer}
```

The above clause defines a group of actors that are used for face-specific processing in an image. This group could be used to restrict the actors to run in private clouds only. Suppose we have previously defined two actor systems (ActorSysPrivate1 and ActorSysPrivate2) that are made of actors restricted to run in a private cloud only. We can group these actor systems into a *Location* definition, as the next clause shows:

```
{Name: PrivateActorSystems, ActorSystems:
ActorSysPrivate1, ActorSysPrivate2}
```

The next clause is a Set-Operation definition for extracting face-specific processing actors by removing the image loader from the face recognition actor system. Note that other conditions, such as selecting the list of actors with "Face" in the actor name attribute, are also supported.

```
{Name: FaceActors, Subject (ActorSystems:ActorSysUser),
Object(Actors:UserImageLoader), Operator (REMOVE)}
```

7.4.7.2 Policy Description

The main goal of writing a policy file is to define required authorization rules on actions among actors. The previously defined grammar allows one to define entities and group or select them, which are prerequisites for defining restriction rules. We now look at using them to express authorization rules and their evaluation ordering.

Each rule definition regulates one action from subject entities to be performed on object entities. Actions include all allowable actions within an actor framework: sending, receiving, migrating, and creating. This allows one to regulate actions, move-around, and communication between the actor components of a mobile hybrid cloud application.

The following are examples of such rules defined for the face recognition application from Section 7.4.5.

```
{Name: ActorSystemIsolation-Rule, Subject(ActorSystems:Actor-
SysUser), Object(ALL), Actions:ALL, Permission:DISALLOWED}
```

The above rule restricts all actions between actors that are in the ActorSysUser actor system and any other actors in the system.

```
{Name: Actor-Gateway-Rule, Subject(Actors:UserFaceRecogni-
zer), Object(Actors:ActorPublicGate, ActorPrivateGate),
Actions:SEND-TO, RECEIVE-FROM, Permission:ALLOWED}
```

The above rule allows the UserFaceRecognizer actor to send/receive messages from public and private gateway actors (assuming that such actors have been defined previously).

```
{Name: UserFaceRecognizer-Rule-Order, Subject(Rules: Actor-Gateway-
Rule), Object(Rules: ActorSystemIsolation-Rule), Order: PRECEDENCE}
```

This is a rule evaluation order definition that allows UserFaceRecognizer to communicate with public/private gateway actors but not with any other actors.

Note that other actions that are supported, such as MIGRATE–TO, CREATE–AT, BE–MIGRATED–FROM, and BE–CREATED–AT–BY, can be used to restrict the creation/migration of private actors (such as the database with faces of known criminals) to public cloud servers in the face-recognition application (from Section 7.4.5).

7.4.7.3 Policy Evaluation

In a mobile hybrid cloud framework with authorization restrictions, every requested action by the subject has to be approved by the authorization framework before being performed on the object. To make a decision, the authorization system has to evaluate the defined policy rules. However, it is possible for different policy rules to contradict each other, as rules are human-defined by different parties, organizations, and developers, at different times, at different levels, and for different purposes. Our framework prioritizes hard policy rules, which are defined at a higher level by the organization, over soft policy rules, which are defined by programmers for individual applications or instances. Prioritizing hard policy restriction rules over soft policy rules resolves any potential conflict between hard and soft policies. In other types of conflicts between rules of the same type, we always prioritize action denials over permissions.

Every authorization rule can be summarized as a five-tuple of the form *<Subject, Object, Action, Sign, Type>*. Here, *Subject* and *Object* are the entities between which the specific action is being restricted. *Sign* can be allowance (+) or prohibition (−), and *Type* covers hard policy (H) or soft policy (S). In order to decide on any requested action, the authorization system has to process rules in a meaningful way from the most prioritized one (usually the most specific rule), to the least prioritized one (the most general one).

7.4.8 Performance- and Energy-Usage-Based Code Offloading

In the previous section, we saw the need to support code mobility to enforce policies for the privacy guarantees of an application. The target offloading goals can affect the component distribution plan in a hybrid cloud environment with multiple public and private cloud spaces in addition to fully parallel application execution. So we next examine application performance and energy usage on mobile devices as target offloading goals and create an offloading decision-making model for the same.

As an alternative to the expensive option of full-VM migration (see Section 7.3.1), we propose a code-offloading mechanism that is more selective, intelligently identifying and migrating only those parts of the mobile application that would most benefit from migration to a more powerful computing platform while keeping communication overhead low. This approach masks the details of the migration process from both users and developers, while providing for a natural partitioning of an actor-based mobile application. The key challenge is that of identifying and selecting the groups of application components (actors) that would be most beneficial to offload based on the current state of the mobile platform and the application, the application behavior, and the primary purpose of the offloading (energy or performance optimization).

7.4.8.1 Offloading for Sequential Execution on a Single Server

There can be wide variation in the goals for the offloading, depending on the usage scenario; they may range from maximizing the application performance, for example, in video games and vision-based applications, to minimizing the energy consumption on the mobile device, for example, in background processes. Regardless of the offloading purpose, however, the implementation costs are highly dependent on the remote platform on which the offloaded components are to be executed. Key factors include both the performance of the target platform and the communication properties, such as latency and bandwidth.

The goals for maximizing application performance [1] and minimizing energy use on mobile devices are shown in Equations 7.1 and 7.2, respectively. First, let

$T_{mobile} \overset{\text{def}}{=}$ execution time on mobile device

$T_{transfer} \overset{\text{def}}{=}$ duration of data transfer

$T_{remote} \overset{\text{def}}{=}$ execution time on remote server

then

$$T_{mobile} > T_{transfer} + T_{remote} \qquad (7.1)$$

Second, let

$E_{active} \overset{\text{def}}{=}$ energy cost on mobile device

$E_{transfer} \overset{\text{def}}{=}$ energy cost of data transfer

$E_{idle} \overset{\text{def}}{=}$ idle energy cost while waiting for result

then

$$E_{active} > E_{transfer} + E_{idle} \qquad (7.2)$$

These equations lead naturally to the pause-offload-resume model [9], which results in sequential execution. For this reason, using CloneCloud [10] or ThinkAir [11] ostensibly to enable opportunistic parallelism results in sequential execution in practice. Furthermore, such models only have to consider a single remote location for offloading. We consider parallelism wherein multiple remote servers work concurrently with mobile devices.

Expanding Equations 7.1 and 7.2, we observed that they are structurally similar and usually result in close decisions, if power consumption on mobile devices for computation, transfer of data to remote servers, and waiting in idle mode are all proportional. That is the case for sequential execution and is the result of assuming that the mobile screen will be on, even in the idle state [9–11].

7.4.8.2 Offloading for Parallel Execution on Hybrid Clouds
Deciding on an optimized offloading plan for parallel applications in a hybrid cloud environment requires consideration of the application type, the available resources at different remote machines, and the effects of offloading on future application behavior.

7.4.8.3 Maximizing Performance
Fully parallel execution refers to both parallel execution on multiple remote locations and simultaneous local and remote execution. As a result, the total application execution time is the maximum time required for any of the mobile or remote spaces to finish executing program code for all of its assigned components. Since local communication between components located on the same machine is relatively fast, we can ignore local communication and

only consider communications between components placed at different locations. The offloading goal can be summarized as maximizing application performance (*MaxAppPerf*) or minimizing application execution time (*MinAppExec*) using:

$$
\max(MaxAppPerf) = \min(MinAppExec) =
$$
$$
\min\left(\max_{0 \leq L \leq M}(ExecAtLoc(L)) + \max_{0 \leq L \leq M}(CommAtLoc(L)) \right) \qquad (7.3)
$$

A mobile application consists of N components, and each component $i \in [1, N]$ is located at $Loc(i, t)$ at time t. Having M different cloud spaces results in $Loc(i, t) \in [0, M]$ where 0 represents the local mobile device and $[1, M]$ corresponds to different cloud spaces. Assuming that we know the application component distribution between the local mobile device and the hybrid cloud spaces at time t_1, our goal is to find the component distribution for the next time interval t_2 such that application performance is maximized.

Thus, different parts of Equation 7.3 can be extended so that the first term $\max_{0 \leq L \leq M}(ExecAtLoc(L))$ captures the maximum (across M different cloud spaces) execution time for all components on each of those locations L. This maximum can be obtained using monitoring and previous profiling for the execution time of each component in its location at time t_2.

Similarly, the second term $\max_{0 \leq L \leq M}(CommAtLoc(L))$ of Equation 7.3 captures the maximum required time for one of the locations to send out all its communications. This maximum can be obtained using the profiled amount of communication between the members of each pair of components during the elasticity manager's running time interval Δ and the location of components across locations in time t_2.

However, not all components of an application are offloadable, so a few constraints must be added to the above optimization problem. As we are considering a hybrid cloud that consists of multiple private and public cloud spaces, application developers or users can specify additional constraints in terms of how different components can be offloaded to different locations. These additional constraints can also address privacy issues in terms of not offloading sensitive or confidential components to public cloud spaces.

7.4.8.4 Minimizing Energy Consumption

Let us now examine the differences in terms of minimizing mobile device energy consumption instead of performance. This goal can be defined as below. Let

$E_{app} \overset{\text{def}}{=} $ application mobile energy consumption

$E_{device} \overset{\text{def}}{=} $ energy saved on mobile device

$E_{remote} \overset{\text{def}}{=} $ total mobile energy saving by remote component execution

$E_{rcom} \overset{\text{def}}{=} $ energy loss because local communication became remote communication

$E_{lcomm} \overset{\text{def}}{=} $ energy saved because remote communication became local communication then

$$\min\left(E_{app}\right) = \max(E_{device}) = \max(E_{remote} - E_{rcomm} + E_{lcomm}) \qquad (7.4)$$

E_{remote} in Equation 7.4 can be further elaborated into

$$\sum_{i=1}^{N} (LocEQ(0, Loc(i, t_1)) * (1 - LocEQ(0, Loc(i, t_2))) * Energy(i)) \qquad (7.5)$$

where $Energy(i)$ is the profiled energy consumption of component i as it runs locally on the mobile device during the time interval Δ, and $LocEQ(l_1, l_2)$ returns 1 if two given locations are identical and 0 otherwise. Note that the first term of the equation considers only components that are currently on the device, and the second term adds the condition that those elements must now be at a remote location. Thus, energy savings are counted only for components that have been migrated from the local device to a remote location. It should be noted again that our goal is to minimize energy consumption at the mobile device and not the total energy. The migration of components between remote locations does not help with that goal and therefore is not considered in the equation.

E_{rcomm} and E_{lcomm} in Equation 7.4 are obtained using the profiled amount of communication between each pair of components and the profiled mobile power during communication with remote servers.

To maximize performance, we can add constraints, for example, that offloading of components to remote locations to save local energy must not affect the performance of the application. In other words, we can specify that energy savings are allowed only as long as a certain level of service performance is maintained. An important observation we made in our fully parallel application model is that the results of our offloading goals are very different for application performance improvement and for energy savings on mobile devices. This is unlike the sequential case in which the models lead to similar configuration results. Therefore, we use the constraints to add restrictions on how much improvement in support of one goal is allowed to negatively affect the other.

7.4.8.5 Energy Monitoring

The difficulty in solving Equation 7.5 arises from $Energy(i)$. As mentioned, $Energy(i)$ is the profiled energy consumption of component i while it is running locally on the mobile device. Evaluating this term requires fine-grained profiling of energy consumption per application component on a mobile device. However, most mobile devices do not provide any tool for direct measurement of the consumed energy. Almost all previous research in this area has relied on external power meters to measure energy consumption. Although use of expensive external power meters works for an experimental setting, we cannot expect end users to carry such devices with themselves to profile the energy consumption of their mobile devices. This is a big challenge in optimizing energy consumption of mobile hybrid cloud applications. Even if the total energy consumption of the mobile device can be measured, there are multiple applications running on a mobile device at any one time. That requires distribution of the total measured energy among those applications. Furthermore, multiple components within our target application are running at any given time, and distributing energy among these components is a challenge. The solution we explore here can scalably profile runtime energy consumption of an application, while treating it as a black box. This approach can detect complex interactions and dependencies between components or actors in an application that affect energy consumption on mobile devices.

We consider mobile applications written using the actor-model-based programming language SALSA [46], which natively supports migration of actors between mobile and cloud platforms. We built the mechanism to profile running applications from an underlying SALSA runtime layer in order to enable attribution of battery drops to subsets of actor types. The mechanism first instruments the SALSA runtime to enable determination of which actors are scheduled in the application at each (predefined) interval of time. Based on the corresponding battery drops in those intervals, a combination of linear regression and hypothesis testing techniques is used to infer battery drop distribution of subsets of actor groups within an execution context.

Note that different subsets of actors would be active in each interval, so if we observed these data for an application from a large number of smartphones, it would then be possible to collect measurements that would help us generate a distribution of battery drop characteristics for different actor types with increasing accuracy. Apart from speeding up the availability of battery drops for subsets of actor types, this crowdsourcing-based approach could handle noise in the sensor readings. We would have to partition the data by execution context, however, which includes hardware context, such as the screen or GPS's being turned on or off, along with software context, such as other applications running on the device. The additional data available would then also have to be used to handle large variability of the context in which different application instances are running, before energy attribution can be done. We leave this crowdsourcing-based monitoring approach as an extension for future work.

7.4.8.6 Security Policies and Energy Monitoring

One use case of the IMCM infrastructure is an authorization system that respects a specified energy policy. Such a policy is enforced through runtime restrictions on the actors that are executing on a mobile cloud platform. Some examples of policies that can be specified include the following:

- Policies that prevent malicious actors from draining batteries on mobile devices to prevent secure actors from carrying out their tasks.

- Energy-consumption-based policies to restrict sending or receiving of messages from abusive actors, and to manage DoS attacks by enforcing maximum-energy-threshold-based restrictions on actor creation within a container.
- Organization-wide policies for abusive actors (based on energy characteristics) as the actor signature. This is useful when runtime actor information is unavailable when such policies are being written.
- Track the energy consumption of an actor over time, in order to detect any large deviations in energy characteristics that may occur because the actor has been compromised.

7.5 Coordinating Actors

Applications on the mobile cloud involve a large amount of concurrent activity. As we described in the last section, dividing a computation into actors isolates its state, enabling security while facilitating parallelism, distribution, and mobility. However, because actors are autonomous and decentralized, not globally controlled, they must explicitly synchronize when they need to coordinate their actions. For example, such synchronization is needed if an action by one actor must precede an action by another.

Synchronization between actors follows a message-passing semantics: When an actor sends a message to another actor, the events leading up to the first message send precede the events that happen after the message has been received. For example, consider a deposit into one bank account that requires a withdrawal from another bank account. These actions must be atomic. One way to implement the atomicity is to require the two actors to explicitly exchange a sequence of messages that implement a two-phase commit protocol.

Synchronization protocols can grow complicated. For example, atomicity may be realized using a two-phase commit protocol, or it may be implemented using a three-phase commit to guard against failure of one of the participants. The atomicity protocol may be optimistic (and undo actions if there is a violation of atomicity), or it may be conservative. Expressing complex protocols via message-passing leads to complex code. Such complexity can introduce software bugs, thus compromising safety, and also make it difficult to monitor a system in order to detect anomalies in the interaction patterns. Moreover, because the same high-level synchronization policy may be implemented using different protocols, the policy may be buried in the code. This means that one has to reason about the code to understand the policy.

7.5.1 Expressing Coordination

A number of methods have been developed to express coordination between actors separately from the functional behavior of an actor. Observe that coordination policies affect the behavior of participant actors locally. Thus, multiple policies can be specified separately and enforced. However, such policies could interfere with each other. Moreover, policies must be enforced by trusted actors and implemented through mediation of the messages between actors, so formal methods are required to support reasoning about coordination.

7.5.1.1 Synchronizers

The oldest framework for declarative expression of coordination behavior is *synchronizers* [3]. A synchronizer is a collection of constraints that enforce an ordering between messages processed by different actors that belong to some interacting group. Such groups may overlap, as an actor may participate in different activities with different groups of actors. When a message does not meet the constraints required for it to be processable, message dispatch is delayed until the state of the system changes (because other messages are processed) and the constraints governing the blocked message are satisfied. Synchronizers are implemented through creation of the appropriate meta-actors, and this process can be automated.

Two types of constraints that provide generality are *disabling constraints* and *atomicity constraints*. Disabling constraints prevent an actor from handling messages that match a given pattern. For example, if all types of messages except an initialization message are disabled, the actor must first receive an initialize message before it processes other messages. Atomicity constraints implement atomic processing of messages by different actors, as in the example discussed earlier.

Programmers declare synchronizers as templates. Similar to classes or actor behaviors, these templates are dynamically instantiated at runtime with concrete values filled in for the parameters. Consequently, synchronizers can have local state that changes with the observed messages. They may also overlap, that is, multiple synchronizers can constrain the same actor. Figure 7.4 shows the effects of a possible synchronizer.

Consider a sliding window protocol. The protocol prevents a sender from sending more than a specified number of messages at a time. More messages can be sent only after some of the messages sent have been processed. The protocol can prevent *denial of service* attacks. It is straightforward to express such a protocol using a synchronizer.

In the code below (Figure 7.5), no more than a given number of messages – as determined by the parameter `size` (representing the size of a "window") – from a specified `sender` to a specified `receiver` may be pending at any one time. Each time one message within the window is processed, the number of acknowledged messages is updated, permitting another message to be sent.

7.5.1.2 Security Issues in Synchronizers

As systems are continually evolving, synchronizers may be dynamically added or removed. As systems scale, the chance that some actor will be compromised or

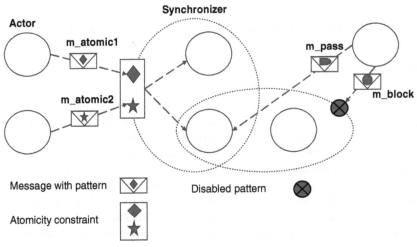

Figure 7.4 Constraints enforced by synchronizers. Synchronizers (dashed ovals) support combinations of atomicity and disabling constraints. Atomicity constraints ensure that a set of messages is dispatched as a whole and without causally intervening events. Messages m_atomic1 and m_atomic2 satisfy the atomicity constraint together and are therefore dispatched together by their synchronizer at the respective target actors. Message m_block matches a disabling pattern in the other (lower) synchronizer and therefore cannot be dispatched. Message m_pass matches no pattern and thus is unconstrained.

be malicious increases. A scalable coordination model must be able to cope with uncooperative actors and gracefully degrade in the presence of failures. It must also guard its reconfiguration mechanisms against abuse. Installation of a synchronizer can lead an actor to wait for a message that never arrives, leading to failure. For example, suppose that an actor A can handle messages of types

```
1   SlidingWindowProtocol (sender, receiver, int size) {
2       int lastFrame;
3       int lastAck;
4
5       disable receiver.message (data, int framen)
6               when lastFrame − lastAck >= size;
7
8       trigger:
9           receiver.message (data, int framen) −>
10                      lastFrame = lastFrame + 1;
11          sender.ack (int ackn) −> lastAck = lastAck + 1;
12  }
```

Figure 7.5 Sliding window synchronization constraint.

```
1   DisablingAttack (a) {
2      disable (a.message1 or a.message2 or ... or a.messageN) when true
3   }
```

Figure 7.6 Disabling attack.

message1, message2, and so on, up to message$N$. A malicious actor M can prevent A from receiving any further messages by installing a synchronizer, as shown in Figure 7.6, that disables all message handlers in A.

To address the problem, we need synchronization constraints to be scoped. A suitable scoping mechanism was introduced in Ref. [47], and the discussion in the rest of this section follows that work. The idea is to restrict the ability of a synchronization constraint to be able to limit messages only if they are coming from actors within a scope. This scoping rule means that messages from outside the scope of a synchronizer can still be received by a target actor, thus preventing the target actor from being permanently locked by a malicious or buggy actor that fails to follow the agreed-upon interaction protocol.

Synchronization constraints are complementary to object capabilities [48]. Object-capability security is a natural model of security in actor systems. An actor's address works like a capability: One cannot send a message to an actor without knowing its address, and these addresses can be guessed. If the address of an actor a_0 is provided to an actor a_1, the recipient a_1 is provided with the capability to send a_0 a message. This provides a level of security: By carefully restricting the distribution of addresses, one can prevent malicious or buggy actors from affecting the behavior of actors with which they should not be able to interact.

In contrast, synchronization capabilities determine the scope of synchronization constraints: They provide actors with the capability to join an interaction and be restrained to follow the synchronizer's constraints. Figure 7.7 shows the scoping effects of synchronization capabilities.

As with object capabilities, we assume that *synchronization capabilities* are unique across the system and cannot be guessed. The distribution of synchronization is similar to that of actor addresses: Actors obtain synchronization capabilities through initialization and introduction by an actor that has the capability. However, synchronization capabilities are "inherited" from the creator: When a new actor is created, it is subject to the same synchronization rules as its creator. This prevents an actor from escaping its synchronization constraints by simply creating a new actor that circumvents the constraints on its messages.

Synchronization capabilities can prevent a buggy or malicious actor from intentionally creating a deadlock through installation of a synchronizer. In the DisablingAttack synchronizer from Figure 7.6, scoping the synchronization constraint results in behavior similar to that of the lower right actor in Figure 7.7:

Scoped synchronizer

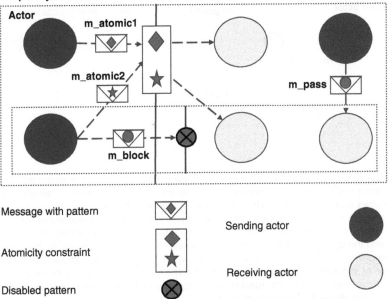

Message with pattern		Sending actor
Atomicity constraint		
		Receiving actor
Disabled pattern		

Figure 7.7 Scoped synchronizers enforcing constraints. Scoped synchronizers (dashed boxes) constrain only messages sent by actors for which they hold the synchronization capability. These actors are shaded in dark gray in the diagram. Their sent messages must satisfy the constraints before they can be dispatched at the recipients (shown in light gray). Since message m_block matches a disabling pattern of the inner synchronizer, it cannot be dispatched. However, the respective synchronizer lacks control over the sender message m_pass, so m_pass can be dispatched despite having the same shape as m_block.

If the synchronizers do not have a synchronization capability that affects an actor's messages, these messages will pass through. Moreover, accidental interference of constraints also becomes less likely [47].

When a message is dispatched, all synchronizers that belong to matching update patterns receive information about the event. For example, consider cooperating resource administrators that share a limited number of resources. When a resource is released by one administrator, it enables another administrator to service requests that require the resource. Synchronizers that are not constraining the sender of a message (and thus that type of message) may still need to receive information about the dispatch of a message by an actor whose other messages are being constrained. It needs that information in order to have a correct view of the state of the target actor, since it needs the information to be able to receive a specific type of message (that may need to be constrained). Although a visible message dispatch provides a consistent view of the system, it can enable malicious actors to spy on other actors, as shown in Figure 7.8. This security issue remains an open problem.

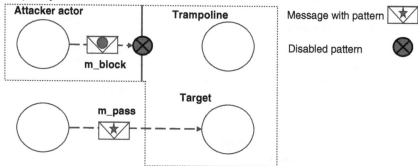

Figure 7.8 Information leak through updates. Scoping only limits the constraining power of synchronizers. To guarantee a consistent view on the system, synchronizers can observe all messages that an actor dispatches, regardless of the synchronization capabilities the synchronizer holds. An attacker actor exploits this fact to gather information about the target actor. First, the attacker creates a trampoline actor and installs a synchronizer on the target and the trampoline. The synchronizer disables the dispatch of message m_block at the trampoline until it observes message m_pass at the target. Then, the attacker sends message m_block to the trampoline. Once the trampoline has dispatched m_block, it bounces a message back to the attacker, providing the attacker with the knowledge that the target has dispatched message m_pass.

7.6 Session Types

One way to reason about the constraints discussed above is in terms of the interface of the actors involved in an interaction. An interface is expressed in terms of abstract data types in programming languages. For example, the type of a *bounded buffer* is the operations the buffer accepts: *put, get, full?* and *empty?*. In sequential languages, such types do not account for when an operation may take place. For example, a bounded buffer may not perform a *put* when it is full, and the buffer may not perform a *get* when it is *empty*. The programmer must explicitly check whether the buffer is *full* or *empty* before performing a *put* or *get* operation, respectively. Such a check does not help us in a concurrent system; while one actor checks (by sending a message) whether the buffer is *full*, another actor may perform a *put* that changes the state to *full*.

Synchronizers provide a high-level specification of interaction protocols. They are related to the evolution of types based on communication between actors. One theoretical framework for capturing interaction protocols is to consider the communication protocols that they represent. Session types [5] describe communication protocols among parallel processes in terms of the evolution of types of participant actors. The general methodology starts with the specification of a *global type,* which captures the permissible sequences of messages that participants may exchange in a given *session.*

Note that a "global type" is so called because it expresses the evolution of type properties of the behavior of a group of actors involved in an interaction. It is not global in the sense of saying something about the behavior of every actor in a system. Based on a global type, a *projection* algorithm is needed to generate the necessary session types for each participant. The session types for participants are known as *end-point types* or *local types*. These local types describe the evolution of the expected behavior of the individual participants in the protocol.

A type checker can compare the projected localized specification against the implementation of each participant's behavior. Conventional session types can be generalized to typing coordination constraints in actor programs, which can then be enforced using, for example, synchronizers.

7.6.1 Session Types for Actors

Typing coordination constraints in actors pose two challenges: first, asynchronous communication leads to delays that require consideration of arbitrary shuffles; second, *parameterized* protocols must be considered. For example, assume that two actors are communicating through a sliding window protocol; the actors agree on the length of the window (i.e., the number of messages that may be buffered) and then proceed to a concurrent exchange of messages. Conventional session types are not suitable for typing interactions such as the sliding window protocol. The reason for this limitation is that their respective type languages depend on other formalisms for type checking (such as typed λ-calculus [49] or System T [50]), and these formalisms do not support a concurrency construct.

To overcome such limitations, we propose a programming language, Lang-A, along with a session type system, System-A, that allows type parameters in novel constructs that capture the inherent asynchrony of actor programs. In particular, the system includes novel *parameterized* typing constructs for expressing asynchrony, concurrency, sequence, choice, and atomicity in protocols. An inference algorithm derives local System-A types from Lang-A programs, and a type system checks those types against conformance criteria.

7.6.1.1 Example: Sliding Window Protocol

Recall the sliding window protocol. Assume that an actor a sends messages of type m to an actor b, which acknowledges every received message with an *ack* message. The protocol determines that at most n messages can be unacknowledged at any given time, so that a ceases sending until it receives another *ack* message. In this example, the window size n is a parameter, which means that we need a way to express the fact that n sending and acknowledging events can be in transit at any given instant in time. The global type of the protocol is as follows:

$$\underbrace{(a \xrightarrow{m} b; b \xrightarrow{ack} a)^* \parallel (a \xrightarrow{m} b; b \xrightarrow{ack} a)^* \parallel \cdots \parallel (a \xrightarrow{m} b; b \xrightarrow{ack} a)^*}_{n \ times}$$

Here $a \xrightarrow{m} b$ means that a sends a message of type m to b. The operator ; is used for sequencing interactions. Operator $\parallel$ is used for the concurrent composition of its left and right arguments, while the Kleene star has the usual semantics of an unbounded – yet finite – number of repetitions.

The above type can be expressed using the notation of Castagna *et al.* [27,51], albeit with a fixed window size n. In System-A, on the other hand, we can parameterize the type in n and statically verify that participants follow the protocol without knowing its runtime value. Using $\left\Vert\right._{i\,=\,1}^{n}$ to denote the concurrent composition of n processes, we obtain the following type:

$$\overset{n}{\underset{i=1}{\Vert}} (a \xrightarrow{m} b; b \xrightarrow{ack} a)^* \qquad (7.6)$$

7.6.2 Global Types

A global type describes a protocol to which the whole system must adhere.

For example, the sliding window protocol specification is a global type, since it describes the behavior of all participants. Table 7.1 presents the grammar that generates the syntactic category $\mathcal{G}$ of global types. The elements of $\mathcal{G}$, which are

Table 7.1 The syntax of global types.

$\mathcal{G} ::=$	$a \xrightarrow{m} b$	(G-Interaction)	$\mid$	$(\mathcal{G})$	(G-Paren)
$\mid$	$\mathcal{G} ; \mathcal{G}$	(G-Seq)	$\mid$	$\overset{n_2}{\underset{i=n_1}{\odot}} \mathcal{G}_i$	(G-Seq-N)
$\mid$	$\mathcal{G} \oplus \mathcal{G}$	(G-Choice)	$\mid$	$\overset{n_2}{\underset{i=n_1}{\oplus}} \mathcal{G}_i$	(G-Choice-N)
$\mid$	$\mathcal{G} \parallel \mathcal{G}$	(G-Paral)	$\mid$	$\overset{n_2}{\underset{i=n_1}{\Vert}} \mathcal{G}_i$	(G-Paral-N)
$\mid$	$\mathcal{G} \otimes \mathcal{G}$	(G-Shuffle)	$\mid$	$\overset{n_2}{\underset{i=n_1}{\otimes}} \mathcal{G}_i$	(G-Shuffle-N)
$\mid$	$\mathcal{G}^n$	(G-Exp)	$\mid$	$\mathcal{G}^*$	(G-KleeneStar)

We use a and b to denote possibly indexed actor names; m to denote possibly indexed message types; variants of n for type parameters; and i for index names. For parsing purposes, we assume operators associate with the right, and that operator precedence $<_p$ is $\otimes <_p \parallel <_p \oplus <_{pi}$.

instances of global types, will be denoted by variations of the variable G. Intuitively, the rules capture the following concepts:

(G-Interaction) denotes the sending and receiving of a message. For instance, $a \xrightarrow{m} b$ means that participant a sends a message of type m to participant b.

(G-Seq) is used for the sequential composition of events.

(G-Choice) denotes exclusive choice between the arguments. For $G_{1,2} \in \mathcal{G}$, $G_1 \oplus G_2$ means that only one of G_1, G_2 will be executed.

(G-Paral) means that the arguments run concurrently. Interleavings are allowed, as long as the order established by the ; operator is respected. For example, $\left(a \xrightarrow{t_1} b; a \xrightarrow{t_2} c\right) \| c \xrightarrow{t_3} b$ means that all interleavings ABC, ACB, CAB are possible, where $A = \left(a \xrightarrow{t_1} b\right)$, $B = \left(a \xrightarrow{t_2} c\right)$, and $C = \left(c \xrightarrow{t_3} b\right)$. Note that B is not allowed to precede A because of how operator ; orders them. (Hence, BAC, BCA, and CBA are not valid interleavings in this example.)

(G-Shuffle) means that both arguments are executed atomically, in an unspecified order. Formally, $G_1 \otimes G_2 = (G_1; G_2) \oplus (G_2; G_1)$ with the = relation denoting semantic equivalence, as detailed in the original work of Charalambides *et al.* [4].

(G-KleeneStar) has the usual semantics of zero or more repetitions of the argument. We assume a finite number of repetitions.

The n-ary versions of the operators express behaviors for which the values of n, n_1, and n_2 are unknown at compile time. Intuitively, the rules (G-Seq-N), (G-Choice-N), (G-Paral-N), and (G-Shuffle-N) apply the respective binary operator $n_2 - n_1$ times, generating a global type for each of the $n_2 - n_1 + 1$ values of i. (G-Exp) denotes the n-fold, sequential repetition of the argument. Note that for known parameter values, these expansions can take place during program compilation.

For parsing purposes, operators associate with the right. In terms of semantics, however, all of the operators are commutative, with the exception of sequencing. Furthermore, all operators are associative, with the exception of shuffling. In particular,

$$\overset{n}{\underset{i=1}{\bigotimes}} G_i \neq (\cdots((G_1 \otimes G_2) \otimes G_3 \cdots) \otimes \cdots \otimes G_n)$$

because the meaning of $\otimes_{i=n_1}^{n_2} G_i$ is that all arguments G_i are executed atomically, but in an unspecified order. Instead, the right-hand side above prevents, for example, G_3 from occurring between G_1 and G_2.

The distinction between the Kleene star and exponentiation is fundamental. The use of G^n means that the protocol conformance checker will have to prove that the system is correct for any fixed value of the parameter n. G^*, on the other hand, refers to an unbounded number of repetitions of G. There is no parameter fixing this number, and it may be different from instance to instance of the

Kleene star and/or among executions of the same program with the same runtime values for its parameters; the Kleene star entails a choice as to when to exit the loop.

We define an *event* to be a single interaction $p_1 \xrightarrow{m} p_2$, and a *trace* to be any finite sequence $e_1 \cdots e_k$ of events. Given a global type $G \in \mathcal{G}$, the set of possible traces $tr(G)$ that it can produce determines the permissible sequences of messages that participants may exchange. We omit the full semantics of global types in this text, and refer the interested reader to the original work of Charalambides *et al.* [4].

7.6.3 Programming Language

Global types by themselves provide no implementation of protocols; implementations are given in the language Lang-A, which is used in the example in Figure 7.9. The figure shows an implementation of the sliding window protocol described earlier. The spawn statement launches n parallel instances of its block argument, one for each value of the provided index expression. Sends and receives coming from different spawned operations can be interleaved in any way possible. In this example, both the sender and the receiver spawn n parallel operations, each consisting of a repeating send/receive pair. This allows any interleaving of sends and receives, as long as no more than n sends are left unacknowledged.

In general, a Lang-A program begins with the declaration of the program parameters, akin to type parameters n, n_1, n_2, and so on, from Table 7.1. Then come message structure definitions, and the code for each actor. Both actor and message definitions can include an optional array syntax after their name. In the case of actors, this syntax declares as many actors as does the array parameter.

```
n : param
```

```
// the sender
actor a = {
  message m : Int
  message ack : Int
  var NotDone : Boolean
  NotDone = true

  spawn( i = 1..n
    while NotDone {
      m = ...
      send(b, m) ;
      recv(b, ack) ;
      NotDone = ...
    })
}
```

```
// the receiver
actor b = {
  message m : Int
  message ack : Int
  var NotDone : Boolean
  NotDone = true

  spawn( i = 1..n
    while NotDone {
      recv(a, m) ;
      ack = ...
      send(a, ack) ;
      NotDone = ...
    })
}
```

Figure 7.9 The sliding window example in Lang-A.

In the case of message structures, the syntax declares as many message types as does the array parameter. This allows the expression of protocols for which both actor names and message types are parameterized, such as $\left\| \begin{smallmatrix} n \\ i = 1 \end{smallmatrix} \alpha_i \xrightarrow{\mu_i} \beta_i \right.$. Lang-A is defined such that there is an almost one-to-one correspondence between the language constructs and the syntax of local types; the precise syntax and semantics are detailed in the work of Charalambides *et al.* [4].

7.6.4 Local Types and Type Checking

A local type specifies the abstract behavior of a single protocol participant, for example, one of the actors in the program in Figure 7.9. This section discusses System-A local types [4,28], whose syntax is shown in Table 7.2. In the grammar,

(L-Send) denotes sending of a message of type m to actor a.

(L-Recv) denotes receiving of a message of type m from actor a.

(L-Seq), (L-Choice), (L-Shuffle), (L-Exp), (L-KleeneStar) describe the same concepts as for the global types.

(L-Paral) is also defined as for global types. As before, the local type $(a!t; a!u) \parallel a?v$ allows three orderings of the events $T = a!t$, $U = a!u$, and $V = a?v$, namely, TUV, TVU, and VTU. As above, the specification $a!t; a!u$ ensures that T happens before U.

Projecting a global type onto each participant in a protocol results in a set of local types, one for each actor. For example, projecting $G = a \xrightarrow{m} b$ onto a,

Table 7.2 The syntax of local types.

$\mathscr{L} ::=$	$(\mathscr{L})$	(L-Paren)	$\mid$	τ	(L-Empty)
	$\mid$ $a!t$	(L-Send)	$\mid$	$a?t$	(L-Recv)
	$\mid$ $\mathscr{L}\,;\,\mathscr{L}$	(L-Seq)	$\mid$	$\overset{n_2}{\underset{i=n_1}{\odot}}\mathscr{L}_i$	(L-Seq-N)
	$\mid$ $\mathscr{L} \oplus \mathscr{L}$	(L-Choice)	$\mid$	$\overset{n_2}{\underset{i=n_1}{\bigoplus}}\mathscr{L}_i$	(L-Choice-N)
	$\mid$ $\mathscr{L} \parallel \mathscr{L}$	(L-Paral)	$\mid$	$\overset{n_2}{\underset{i=n_1}{\parallel}}\mathscr{L}_i$	(L-Paral-N)
	$\mid$ $\mathscr{L} \otimes \mathscr{L}$	(L-Shuffle)	$\mid$	$\overset{n_2}{\underset{i=n_1}{\otimes}}\mathscr{L}_i$	(L-Shuffle-N)
	$\mid$ $\mathscr{L}^n$	(L-Exp)	$\mid$	$\mathscr{L}^*$	(L-KleeneStar)

As in the syntax of global types, we use a to denote possibly indexed actor names; m for possibly indexed message types; variants of n for type parameters; and i for index names. For parsing purposes, we assume operators associate with the right, and that operator precedence $<_p$ is $\otimes <_p k <_p \oplus <_{p^;}$.

written $G \triangleright a$, gives the local type $b!m$. Similarly, $G \triangleright b = a?m$. These can then be checked against the types inferred from the Lang-A code.

Some (local) type inference rules are shown in Table 7.3. Rule (Inf-Skip) assigns the type τ to the skip construct and to all value computations, since those do not involve interactions between processes. Rules (Inf-Send) and (Inf-Recv) assign the (L-Send) and (L-Recv) constructs to send and receive actions, respectively. Sequencing is straightforward, but typing the spawn construct is a little more involved: The index i as well as the range $c_1 \cdots c_2$ transfers into the inferred type, which includes the parameterized construct $\|$.

The type-checking methodology requires the comparison of local types projected from the global type with those inferred from the program code. In order to perform the comparison on a syntactic level, local types are reduced to a *normal form*, in a semantics-preserving manner. For a detailed discussion of the projection function, the full list of type inference rules, and the semantics of local types and their relation to the semantics of Lang-A programs, we refer the interested reader to the original work of Charalambides *et al.* [4].

7.6.5 Realization of Global Types

A global type must satisfy certain properties in order to be *projectable*. Applying the projection function to a projectable global type will result in local types for the participants whose combined behavior is consistent with the global type.

First, to be projectable, the sequential constructs of a global type must retain their sequential semantics after projection. For example, the type $a \xrightarrow{m} b; c \xrightarrow{m'} d$ does not preserve said semantics after projection, because c has to send m' after b receives m, and programming this sequence requires interactions not captured in the above type.

Table 7.3 Rules for inferring local types from Lang-A programs.

(Inf-Skip)	(Inf-Compute)
$\Gamma \vdash \texttt{skip} : \tau$	$\Gamma \vdash \text{Computation} : \tau$
(Inf-Send)	(Inf-Recv)
$\dfrac{\Pi \vdash a : actor \quad S \vdash x : m}{\Gamma \vdash \texttt{send}(a, x) : a!m}$	$\dfrac{\Pi \vdash a : actor \quad S \vdash x : m}{\Gamma \vdash \texttt{recv}(a, x) : a?m}$
(Inf-Seq)	(Inf-Spawn-N)
$\dfrac{\Gamma \vdash R_1 : L_1 \quad \Gamma \vdash R_2 : L_2}{\Gamma \vdash R_1; R_2 : (L_1; L_2)}$	$\dfrac{S \vdash c_1 : cons \quad S \vdash c_2 : cons \quad \Gamma \vdash R[i/k] : L_i \quad k \in fv(R)}{\Gamma \vdash \texttt{spawn}(k = c_1..c_2\, R) : \left(\|_{i=c_1}^{c_2} L_i\right)}$

The environments S and Π contain sorts of variables and actor names, respectively. Γ assigns local types $L \in \mathcal{L}$ to (sequences of) statements R. We omit brace literals from the program syntax when the meaning is clear from the context. We use variants of c for both constants and type parameters. The variables a, b, m, and so on are as in Section 7.6.2.

In addition, a global type $G_1 \oplus G_2$ is projectable if all participants can recognize which branch of the choice operator they need to take during execution. Similar requirements apply to $G_1 \otimes G_2$, for which it must be possible for the constituents to be sequenced both ways, and for the participants to distinguish the two possible orderings.

Concurrent composition requires special care to prevent actions in one concurrent branch from affecting choices made on another, while for the Kleene star we need to ensure that the entry and exit conditions to the starred type can be identified by all participants.

If a global type G is projectable according to the criteria sketched above, the projection function generates local types that are functionally consistent with the global type. We use Δ_G to denote the set of local types that result from the projection of G onto each one of the participants. That is, $\Delta_G = \{p : G \triangleright p\}_{p \in \Pi}$ where Π is the set of participating actors. The set of traces $tr(\Delta_G)$ producible by Δ_G is the union of the sets of traces producible by the local types in Δ_G. Now, let PR be the set of projectable global types. The key correctness property is then that $G \in PR \Rightarrow tr(G) = tr(\Delta_G)$. The proof is by induction on the structure of global types [4].

7.7 The Future

Our mobile cloud framework can enable effective balancing of resource use with performance while preserving security. A particular concern is energy management in mobile devices. Further research is needed in techniques to infer the energy consumed by a specific application on a mobile device. While the execution times of different components can be individually recorded using the system clock, a mobile device reports only lump-sum energy consumption. Figuring out the behavior thus requires inferential techniques based on a larger pool of data.

Monitoring techniques that infer coordination constraints and session types are another area of research. This would allow patterns of interaction in a running system to be inferred. System assurance will be enhanced by flagging suspicious deviations as well as preventing harmful actions. Moreover, the current notations to represent session types are formal in nature and not suitable for use by system developers. A friendly interface could help programmers visualize the interaction behaviors as well as reduce errors in the system. Moreover, session types may help detect information leaks, as certain sequences may reveal more information than a single message interaction would. In addition, statistical sampling can help detect violations of quantitative coordination constraints when not just information sequences, but the sum total of information revealed from a very large number of sources needs to be constrained.

The challenge of addressing the impact of failures and faults on interaction patterns needs to be studied further. Such failures are not explicitly incorporated either into the language of synchronizers or into session types. Adding support for dynamic process creation is an important direction for future work in session types for actor systems. In its current form, System-A cannot express actor creation as a behavior, and global types assume that all participants already exist. Matching a created actor with its subsequent use in a type requires an extra step that is not obvious. Furthermore, System-A omits support for session delegation, and does not deal with issues of progress. In addition, it does not consider overlapping indexed names when they are nested in multiple operators. That omission disallows some cases, such as all-to-all communication.

Acknowledgments

The authors would like to thank the members of the Open Systems Laboratory (OSL) group at the University of Illinois at Urbana-Champaign for their valuable discussion and insight. This chapter was made possible in part by sponsorship from the Air Force Research Laboratory and the Air Force Office of Scientific Research under agreement FA8750-11-2-0084, and the material is based upon work supported by the National Science Foundation under grants NSF CCF 14-38982 and NSF CCF 16-17401. Any opinions, findings, and conclusions or recommendations expressed in this material are those of the authors and do not necessarily reflect the views of the National Science Foundation.

References

1 Kumar, K., Liu, J., Lu, Y.-H., and Bhargava, B. (2013) A survey of computation offloading for mobile systems. Mobile Networks and Applications, **18** (1), 129–140.

2 Agha, G. and Hewitt, C. (1985) Concurrent programming using actors: exploiting large-scale parallelism, in *Foundations of Software Technology and Theoretical Computer Science, FSTTCS 1985* (ed. S.N. Maheshwari), Lecture Notes in Computer Science, vol. 206, Springer, Berlin, pp. 19–41.

3 Frølund, S. (1996) *Coordinating Distributed Objects: An Actor-Based Approach to Synchronization*, MIT Press.

4 Charalambides, M., Dinges, P., and Agha, G. (2016) Parameterized, concurrent session types for asynchronous multi-actor interactions. *Science of Computer Programming*, **115** (C), 100–126.

5 Honda, K., Vasconcelos, V.T., and Kubo, M. (1998) Language primitives and type discipline for structured communication-based programming, in

Proceedings of the 7th European Symposium on Programming: Programming Languages and Systems (ESOP) (ed. C. Hankin), Lecture Notes in Computer Science, vol. 1381, Springer, Berlin pp. 122–138.

6 Rahman, M., Gao, J., and Tsai, W.-T. (2013) Energy saving in mobile cloud computing, in Proceedings of the IEEE International Conference on Cloud Engineering (IC2E), pp. 285–291.

7 Satyanarayanan, M., Bahl, P., Caceres, R., and Davies, N. (2009) The case for VM-based cloudlets in mobile computing. *IEEE Pervasive Computing*, **8** (4), 14–23.

8 Imai, S., Chestna, T., and Varela, C.A. (2012) Elastic scalable cloud computing using application-level migration, in Proceedings of the IEEE 5th International Conference on Utility and Cloud Computing, pp. 91–98.

9 Cuervo, E., Balasubramanian, A., Cho, D.-k., Wolman, A., Saroiu, S., Chandra, R., and Bahl, P. (2010) MAUI: making smartphones last longer with code offload, in Proceedings of the 8th International Conference on Mobile Systems, Applications, and Services, pp. 49–62.

10 Chun, B.-G., Ihm, S., Maniatis, P., Naik, M., and Patti, A. (2011) CloneCloud: elastic execution between mobile device and cloud, in Proceedings of the 6th European Conference on Computer Systems, pp. 301–314.

11 Kosta, S., Aucinas, A., Hui, P., Mortier, R., and Zhang, X. (2012) ThinkAir: dynamic resource allocation and parallel execution in the cloud for mobile code offloading, in Proceedings of the IEEE INFOCOM, pp. 945–953.

12 Shifteh Far, S. (2015) A flexible fine-grained adaptive framework for parallel mobile hybrid cloud applications, Ph.D. dissertation, University of Illinois at Urbana-Champaign.

13 Varela, C.A. and Agha, G. (1999) A hierarchical model for coordination of concurrent activities, in *Proceedings of the 3rd International Conference on Coordination Languages and Models: COORDINATION '99* (eds. P. Ciancarini and A.L. Wolf), Lecture Notes in Computer Science, vol. 1594, Springer, Berlin, pp. 166–182.

14 Astley, M., and Agha, G.A. (1998) Customization and composition of distributed objects: middleware abstractions for policy management, in Proceedings of the 6th ACM SIGSOFT International Symposium on Foundations of Software Engineering (FSE-6), pp. 1–9.

15 Sturman, D.C. (1996) Modular specification of interaction policies in distributed computing, Ph.D. dissertation, University of Illinois at Urbana-Champaign.

16 Meseguer, J. and Talcott, C. (2002) Semantic models for distributed object reflection, in *16th European Conference on Object-Oriented Programming (ECOOP 2002)* (ed. B. Magnusson), Lecture Notes in Computer Science, vol. 2374, Springer, Berlin pp. 1–36.

17 Venkatasubramanian, N. and Talcott, C. (1995) Reasoning about meta level activities in open distributed systems, in Proceedings of the 14th Annual ACM Symposium on Principles, PODC, pp. 144–152.

18 Ren, S., Yu, Y., Chen, N., Marth, K., Poirot, P.-E., and Shen, L. (2006) Actors, roles and coordinators: a coordination model for open distributed and embedded systems, in *Proceedings of the 8th International Conference on Coordination Models and Languages: COORDINATION 2006*, Lecture Notes in Computer Science, vol. 4038, Springer, pp. 247–265.

19 Milner, R., Parrow, J., and Walker, D. (1992) A calculus of mobile processes, I. *Information and Computation*, **100** (1), 1–40.

20 Honda, K. (1993) Types for dyadic interaction, in *Proceedings of the 4th International Conference Concurrency Theory (CONCUR)* (ed. E. Best), Lecture Notes in Computer Science, vol. 715, Springer, Berlin, pp. 509–523.

21 W3C: The Web Services Choreography Description Language, Version 1.0, 2005. Available at http://www.w3.org/TR/ws-cdl-10/.

22 Carbone, M., Honda, K., and Yoshida, N. (2007) Structured communication-centred programming for web services, in *Proceedings of the 16th European Symposium on Programming (ESOP)* (ed. R. De Nicola), Lecture Notes in Computer Science, vol. 4421, Springer, pp. 2–17.

23 Honda, K., Yoshida, N., and Carbone, M. (2008) Multiparty asynchronous session types, in Proceedings of the 35th Annual ACM SIGPLAN-SIGACT Symposium on Principles of Programming Languages (POPL), pp. 273–284.

24 Gay, S. and Hole, M. (2005) Subtyping for session types in the pi calculus. *Acta Informatica*, **42** (2–3), 191–225.

25 Gay, S.J. and Vasconcelos, V.T. (2010) Linear type theory for asynchronous session types. *Journal of Functional Programming*, **20** (1), 19–50.

26 Yoshida, N., Deniélou, P.-M., Bejleri, A., and Hu, R. (2010) Parameterised multiparty session types, in *International Conference on Foundations of Software Science and Computational Structures (FoSSaCS)* (ed. L. Ong), Lecture Notes in Computer Science, vol. 6014, Springer, Berlin, pp. 128–145.

27 Castagna, G., Dezani-Ciancaglini, M., and Padovani, L. (2011) On global types and multi-party sessions, in *Formal Techniques for Distributed Systems* (eds. R. Bruni and J. Dingel), Lecture Notes in Computer Science, vol. 6722, Springer, Berlin, pp. 1–28.

28 Charalambides, M., Dinges, P., and Agha, G. (2012) Parameterized concurrent multi-party session types, in Proceedings of the 11th International Workshop on Foundations of Coordination Languages and Self Adaptation (FOCLASA), Newcastle, UK, Sep. 8, 2012, *Electronic Proceedings in Theoretical Computer Science (EPTCS)*, vol. 91, pp. 16–30. Available at http://eptcs.web.cse.unsw.edu.au/content.cgi?FOCLASA12.

29 Bocchi, L., Chen, T.-C., Demangeon, R., Honda, K., and Yoshida, N. (2017) Monitoring networks through multiparty session types, in Theoretical Computer Science, vol. 669, pp. 33–58.

30 Neykova, R., Bocchi, L., and Yoshida, N. (2017) Timed runtime monitoring for multiparty conversations. *Formal Aspects of Computing*, **29** (5), 877–910.

31 Honda, K., Mukhamedov, A., Brown, G., Chen, T.-C., and Yoshida, N. (2011) Scribbling interactions with a formal foundation, in *Proceedings of the Distributed Computing and Internet Technology (ICDCIT 2011)* (eds. R. Natarajan and A. Ojo), Lecture Notes in Computer Science, vol. 6536, Springer, Berlin, pp. 55–75.

32 Noble, B.D., Satyanarayanan, M., Narayanan, D., Tilton, J.E., Flinn, J., and Walker, K.R. (1997) Agile application-aware adaptation for mobility. *ACM SIGOPS Operating Systems Review*, **31** (5), 276–287.

33 Moinzadeh, P. (2013) I-AdMiN: a framework for deriving adaptive service configuration in wireless smart sensor networks, Ph.D. dissertation, University of Illinois at Urbana-Champaign.

34 Oliner, A.J., Iyer, A.P., Stoica, I., Lagerspetz, E., and Tarkoma, S. (2013) Carat: collaborative energy diagnosis for mobile devices, in Proceedings of the 11th ACM Conf. Embedded Networked Sensor Systems, Article No. 10.

35 Subashini, S. and Kavitha, V. (2011) A survey on security issues in service delivery models of cloud computing. *Journal of Network and Computer Applications*, **34** (1), 1–11.

36 Chow, R., Golle, P., Jakobsson, M., Shi, E., Staddon, J., Masuoka, R., and Molina, J. (2009) Controlling data in the cloud: outsourcing computation without outsourcing control, in Proceedings of the ACM Workshop on Cloud Computing Security, pp. 85–90.

37 Grewal, R.K. and Pateriya, P.K. (2013) A rule-based approach for effective resource provisioning in hybrid cloud environment, in *New Paradigms in Internet Computing* (eds. S. Patnaik *et al.*), Advances in Intelligent Systems and Computing, vol. 203, Springer, Berlin, pp. 41–57.

38 Kumar, K. and Lu, Y.-H. (2010) Cloud computing for mobile users: can offloading computation save energy? *Computer*, **43** (4), 51–56.

39 Zhang, X., Schiffman, J., Gibbs, S., Kunjithapatham, A., and Jeong, S. (2009) Securing elastic applications on mobile devices for cloud computing, in Proceedings of the 2009 ACM Workshop on Cloud Computing Security (CCSW'09), Chicago, IL, pp. 127–134.

40 Chow, R., Jakobsson, M., Masuoka, R., Molina, J., Niu, Y., Shi, E., and Song, Z. (2010) Authentication in the clouds: a framework and its application to mobile users, in Proceedings of the ACM Cloud Computing Security Workshop (CCSW'10), Chicago, IL, pp. 1–6.

41 Huang, D., Zhang, X., Kang, M., and Luo, J. (2010) MobiCloud: building secure cloud framework for mobile computing and communication, in Proceedings of the 5th IEEE International Symposium on Service Oriented System Engineering, pp. 27–34.

42 Khan, A.N., Kiah, M.L.M., Khan, S.U., and Madani, S.A. (2013) Towards secure mobile cloud computing: a survey. *Future Generation Computer Systems*, **29** (5), 1278–1299.

43 Yu, X. and Wen, Q. (2012) Design of security solution to mobile cloud storage, in *Knowledge Discovery and Data Mining* (ed. H. Tan), Advances in Intelligent and Soft Computing, vol. 135, Springer, Berlin, pp. 255–263.

44 Jansen, W. and Grance, T. (2011) Guidelines on Security and Privacy in Public Cloud Computing. National Institute of Standards and Technology (NIST), U.S. Department of Commerce, Special Publication 800-144, Computer Security Division, Information Technology Laboratory, NIST, Gaithersburg, MD.

45 Damiani, E., De Capitani di Vimercati, S., Paraboschi, S., and Samarati, P. (2002) A fine-grained access control system for XML documents. *ACM Transactions on Information and System Security*, **5** (2), 169–202.

46 Varela, C. and Agha, G. (2001) Programming dynamically reconfigurable open systems with SALSA. *ACM SIGPLAN Notices*, **36** (12), 20–34.

47 Dinges, P. and Agha, G. (2012) Scoped synchronization constraints for large scale actor systems, in *Proceedings of the International Conference on Coordination Languages and Models: COORDINATION 2012* (ed. M. Sirjani), Lecture Notes in Computer Science, vol. 7274, Springer, Berlin pp. 89–103.

48 Miller, M.S. (2006) Robust composition: towards a unified approach to access control and concurrency control, Ph.D. dissertation, Johns Hopkins University.

49 Barendregt, H. (1992) Lambda calculi with types, in *Handbook of Logic in Computer Science, Vol. II: Background: Computational Structures* (eds. S. Abramsky, D.M. Gabbay, and T.S.E. Maibaum), Oxford University Press, pp. 117–309.

50 Gödel, K. (1958) Über eine bisher noch nicht benützte Erweiterung des finiten Standpunktes. *Dialectica*, **12** (3–4), 280–287.

51 Castagna, G., Dezani-Ciancaglini, M., and Padovani, L. (2012) On global types and multi-party sessions. *Logical Methods in Computer Science*, **8** (1), Available at https://lmcs.episciences.org/773.

8

Certifications Past and Future: A Future Model for Assigning Certifications that Incorporate Lessons Learned from Past Practices

Masooda Bashir,[1] Carlo Di Giulio,[2,3] and Charles A. Kamhoua[4]

[1]*School of Information Sciences, University of Illinois at Urbana-Champaign, Champaign, IL, USA*
[2]*Information Trust Institute, University of Illinois at Urbana-Champaign, Urbana, IL, USA*
[3]*European Union Center, University of Illinois at Urbana-Champaign, Champaign, IL, USA*
[4]*U.S. Army Research Laboratory, Network Sciences Division, Network Security Branch, Adelphi, MD, USA*

Security certifications are widely used to demonstrate compliance with privacy and security principles, but over the last few years, new technologies and services – such as cloud computing applications – have brought new threats to the security of information, making existing standards weak or ineffective.

Three of the most highly regarded information technology security certifications used to assess cloud security are ISO/IEC 27001, SOC 2, and FedRAMP. ISO and SOC 2 have been used worldwide since 2005 and 2011, respectively, to build and maintain information security management systems or controls relevant to confidentiality, integrity, availability, security, and privacy within a service organization; FedRAMP was created in 2011 to meet the specific needs of the U.S. government in migrating its data on cloud environments.

This chapter describes the evolution of these three security standards and the improvements made to them over time to cope with new threats, and focuses on their adequacy and completeness by comparing them to each other. Understanding their evolution, resilience, and adequacy sheds light on their weaknesses and thus suggests improvements needed to keep pace with technological innovation.

8.1 Introduction

As cloud computing becomes more and more pervasive in our lives, and cloud services evolve to be easily available, flexible, and usable, their presence is accompanied by concerns for privacy and security. Cloud computing consists in "ubiquitous, convenient, on-demand network access to a shared pool of configurable computing resources" [1] built on one of three service models working on infrastructure maintained by a cloud service provider (CSP). The first model

Assured Cloud Computing, First Edition. Edited by Roy H. Campbell, Charles A. Kamhoua, and Kevin A. Kwiat.

is called *software as a service (SaaS)* and allows access to predefined software applications managed by the CSP. The second is *platform as a service (PaaS)*, which allows users to deploy supported software applications on the CSP's infrastructure. The third is *infrastructure as a service (IaaS)*, which allows greater flexibility and deployment of arbitrary software [1].

The great success of cloud computing is based in its flexibility and the unprecedented possibilities offered by access to scalable, potentially unlimited storage and computing resources at a relatively low cost. The cloud represents a completely new approach to provisioning of and access to information technology resources. On the one hand, CSPs can leverage economies of scale, and the more their services are accessed and used and their capabilities expanded, the higher their performance or the lower their price for the same service. On the other hand, users can rely on commercial services for processing their business-related information, thus completely or partially removing their costs for maintenance and configuration of hardware and software; further, the cheaper the service, the more convenient it becomes to abandon the old "on-premises" model [2,3]. These characteristics were determinant for the U.S. federal government's adoption of a coordinated set of initiatives aimed to reallocate investment in information technology (IT)[1] from the building and maintenance of on-premises infrastructure to the on-demand access to remote infrastructure, platform, or software resources [5].

Cloud computing represents a consistently growing trend in storage of information and access to computational resources, and its use in industry and government, as well as at a customer level, is not likely to decrease in the next few years. A study by Gartner [6] projected 17.2% growth of investments in public cloud services in 2016 compared to the previous year, including an increase of over $208 billion, and an almost 43% increase in investments in IaaS worldwide.

Yet with higher performance and lower costs, the cloud also brings new security vulnerabilities, as the use of shared infrastructures raises the risk of unauthorized access to information, disclosure, and data loss. We are seeing new forms of exploitation of cloud-specific vulnerabilities, such as "side channel" attacks. The presence of multiple tenants on the same infrastructure makes cloud systems behave differently from noncloud systems, and by figuring out this behavior – for example, by analyzing CPU utilization – a malicious attacker residing on the system might be able to infer the information processed in the system by another tenant [7–10]. Similarly, by compromising the hypervisor – the shared platform that governs the functioning of single virtual machines

---

1 The total IT expenditures of the U.S. government were less than $50 billion in 2001, with a compound annual growth rate (CAGR) of 7.1% until 2009. With the adoption of cost-saving initiatives such as the 2012 Cloud Computing Strategy, the CAGR has been lowered to 1.8% during 2009–2017, with 8.2% of $89.9 billion reserved for cloud services [4].

(VMs) – it is possible to compromise the VMs installed on the shared infrastructure.[2]

Such new vulnerabilities are evolving alongside the known threats and issues that are common to all IT systems; insecure or malicious software, for example, can result in damage regardless of whether there is shared infrastructure. Thus, the need for cloud-specific countermeasures is just one element of the modern cybercrime and security landscape that is forcing governments and industry to pursue new techniques to assure confidentiality, integrity, and availability (C-I-A) of information.[3] Among these is the adoption of security standards.

Standards are a mechanism historically used to guarantee the characteristics of products and services and meet users' expectations. If vendors state that their products comply with some particular standard, users can feel assured that the products will possess certain features and meet certain specifications.

By being compliant with a security standard, a CSP shows its commitment to maintaining high levels of security, with an immediate benefit for their clients and a considerable return in building a reputation for trustworthiness. Thus, CSPs are willing to undergo expensive audits conducted by third parties to certify their compliance with certain standards and gain in credibility.

8.1.1 What Is a Standard?

Standards are the result of agreed-upon measures that act as facilitators in everyday life. Standards describe commonly understood concepts, such as the standard time, standards of living, the gold standard, cultural standards, or standards in education.

Through standardization, we achieve "uniformities across time and space . . . [and we] make things work together over distance or heterogeneous metrics" [12]. In other words, by using a standard we assure a common reference across countries and continents, and consistency over the years in describing the world around us. For example, the use of customary units of volume in the United States allows producers and consumers to have a common reference on the capacity of a bottle of water. It creates the conditions under which producers can sell their bottled water across the nation, and users can buy it knowing exactly the quantity they are paying for.

Standards can be derived from unwritten social norms or can result from negotiation and agreement among parties. More often, in the case of particularly complex or specific matters, standards are promulgated by government or nongovernmental organizations and can be formalized in normative documents

---

2 See Chapter 4 for a more thorough discussion of this category of attacks.

3 According to a Verizon [11] report on security incidents in 82 countries, 64,199 incidents took place in 2015, of which more than 47,000 targeted governments or government agencies.

with detailed specifications of the requirements to be satisfied. For instance, the Environmental Protection Agency (EPA) specifies the allowed level of contaminants in drinking water, thus setting a standard for those who sell or distribute drinking water in the United States. In such cases, the creation of standards is assigned to experts with a thorough understanding of the issue. As in the example of drinking water, a simple agreement among users and distributors on the characteristics of water is not sufficient to assure crucial qualities, such as safety for human consumption. Experts in the fields of chemistry, biology, and medicine must be involved to determine what chemicals in what quantities should be allowed to assure that contaminants are minimized as needed and that the end result is acceptable to users.

Standards can often act as authoritative rules [12] and be as effective as formal legislative acts or social norms in regulating our society [13]. Regulation through standards can be a task for either governments or nongovernment standards organizations, which are expert groups devoted to creating standards in a multitude of fields. For example, since it was founded in 1947, the International Organization for Standardization has issued more than 21,000 publications [14] in fields ranging from information security to occupational health and safety to food safety management, among many others [14]. Another example is the National Institute of Standards and Technology (NIST), which is part of the U.S. Department of Commerce. NIST is dedicated to measurement and standardization in science and engineering [15].

NIST's definition of standards [16] encompasses two categories that are based on the standards' function. *Technical standards* clarify the specifics of products or materials and detail the required procedures for managing them, while *performance standards* define results but do not specify the procedures to obtain them.

An everyday example of a *technical* standard can be seen in the design of power sockets and plugs. In the United States, since 1917, people can count on using the same type of power socket for the great majority of their home appliances [17], and can expect to move to any other location within the United States and be able to plug in their devices without the use of adaptors. However, Americans traveling to most countries outside the United States will need adaptor plugs in order to use their appliances, and sometimes even the adaptors are insufficient, since differences in voltage make it impossible to use some appliances in some countries. To allow interoperability of power sources and appliances, a technical standard prescribes a specific design for sockets and plugs, as well as a common voltage with which home appliances should work. Producers of electricity-powered devices such as dishwashers and refrigerators need to follow specific requirements prescribed for their products in technical standards; consumers, on the other hand, can expect that the devices they buy are compliant with standards without having to verify the operability of the device in their homes.

An example of *performance* standards, on the other hand, would be the emission standards that regulate the pollutants produced by engine-powered vehicles. Governments set goals, which are represented by quantitative values for the maximum acceptable emissions of greenhouse gases or other pollutants, but they do not specify what technique or methods should be adopted to achieve these goals. Manufacturers are given the autonomy to decide the best solution to adopt, so long as their vehicles do not exceed the threshold imposed by the government. When purchasing a vehicle, users can expect it to produce emissions below the limit prescribed in the standard, without further verification.

Another distinction among standards, based on NIST's definition [16], depends on their source. Standards are "voluntary" if not imposed by laws or regulations and/or "nongovernmental" if they are not created and enforced by authorities [16]. While a standard can be imposed by government (e.g., the mentioned emission standards), it could instead be adopted on an organization's own initiative, generally following a set of rules suggested by standardization bodies. An example of a voluntary standard is the safety standards used in the food industry. Although not always imposed by law, these standards are adopted to guarantee higher quality of the products and to reassure consumers on the practices used in producing and processing their food. As with technical and performance standards, the purpose of adhering to a voluntary or non-governmental standard is to induce trust in users and reassure them on the quality of a service or product, avoiding case-by-case verification.

The distinctions explained above illustrate the characteristics and goals of standards. The examples presented thus far hint at how standardization is pervasive in our lives and covers the most disparate aspects of the production and provisioning of services and goods. Information technology and cloud computing, just like any other product, follow the same general principles about standardization. Users' expectations, however, are not limited simply to interoperability or safety and may involve aspects such as confidentiality, integrity, and availability of information processed in IT applications. At the same time, the techniques adopted to meet the requirements of IT security standards are not based solely on either technical measures or performance but rather a combination of both.

8.1.2 Standards and Cloud Computing

To understand the use of standards in cloud computing, we must consider how IT security is deeply affected by the rapidly evolving technologies upon which it builds and the equally rapidly evolving cyberthreats. New technologies often mean new threats and with them the necessity of updated security techniques. Still, users expect a certain level of security in accessing applications and services, and it is the provider's responsibility to meet those expectations.

In this scenario, security standards are an essential tool. By imposing measures and procedures aimed at increasing information assurance, a standard sets a security baseline. By complying with the standard, a CSP can make sure to meet that baseline protection against threats and vulnerabilities. In cloud computing, not only do security and quality of service depend on technical precautions, but careful consideration of the internal procedures adopted by the CSP is also absolutely essential. In spite of the implementation of cutting-edge technical measures, attackers can leverage weaknesses in the information management process to compromise the confidentiality, integrity, and availability of information being managed and processed in the cloud. Technical precautions are necessary and must be balanced with equal attention to the usability of applications while guaranteeing the security of information to avoid unacceptable trade-offs.[4] Still, poor-quality internal processes and procedures can negate technical efforts, producing flaws that are easily exploitable by malicious attackers and generating data losses or disclosure.

If a CSP lacks clear policies on the use of workstations or mobile devices, that constitutes a procedural flaw that may open up backdoor access to cloud resources despite the presence of technical security measures. For example, mandatory logouts from individual users' accounts offer an important form of security protection; if employees are not required to log out upon leaving their workstations unattended, their accounts could easily be exploited by other employees to gain access to resources for which the malicious employees might not have the necessary authorization level. Although technical measures may have been implemented, such as least-privilege access rules based on user identity and two-factor authentication to the workstation, a lack of attention to internal procedures can still create a security flaw. (We assume that physical access control procedures would be implemented correctly, limiting the pool of potential attackers to employees who work for the same organization.) For a second example, the absence of password policies or even lock-out screen policies for portable devices (such as smartphones or tablets) assigned to a CSP employee may have even worse consequences. For instance, a mobile device that is accessible off-premises, and can connect to corporate networks and store or access information controlled by the organization, can be targeted by malicious attackers to compromise information controlled by the CSP. A number of different attacks can be performed on mobile devices, for example, enabling remote code execution and privilege escalation (like the so-called "Stagefright" exploit) [18,19], or using multimedia messaging services (MMS) to target Android kernel vulnerabilities [20]. Mobile-specific vulnerabilities can easily be exploited by a malicious attacker if access to the device is not properly protected.

Those two examples suggest the importance of well-designed internal procedures to protect against undesired access and violation of security measures. To

---

4 An example is the implementation of monitoring and detection techniques that detect malicious attackers while maintaining high performance and availability of services. See Chapter 5 for a discussion of detection, security, and resilience in cloud computing systems.

assure adequate protection, standards must include security measures encompassing procedures and policies, not just the technical aspects. Compliance with a standard can be instrumental in reassuring users about the security measures implemented by the organization. Furthermore, in a commercial context in which the provider is a private vendor, the ability to show users that the organization complies with security measures can be as important as the implementation of the measures itself. Demonstrating higher-quality products or services than the competitors can make all the difference in the commercial success of a CSP, and users have great interest in verifying that their vendors are implementing strong security measures and best practices.

However, it is not always possible to offer sufficient insights on IT systems and procedures to demonstrate adequate implementation of security measures. In the auditing of a CSP, a number of challenges make it almost impossible for a single user to perform a full assessment. To prevent accidental or malicious damage, for example, access to the CSP's premises must be restricted to a few authorized subjects, and if auditors are admitted, they must be subject to specific rules.

The solution to the impasse is to rely on trusted third parties to perform the audit and assign a "quality label" certifying the CSP's full compliance with the best practices or standards prescribed for the service provider. *Standards* may refer to a quality or qualification, such as those acquired with a certification [13], and it is in this context that the terms *standard* and *certification* become intertwined. Security audits are conducted by professional auditors with specialized expertise in assessing service providers and verifying compliance with the requirements prescribed in a standard. Different standards require dedicated auditors (although auditors are often authorized to perform assessments for multiple standards) and can result in a certification, authorization, or report that can be used to attest compliance with a standard or framework.

8.2 Vision: Using Cloud Technology in Missions

The federal cloud has not been immune to the need to standardize. Following the release of a 25-point strategy for IT management at the federal level, and in response to the need to reduce IT expenditures while maintaining high security standards, the U.S. government created its own assurance standard: the Federal Risk Authorization Management Program (FedRAMP), which was published in 2011. It is based on a set of security controls organized in three baselines: low, moderate, and high.[5]

FedRAMP prescribes a set of prerequisites that CSPs must meet to offer cloud services to federal agencies and builds on an evaluation and auditing process that culminates in an Authorization to Operate (ATO) for the CSPs. Some agencies,

---

5 For more details on FedRAMP, see Section 8.3.1.

however, have more stringent requirements. That is the case with the Department of Defense (DoD), which – on top of demanding a preliminary FedRAMP authorization – requires further controls that depend on the security impact level of the data transmitted to the cloud provider. The DoD distinguishes six impact levels: Levels 1 and 2 are for unclassified, uncontrolled information; levels 3 and 4 are reserved for unclassified, controlled information; level 5 is for unclassified, controlled information with higher sensitivity than level 4; and level 6 is dedicated information ranging from secret to top secret. For the purposes of authorizing the provisioning of services to the DoD, the Defense Information Systems Agency (DISA) has published a set of guidelines [21] indicating how, once a FedRAMP authorization has been achieved, DISA will perform further verification of additional requirements on the ground before authorizing the CSP. This authorization process (referred to by DISA as *FedRAMP+*) has one exception, which is the authorization for impact levels 1 and 2; for them, a FedRAMP authorization is sufficient, and no further verification is required in order to obtain a DoD authorization. Since the 2016 publication of a high security baseline for FedRAMP authorization, the FedRAMP governing board and DoD have been evaluating the possibility of automatic access to DoD authorization up to impact level 4 for CSPs with a FedRAMP high ATO [22]. If we consider the importance of unclassified, controlled information, it is easy to see how wrong calibration of the measures imposed in FedRAMP can result in serious consequences for the agency relying on the authorization process.

Information covered by assurance level 4 includes, but is not limited to, personally identifiable information (PII), including Social Security numbers and medical records of employees; information regarding ongoing investigations; and information disclosing infrastructure vulnerabilities, design, or construction information [23]. Loss or disclosure of such controlled unclassified information may have serious effects, so it must be protected adequately.

The purpose of FedRAMP is to set a baseline of security requirements for (i.e., to standardize) cloud services for the federal government. It benefits federal agencies by streamlining the authorization process so that the agencies are not forced to do case-by-case evaluations of the security measures implemented by CSPs. By referring to a standardized authorization assessment and procedure, federal agencies can save time and resources by requesting a product or service that is compliant with well-known and verifiable requirements.

8.3 State of the Art

In the United States, FedRAMP was created to build trust in cloud security following a period of increased attention to cloud computing from the federal government. To reduce high expenditures on IT services at a federal level,[6] at the end of 2010, the Office of Management and Budget (OMB) released the "25 Point Implementation Plan to Reform Federal Information Technology Management" [24]. The Implementation Plan was followed a few months later by the "Cloud Computing Strategy" [5], which set a roadmap for migrating federal IT services to cloud environments. The first release of FedRAMP was at the end of 2011; its structure is similar to that of the existing Federal Information Security Management Act (FISMA), with which all federal IT systems must comply, including on-premises infrastructures. Upon its release, FedRAMP joined other existing frameworks for IT security auditing, assessment, and certification requirements.

Two of the most commonly used standards in the IT industry were designed to demonstrate strong protection of privacy and security in IT systems; they include Service Organization Control (SOC) audits, which are based on the Trust Services Principles and Criteria (TSPC) developed by the American Institute of Certified Public Accountants (AICPA), and the ISO/IEC 27001 certification, which was built on criteria published by the International Organization for Standardization (ISO) jointly with the International Electrotechnical Commission (IEC).

ISO/IEC certifications represent a well-established international benchmark for a variety of standards in different fields. One of the prevalent frameworks for the regulation of IT best practices is the ISO/IEC 27001. This standard has regularly evolved to keep pace with technical innovation by prescribing security techniques for information security management systems. Another set of standards is similar to ISO/IEC certifications and FedRAMP, as it is used to attest the security of IT systems: the AICPA SOC reports.

Although the conclusions of AICPA's SOC reports cannot technically be defined as "certifications," they share some of the characteristics and goals of the certifications and authorization standards that were highlighted in previous sections of this chapter. The specific controls performed in SOC audits may vary, touching upon slightly different things in different situations. Thus, the result of an independent assessment is more properly defined as a "report," as it is based on a selection of controls determined on a case-by-case basis. Even so, assessment criteria are both well-defined and mandatory in SOC reports, and that can be used to increase the confidence of customers or stakeholders in the security measures implemented by an organization [25] and reassure them that the service will meet their expectations. To that extent, we will consider SOC reports alongside other frameworks, such as FedRAMP and ISO/IEC 27001, that share the same underlying purpose, and we will group them together in the category of standards and certifications. Although they do not define in detail the controls to adopt, but rather establish guidelines, the technical and procedural criteria that

---

6 In 2011, the total federal IT expenditures were $75.4 billion [2].

are used for SOC reports must be used to assess the service organization and provide auditors with detailed guidance, as the criteria specify not only a goal (the security of information) but also the procedure or methods to achieve it. Among the three possible SOC reports, the most specific and complete is SOC 2, which has the goal of "Reporting on Controls at a Service Organization Relevant to Security, Availability, Processing Integrity, Confidentiality or Privacy" [25] and is conducted by an independent auditor who will attempt to verify that the organization meets the criteria of the TSPC.

In TSPC, and consequently in SOC 2 reports, requirements and controls are thorough and appear to be comparable to the ISO/IEC 27001 provisions and NIST provisions for FedRAMP. Although the similarities of the three standards can easily be seen, no formal effort has been made to enable mutual recognition among them.

8.3.1 The Federal Risk Authorization Management Program

The effort to bring FedRAMP into being started in 2009 as a result of the Federal Cloud Computing Initiative [2,26] and the cooperation of multiple agencies,[7] local governments, academia, and nongovernmental organizations [27], which worked together to create a standardized approach for assessment and authorization procedures for providers of cloud services to the U.S. government [27]. The FedRAMP effort not only created a security framework for cloud services in compliance with FISMA [28], but it also moved to a "do once, use many times" approach, which allows federal agencies to leverage existing authorization processes instead of creating a new assessment procedure every time an assessment is needed; thus, it helps reduce federal agencies' IT expenditures in alignment with the "25 Point Implementation Plan" [24]. FedRAMP was officially initiated in December 2011 with the publication of the OMB's "Security Authorization of Information Systems in Cloud Computing Environments: Memorandum for Chief Information Officers" [27], and it released its first authorization in May 2013.[8]

FedRAMP assessment follows the Risk Management Framework suggested in NIST 800-37 [30]. This framework specifies that assessments should be conducted in four steps: (i) In the initial step, the auditor collects information on the system; (ii) then, the auditor performs an assessment of security requirements; (iii) next is the authorization phase, at which point the CSP is allowed to supply

---

7 Agencies and federal working bodies involved in the collaboration include the Office of Management and Budget (OMB), the National Institute of Standards and Technology (NIST), the General Services Administration (GSA), the Department of Defense (DoD), the Department of Homeland Security (DHS), the U.S. Chief Information Officers Council (CIO Council), and the Information Security and Identity Management Committee (ISIMC) [27].

8 The first CSP to receive an official FedRAMP ATO was Amazon Web Services, on May 1, 2013 [29].

its services to the designated agency; and (iv) finally, there is a continuous monitoring phase, intended to guarantee that the same level of security will be maintained after the initial assessment has been performed.

The FedRAMP authorization process offers to CSPs three different paths to obtaining a FedRAMP Authorization to Operate (ATO), and it makes it possible for multiple agencies to leverage the same authorization. For all three options, the CSP is required to undergo an assessment procedure and submit the results of the assessment to the FedRAMP Program Management Office (PMO), which will add the authorized system to the list of compliant systems [28]. The assessment must be conducted by an independent auditor that is among those listed as FedRAMP Third-Party Assessment Organizations (3PAO). Organizations wishing to become 3PAOs must themselves be certified by the American Association for Laboratory Accreditation against the ISO/IEC 17020:2012 requirements[9] [32]. The independent auditor performs the necessary assessment of the CSP in terms of the FedRAMP controls, attesting to the CSP's compliance. In the first possible path to obtaining an ATO, the FedRAMP Joint Authorization Board (JAB) and the CSP work together directly. At the end of the assessment, the CSP will be considered "FedRAMP ready," meaning that all the requirements are satisfied and federal agencies can issue their own ATOs with no further assessments or reviews. In the second option, the CSP works on its own initiative to comply with FedRAMP in anticipation of future authorization. The results of the assessment are submitted to the FedRAMP PMO, which reviews the results and marks the CSP as "FedRAMP ready." The third possible path is initiated at the request of an agency issuing the ATO, which works together with the CSP to comply with the FedRAMP requirements. Also in this case, after the 3PAO's assessment of the CSP is complete, the request for the ATO must be submitted to the FedRAMP PMO [28].

Following the same approach adopted in FISMA, the FedRAMP assessment process includes a selection of controls derived from NIST Special Publication (SP) 800-53. This Special Publication is part of the 800 series, which "reports on ITL's research, guidelines, and outreach efforts in information system security, and its collaborative activities with industry, government, and academic organizations" [33]. SP 800-53 includes technical and procedural controls aimed at protecting privacy and security in federal information systems and organizations. It is organized into 18 control families, each of which is made of controls and control enhancements. The goal of each control is to provide specific guidance to service organizations on increasing privacy and security, while the enhancements prescribe further measures for improving the protection assured by the control from which they descend. SP 800-53 is structured according to

---

9 ISO/IEC 17020 is entitled "Conformity Assessment: Requirements for the Operation of Various Types of Bodies Performing Inspection" and promotes the impartiality and consistency of audits and inspections [31].

three distinct baselines, which were chosen based on the impact levels described in Federal Information Processing Standard (FIPS) 199 and the security control selection process specified in FIPS 200. The three baselines are a *low* baseline, which includes controls for systems that process information for which a loss of C-I-A would have limited impact; a *moderate* baseline, which relates to information for which a loss of C-I-A may result in serious damage to the organization; and a *high* baseline, which covers information for which a loss of C-I-A could have severe or catastrophic consequences for the organization and its activities [34]. FedRAMP follows an identical organization and builds on the same three impact levels.

In 2011, the year of its first release, FedRAMP considered only low and moderate security baselines for a total, respectively, of 116 and 297 controls and enhancements divided into 17 families, as selected from the 810 controls and control enhancements detailed in NIST SP 800-53. In 2014, following a new review of NIST SP 800-53, FedRAMP was updated to include more controls and enhancements, reaching a total of 325 for the moderate baseline and 125 for the low baseline. A high baseline was added to the FedRAMP program in July 2016, bringing an additional 96 controls and enhancements. The introduction of the high baseline allows federal agencies to move to cloud systems 20% of federal information that by itself accounts for 50% of the total federal spending on IT [35].

As of May 2017, 81 products have been authorized, 4 at a high level; 62 products are in the process of being approved by the FedRAMP PMO, and another 10 are ready for authorization, although they have not yet been authorized by any agency. The total number of 3PAOs is 43, but only 18 of them have performed at least 1 FedRAMP assessment in the 6 years of the program [36].

8.3.2 SOC Reports and TSPC

The American Institute of Certified Public Accountants (AICPA) has worked since 1887[10] on activities that include, but are not limited to, advocacy, certification, standardization, communication, and education in the accounting profession [37]. AICPA counts more than 418,000 members, most of whom are involved in public and management accounting (representing 47 and 34% of the voting members, respectively), and is active in 143 countries worldwide [37].

One of AICPA's standards, the Statement on Auditing Standards (SAS) No. 70, was released in 1992 to support examination of organizations' financial statements and was widely adopted by service organizations to demonstrate compliance with security and privacy best practices [38]. A SAS 70 audit

---

10 AICPA was originally called the American Association of Public Accountants, and became AICPA in 1936 [37].

consisted of an assessment by external auditors of a service organization's procedures, policies, and security controls, and concluded with reports that documented observations on possible security shortcomings and subsequent improvements [39].

SAS 70, however, was conceived to achieve compliance with financial regulations such as the Sarbanes–Oxley Act in the United States, and its adoption as an attestation of compliance with security and privacy best practices was essentially a misuse. Furthermore, SAS 70 reports were conceived as an auditor (on the financial statements of the service organization) to auditor (on the financial statements of the user organization) document but evolved over the years to include a marketing element, as the reports were made available to prospective clients to demonstrate the use of security controls in the organization [39]. The inclusion of the marketing element led AICPA to revisit the standard as part of the Clarity Project conducted in 2011 by the Auditing Standards Board, which created distinctions between different reports according to the purposes they served, and created the Service Organizations Controls (SOC) reports [40].

Audits are carried out by independent third parties and produce three main types of SOC reports: SOC 1, SOC 2, and SOC 3.

SOC 1 reports serve the original purpose of SAS 70 reports, as they address controls that are relevant for the audited entity's financial statements and can be differentiated into Type I reports, which address the suitability of the design of the controls chosen to perform the assessment according to the control objectives set by the management, and Type II reports, which examine the suitability of the design *and operating effectiveness* of those controls. The main difference between the two types is the duration of time covered: A Type I report is limited to a specific moment in time and attests that an organization has selected a number of controls to be adopted, while Type II reports attest implementation of those controls over a longer period of time.

While SOC 1 is focused on financial statements, the SOC 2 and SOC 3 reporting standards attest to the security and privacy measures utilized in the service organization. SOC 2 and SOC 3 reports are released following an assessment that verifies compliance with the list of Trust Services Principles and Criteria (TSPC) published by AICPA [25], which were designed to help organizations promote the C-I-A principles. SOC 2 reports are thorough and are available as either Type I or Type II, with the same distinction and qualities used for SOC 1 reports. Although SOC 2 reports are used to reassure customers and stakeholders on the implementation of security and privacy measures within an organization, they are not public but can be made available upon request. SOC 3 reports are similar in scope to SOC 2 reports, but consist of brief statements rather than complete analyses of security measures' compliance with the TSPC; they are generally available to the public, and thus are suitable for the marketing applications that (improperly) characterized SAS 70 until it was repealed in

2011 [25]. To issue a SOC report, auditors must be certified by AICPA as Certified Public Accountants (CPAs).

The first version of the TSPC that is relevant to the SOC reports was released in 2009; thorough revisions were done in 2014 and 2016.[11] The TSPC documents are structured in multiple sections, of which section TSP 100 specifies "Trust Services Principles and Criteria for Security, Availability, Processing Integrity, Confidentiality, and Privacy." That section is used as the basis for the SOC 2 and SOC 3 assessments and reports.

TSPC 2009 presents 117 criteria organized into four groups based on the principles that they protect: *confidentiality, integrity, availability,* and the more general principle of *security*. Within each group, the criteria are split into four categories: (i) policies, which oversee the creation of relevant documentation to protect the mentioned principles; (ii) communications, which aim to make sure that policies are shared and approved across the organization; (iii) procedures, which establish formal mechanisms for implementing the necessary measures to help the organization meet its security, availability, integrity, and confidentiality objectives; and (iv) monitoring, which relates to enforcement of measures, policies, and procedures across all levels of the organization. Privacy criteria are not included in the TSPC, which for their definition refer to the more extensive list of Generally Accepted Privacy Principles (GAPP). GAPP was created in 2009 as a collaborative effort of AICPA and the Canadian Institute of Chartered Accountants (CICA).

The TSPC were heavily revised in 2014; the content was radically reorganized into 4 groups and 47 criteria. The great reduction in the number of criteria reflects the creation of a new group made of common criteria that were applicable to all of the C-I-A principles, thus reducing repetition in the document and simplifying assessment activities for organizations and auditors. The remaining content was reorganized into three other groups that contained additional criteria specific to each of the C-I-A principles. Like the 2009 release, the 2014 version of TSPC refers to GAPP for privacy criteria.

In 2016, further changes were made to the TSPC criteria. General improvements were made to the definition of the scope of the criteria, and some syntactical changes optimized the content and clarity, but the most relevant innovation in the 2016 issue was the introduction of additional privacy controls in a dedicated group. TSPC 2016 includes 44 general criteria (counting both those that are common to all principles and those that are C-I-A-specific) and 20 additional privacy criteria, all organized in 5 groups. The privacy-specific section introduced with TSPC 2016 supersedes GAPP.

---

11 AICPA's first release of the TSPC was in 2006. However, since this chapter only looks at the impact of SOC reports, which were introduced in 2011 and do not refer to versions of TSPC earlier than 2009s, we limit our analysis to the latest three versions of the TSPC.

8.3.3 ISO/IEC 27001

ISO/IEC 27000 is a large family of standards that is the result of a joint effort between the International Organization for Standardization (ISO) and the International Electrotechnical Commission (IEC). In 1987, ISO and IEC together created the Joint Technical Committee 1 (JTC1) to cover the field of "Information Technology." Later, in 1989, they collaborated on the Sub-committee 27, which developed 151 published standards specifying security techniques in IT environments [41]. Among these standards, The ISO/IEC 27001 Information Security Management standard prescribes requirements and procedures for the implementation, maintenance, and improvement of Information Security Management Systems (ISMSes). The standard addresses measures that are common to all ISMSes and does not differentiate among different types of organization undergoing the certification process [42].

Today, ISO/IEC 27001 is the basis of more than 27,000 certifications across five continents worldwide; indeed, there has been a consistent increase in its adoption. For example, there was a 20% growth in ISO/IEC 27001 certifications in 2015 compared to the previous year, with the North American region registering the highest increase, 78%. In absolute numbers, the highest concentration of certifications is in East Asia and the Pacific, followed by Europe, Central and South Asia, and North America [43]. (Together, East Asia, the Pacific, and Europe account for more than 22,000 certifications.)

The original precursor of ISO/IEC 27001 was called British Standard 7799, which was created in 1995 and revised in 1998 to become ISO 17799. ISO and IEC worked together in 2005 to convert the ISO 17799 into ISO/IEC 27001 and ISO/IEC 27002 [44]. ISO/IEC 27001 is organized in general clauses that provide guidance to auditors and organizations, plus more specific controls to be implemented to achieve compliance with the standard. The controls were derived from those in ISO/IEC 27002, which was created at the same time as ISO/IEC 27001, and were selected for their relevance in the creation and maintenance of an ISMS.

The first release of ISO/IEC 27001, in 2005, included 133 controls in 11 control groups. The standard was revised in 2013 to provide enhanced flexibility and interoperability with other standards and to update the standard to keep pace with the evolution of technology, including protections for new challenges such as identity theft and mobile security [45]. After that review, the number of controls was reduced to 114 across 18 categories. In ISO/IEC 27001:2013, the P-D-C-A (Plan, Do, Check, Act) approach is still considered a best practice, and thus is regularly adopted despite its no longer being required by any explicit provision [46]. (The 2005 version considered the approach mandatory.) The first step for the organization undergoing the certification process, *Plan*, requires consideration of goals and objectives and careful planning of the controls to be enacted; the second step, *Do*, draws from the controls and security measures identified in the first phase, to proceed to their implementation; the third phase, *Check*, is aimed at overseeing the controls implemented in step two, evaluating their effectiveness; and the fourth and last phase, *Act*, works on the flaws identified in the previous step, making further adjustments where required. The organization obtains a certification after the successful completion of the P-D-C-A cycle by an accredited auditor who is following the criteria and controls in ISO/IEC 27001; periodic controls and independent audits are required to maintain the certified status.

Auditing organizations that perform ISO/IEC 27001 assessments must be certified against both ISO/IEC 17021[12] and ISO/IEC 27006.

8.3.4 Main Differences among the Standards

The three standards, ISO/IEC 27001, SOC, and FedRAMP, have important similarities, including similar goals. Security and privacy are among the principal objectives of all three standards, and all three are structured on a given number of criteria and controls that are to be assessed by independent and accredited third parties. Yet some differences among them can be identified, giving a more accurate idea of their range and scope.

ISO/IEC 27001, SOC, and FedRAMP are all grounded in careful evaluation of management processes and targeted implementation of controls to guarantee the integrity and security of information [44]. All of them examine infrastructure, internal procedures, and risk management. FedRAMP, however, focuses on cloud services offered by private outsourcers to the federal government. Although the most recent versions of ISO and SOC include elements useful for cloud security assessments, they have a more general scope and require service providers to guarantee the C-I-A principles regardless of what type of service they offer; thus, their security assessment tools are available for use by any kind of ISMS or service organization. The difference in scope is connected to the type of guidance and the number of controls required in the three different standards. Further, FedRAMP requires at least twice as many controls as the other standards, reflecting the fact that FedRAMP's controls are more specific and detailed, while those in ISO/IEC 27001 and SOC/TSPC are more general. SOC 2 and ISO/IEC 27001 do not specify how to implement the criteria and controls that they require, showing higher flexibility than FedRAMP.

Perhaps the most notable differences are in geographical distribution and diffusion. FedRAMP is limited to U.S. federal agencies and enforced by U.S. auditors; ISO/IEC 27001 and SOC are international standards with global

---

12 The full rubric of the ISO/IEC 17021 is "Conformity assessment – Requirements for Bodies Providing Audit and Certification of Management Systems," and ISO/IEC 27006 is titled "Information Technology – Security Techniques – Requirements for Bodies Providing Audit and Certification of Information Security Management Systems."

applicability and auditability. Furthermore, the United States – with its 1247 total certificates, according to the most recent data available (from 2015) [43] – is among the 5 countries with the highest numbers of ISO/IEC 27001 certificates.[13] It has far more ISO/IEC certifications than FedRAMP authorizations, the count of which stood at 348 authorizations among 81 providers as of May 2017. In some cases, the same CSP has received multiple ATOs, each issued for a different product [36]. Because TSPC and SOC do not result in formal certifications, but rather in personalized reports, no current data are available on their worldwide implementation.

Finally, the times of the three standards' releases and the gaps between their updates show visible divergence. Whereas ISO/IEC's only review of the 27001 standard was in 2013 (8 years after its first publication), SOC and FedRAMP have received more frequent updates (after 2 and 5 years and after 4 years, respectively).

8.3.5 Other Existing Frameworks

The landscape of security standards tailored or adaptable for cloud computing systems does not consist only of FedRAMP, ISO/IEC 27001, and SOC 2. The other existing frameworks, however, are of less interest for our purposes. Some are data-specific, meaning that they focus on particular qualities of the target system that are related to the type of data processed by the system, such as financial data. Some are not supported by external audits that prove compliance with the provisions in the standard. In addition, the popularity of some of them is not significant compared to that of the three standards analyzed in this chapter. Even so, in order to provide a more comprehensive overview of the IT security certification landscape and note some similarities in other frameworks, in the following section we briefly note three other standards that have attracted some interest for cloud computing.

8.3.5.1 PCI-DSS

The Payment Card Industry (PCI) Security Standard Council was founded in 2006 to help anyone involved with credit card payments, such as merchants or financial institutions, with the implementation of standards for security policies and technology [47]. The PCI Data Security Standard (PCI-DSS) creates a structure for developing a payment card data security process [47]. PCI-DSS is built on a three-step process (Assess, Remediate, Report) and includes specific controls and directions on how to implement these controls to cover security procedures, access control, and software development [48,49]. Organizations that process cardholder data are required to comply with the standard, and their

---

13 The first is Japan (8240 certificates), followed by the United Kingdom (2790), India (2490), and China (2469).

compliance must be attested by third-party Qualified Security Assessors (QSA) authorized by PCI.

8.3.5.2   C5

The Bundesamt für Sicherheit in der Informationstechnik (BSI–German Information Security Office) presented its Cloud Computing Compliance Control Catalogue (C5) in February 2016. The C5 is a set of cloud-specific security and privacy controls, largely based on ISO/IEC 27001 and the SOC reporting standard [13,50]. C5 was intended merely as a guideline for CSPs, and as a supplement to other certifications. However, C5 can also be used as a standalone framework to verify the adoption of technical and procedural measures promoting cloud assurance. Like other frameworks, C5 relies on security controls organized into control domains. It prescribes a set of basic requirements that clouds are required to meet to be considered compliant. Additional requirements are specified to assure higher security levels, where applicable.

8.3.5.3   STAR

The Cloud Security Alliance (CSA) is a nongovernmental organization whose core mission is to develop best practices for cloud assurance and to promote awareness of best practices in cloud environments [51]. Its activities include designing security principles and attesting to their implementation in cloud systems. In 2013, CSA released a security assurance program known as the Security, Trust & Assurance Registry (STAR). The goal of STAR is to offer a cloud-specific certification built on the principles outlined in ISO/IEC 27001, with enhanced controls from other existing schemes (e.g., FedRAMP and SOC 2). Although STAR has not attained the popularity of other standards – only 44 certifications had been issued as of July 2016, it has attracted interest as a self-assessment framework; as of July 2016, 187 self-assessments have been performed [52]. STAR's assurance framework is built on CSA's Cloud Control Matrix (CCM), which is a list of 133 controls and criteria organized into 16 "families" that encompass both technical and procedural security measures. The CCM was first released in 2010 and has been regularly updated over the years; it was last revised in June 2016 [53].

8.3.6   What Protections Do Standards Offer against Vulnerabilities in the Cloud?

Standards set a baseline for protection against vulnerabilities in the cloud and aim to reassure users about the privacy protections and security offered by the service that processes their data. Users' main concerns are the common threats, issues, and vulnerabilities that affect cloud environments. Thus, the adequacy of a standard is directly connected to the protections its proposed measures offer against those issues and vulnerabilities. Depending on the type of issue that could affect the cloud and the type of information processed in the system, the potential consequences for the C-I-A of information processed in the cloud may vary, as may the need for controls that protect against some threats [34]. For example, denial-of-service (DoS) attacks, which are often used as "red herrings" to conceal other forms of attack [54] and can cause serious inconvenience to users and even heavy costs for businesses, can compromise the availability of information for only a limited amount of time. On the other side of the spectrum, side-channel attacks do not have significant consequences for the availability and integrity of information, but they do compromise confidentiality. An ideal security framework should consider those trade-offs to assure higher protection against vulnerabilities when it is more necessary, while optimizing resources and effort. Thus, to maximize the value of both an assessment and the implemented security measures, when the initial objectives of an assessment are being set, the mission of the organization should be identified and the necessary controls should be selected according to the threats, issues, or vulnerabilities of greatest importance in the specific scenario [55]. That approach is a good fit for IT security assessments of organizations that manage their processes and infrastructure on their own premises. However, it might be less effective for assessing vendors of cloud services, as the services of the CSP may be accessible to a wide range of organizations and users (e.g., in health care, the financial sector, or manufacturing) that have different security needs. Industry-based certifications exist and are often built on requirements imposed by sectoral laws (e.g., SOC 1 aims to fulfill the requirements of the Sarbanes–Oxley Act). The broad context in which CSPs operate, however, and their consequent need to comply with multiple standards simultaneously, calls into question the desirability of having multiple concurrent standards.[14] Instead, it suggests the need for a more comprehensive and complete framework that is equally aware of the confidentiality, integrity, and availability of information.

It is necessary to ensure the completeness of a standard, including its ability to protect against issues and vulnerabilities; but that may not be enough. The large variety of vulnerabilities in cloud computing, including both cloud-specific issues and issues independent of the shared infrastructure, must be covered. At the same time, the study of new vulnerabilities is constantly advancing, and work in academia and industry as well as nongovernmental organizations is contributing to the identification of potential threats, the building of awareness, and the improvement of countermeasures. Thus, there is tension between the need to standardize protection methods and controls and the constant discovery of new issues that must be included in the standardized protections.

---

14 The issue of overlapping security standards and controls in cloud services is not new in the IT industry. Adobe [56] and Microsoft [57], for instance, have analyzed the issue and identified considerable overlap among existing regulations and frameworks.

Thus, the study of vulnerabilities must be closely connected to the mainte-nance of security standards in order to keep the standards effective over time. It is necessary to determine how a standard reacts to new threats, whether the measures and controls that it imposes are adequate and effective, and whether the baseline protection is maintained, despite the ever-changing threat land-scape. Resilience is thus a fundamental quality of security standards in cloud computing systems, especially in light of the unpredictability of vulnerabilities generated by new threats. That connection, however, is not prominent in the literature, and studies of new threats and issues conducted in academia and industry typically address threats on a case-by-case basis [58].

The work of the CSA represents a partial exception to that trend. It has been conducting an ongoing study of cloud issues since 2010, and it regularly updates its published list of issues in cloud computing. Its first major publication, "Top Threats in Cloud Computing" [59], was followed by "The Notorious Nine: Cloud Computing Top Threats in 2013" [60], and then "The Treacherous Twelve: Cloud Computing Top Threats in 2016" [54]. CSA's studies are based on surveys of experts who are asked to identify and rank in order of severity the most common issues in cloud computing, such as data breaches, advanced persistent threats (APTs), malicious insiders, or account hijacking. What differentiates CSA's work is that although it is aimed primarily at providing guidance to cloud providers and other stakeholders in evaluating the adequacy of their systems, the studies have direct references to CSA's CCM, which serves as the basis for the STAR certification. Thus, there is a direct connection between the controls in the Matrix (i.e., the standard) and the issues highlighted in the study. In their most recent publication (from February 2016) [54], the Cloud Security Alliance featured 12 issues in cloud computing, for which it gave direct reference to the relevant controls in the CCM. Out of the 133 controls in the Matrix, 82 were noted as offering protection against 1 or more of the "Treacherous Twelve" issues, while the remaining 51 controls do not have a direct connection with any of them.

The work of the CSA is at the foundation of our study, since it helps make a connection between the three security standards analyzed in this chapter, and the common cloud-specific threats and issues. In the following sections, we consider the effectiveness of AICPA SOC 2 (which refers to TSPC), ISO/IEC 27001, and FedRAMP, and compare their performance with the requirements imposed by a third-party framework: the CSA CCM v.3.0.1. We assess the number of controls in the CCM satisfied by each of the four standards and evaluate the relevance of the missing controls by assessing the severity of the threats with which they cannot cope.

8.4 Comparison among Standards

The existence of multiple standards with similar characteristics, all covering security in IT environments (including the cloud), raises a number of questions on their necessity and function. Are there benefits of having multiple frameworks, when the final goal is to achieve the best possible security – or at least fulfill a baseline – and reassure users on the quality of the service that is being offered?

The adequacy of each standard in addressing current threats is another point that requires attention. How does a standard perform in the face of the current, dynamic threat landscape? Are existing standards resilient to the appearance of new issues and vulnerabilities? Can we observe any trends in the protection they offer that can be used to improve their adequacy and resilience?

Standards can be different, as they are often framed around divergent principles and strategies to assure confidentiality, integrity, and availability of information. A standard can be more oriented toward documentation of processes and internal procedures in information management, or have a stronger emphasis on the adoption of technical measures. But with regard to how good a standard is at fulfilling the C-I-A principles, it is necessary to evaluate the standard's ability to offer baseline protection against issues and vulnerabilities. A differentiation in principles can be drawn and different solutions adopted, but the process integrity or the security of applications must always be the final goal.

For example, two standards can approach the same problem from two different perspectives. In regulating physical access control, one standard might be broad and demand that an organization issue control policies that follow a least-privilege rule that requires users to authenticate whenever they transition between physical environments, such as moving from a service room to a data center. Another standard could prescribe the use of strong (two-factor) authentication to verify identity upon physical entry of premises but not mention what the control policy should be or how frequent controls should be (e.g., only upon the first access to the premises or upon any transition between two environments).

In that example, there are two possible flaws: one derives from insufficient technical measures, and the other from lack of policies. If strict policies are implemented, but no attention is given to the technical means for authentication of visitors, the general level of security upon the first access to the premises, as well as in the transition between environments, will be low. Thus, once the authentication system has been compromised, the use of strong policies to separate the environments and define access levels will become useless. On the other hand, if authentication methods are strong enough, it is harder to compromise the system, thus making accesses to the premises more secure. However, a lack of clear access privilege policies may allow uncontrolled transitions between environments and unauthorized accesses to resources. For instance, persons authorized to enter a service area (e.g., air conditioning maintenance personnel) could easily access the data center if the environments are not segregated and access control is not implemented at every level.

To be complete and adequate to address threats and vulnerabilities, standards could use a technical or policy-oriented approach, but either way they must compensate for possible flaws caused by the specific approach they choose. To form a clear vision of such considerations, we can compare standards to each other to highlight their differences and weaknesses. However, a direct comparison, in which we gauge how the controls in one standard match with the provisions of another, might lack objectivity and therefore do a poor job of addressing completeness and adequacy.

Direct comparisons of AICPA TSPC (which serves as the basis of SOC 2), NIST SP 800-53 (the basis for FedRAMP), and ISO/IEC 27001 exist, but only a very limited number of studies compare more than two standards to each other [61–63]. NIST, for instance, released in its last version of the SP 800-53 a direct comparison with ISO/IEC 27001. That comparison limits its observations to the annex in ISO, excluding the introductory clauses that give further specification and thus limiting the completeness of the comparison. Furthermore, the table in NIST SP 800-53 is merely an analysis of how the two sets of controls can match each other, regardless of the threats or vulnerabilities addressed by each control or its effectiveness. Similar comparisons between the TSPC and NIST SP 800-53, and between TSPC and ISO/IEC 27001, have been published by AICPA.

To assess the effectiveness and completeness of the three standards, a more nuanced analysis is required. First of all, they must be observed for their adequacy in addressing cloud issues independent of the adopted approach, which could have a technical emphasis or be more oriented toward the completeness of policies and procedures. Second, the evaluation cannot be built on a reciprocal comparison that uses one of the standards as a reference and observes how the others satisfy its requirements. Third, the observations cannot be limited to the latest version of each standard but must include consideration of how the standard has adapted to the threat environment.

8.4.1 Strategy for Comparing Standards

To analyze the differences among the security and privacy measures in FedRAMP, ISO/IEC 27001, and SOC 2, and to perform a study of them over time, we compare the provisions specified in the clauses and controls upon which the standards are based in their available versions, which include NIST SP 800-53 in its 2012 and 2015 releases; AICPA TSPC 2009, 2014, and 2016; and ISO/IEC 27001:2005 and 2013. Since FedRAMP is based on three possible security baselines (low, moderate, and high), we chose to look at the highest baseline that is common to the two releases of the standard, namely, the moderate baseline, and to look only at the controls in the NIST publication that are relevant for FedRAMP authorization. We also include all the possible AICPA TSPC controls that cover security, availability, processing integrity, confidentiality, and privacy, and limit our study to the three versions used for SOC 2 and SOC 3 reports.[15]

For the comparison, we chose to use a third-party framework that was specifically created for cloud computing, has been updated to encompass protection against newer threats, and is organized in a structure similar to that of FedRAMP, ISO/IEC 27001, and TSPC. The CSA Cloud Control Matrix version 3.0.1 satisfies all those requirements, as it is built on cloud-specific security and privacy controls and is constantly being updated to provide protection against issues and vulnerabilities that target cloud environments. In the following, as we match the controls and requirements of the three standards to the CCM, we will highlight the controls in the CCM that are covered by the three standards, and others that are not sufficiently covered by the three standards. To complete this task, we will rely on the matching made by CSA between the CCM and TSPC 2009 and 2014, FedRAMP rev. 3, and ISO/IEC 27001 [53]; we will partially integrate the matching presented by CSA for FedRAMP rev. 4 [64]; and we will provide the entire matching for TSPC 2016 by building on the same criteria followed by CSA for the matching of its previous versions.

In addition, we will take advantage of CSA's studies of issues, threats, and vulnerabilities in cloud environments and their direct reference to the CCM to select the controls that have a direct connection to any of the "Treacherous Twelve" issues identified in CSA's latest study, from 2016 [54]. As a result of our selection and the matching of the three standards to the CCM, we can verify the impact of the standards on current issues in cloud environments.

8.4.2 Patterns, Anomalies, and Discoveries

The three standards were introduced over a span of 11 years, starting in 2005 with ISO/IEC 27001. Despite being outdated by the appearance of new threats and vulnerabilities, the 2005 ISO/IEC 27001 standard is similar to the other frameworks in terms of its adequacy and comprehensiveness (see Figure 8.1). Its newer version, however, stands out for completeness; the number of matches increased from 91 in 2005 to 130 in 2013. In general, however, completeness does not correlate with the newness of the standard. The completeness of the TSPC-based SOC 2, for example, shows a fluctuation over the 7 years from 2009 (the year of its first release) to 2016; the number of matched controls went from 93 in 2009 to 85 in 2014, and up to 91 in 2016. Finally, FedRAMP showed significant improvement, going from 88 matches in the 2012 version to 103 in the version released in 2015.

---

15 As discussed in Section 8.3.2, SOC reports debuted in 2011 and superseded the existing SAS 70, whose scope (unlike that of SOC 2 and SOC 3) was limited to financial statements.

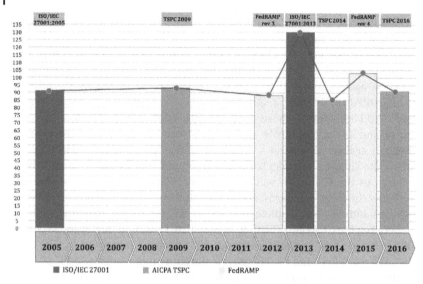

Figure 8.1 Matching controls in each standard over time.

The great improvement in ISO/IEC 27001 is undoubtedly due to the 2013 introduction of controls addressing mobile security, which was not considered in 2005. In the CCM, the control family that includes controls for mobile security contains 20 controls, which specify policies and technical measures. For example, CCM requires "Anti-malware awareness training, specific to mobile devices" for the employees of the CSP, or clear policies defining limits and protection in bring-your-own-device use for the CSP's employees. At the same time, more technical measures include the adoption of automatic lockout screens to protect devices from unwanted access [53]. None of these controls were included in the older version of ISO/IEC 27001. The area of mobile security represents a major divide between ISO/IEC 27001 and all the other standards in any of the versions we observed. FedRAMP rev. 4 shows slight improvement in this area relative to its older version; the introduction of more specific measures and controls for mobile devices[16] has improved its coverage, reducing the number of omitted controls from 20 to 14.

Similarly, the entire control family that protects the interoperability and portability of cloud applications and data is fully covered in ISO/IEC 27001:2013, but missing from all the other standards in our analysis. This

---

16 The introduction of more specific measures for mobile security in FedRAMP must be attributed primarily to the update of NIST SP 800-53. In the newer version, FedRAMP makes explicit mention of mobile devices in some of the controls (e.g., the control SC-8, which deals with protecting communication and integrity of information).

control family consists of five controls aimed at facilitating migration of applications and data and access to information by the tenant through the use of open standards and mutually agreed-upon procedures.

The absence of this control area, however, must be evaluated with respect to the function of each particular standard and its goals. In the case of FedRAMP, which is required only for CSPs that provide services to federal agencies, the applicability of federal laws affects the impact of any omission. Since the U.S. government and NIST have addressed the issue of portability and inter-operability in a separate document, NIST SP 500-293, the omission of this control in FedRAMP is not particularly startling.

For TSPC, the introduction of a more homogeneous set of privacy criteria and a slight reorganization of the content in 2016 clearly improved its effectiveness and completeness in comparison with the preceding version. The same cannot be said for the drastic reorganization done in 2014, which worsened the performance, that is, increased the number of omitted controls.

The scenario changes if we introduce another variable into our analysis, namely, the impact of the omitted controls with respect to the current issues and vulnerabilities in cloud computing environments. When we select only the controls in the CCM that are connected to at least one of the "Treacherous Twelve," we notice a considerable increase in the percentage of controls of the considered standards that match the CCM, counted on a total of 82 relevant controls (see Figure 8.2). The average percentage increase is close to 14%, with the greatest improvement being for FedRAMP rev. 3, for which the number of matching controls moved from 66 to 85%. ISO/IEC 27001:2013 is the only standard showing a slight decrease of 1%.

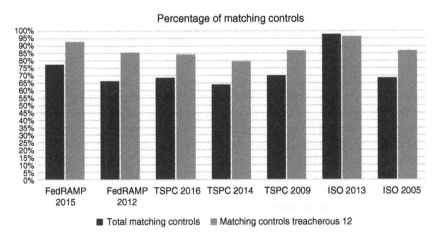

Figure 8.2 Total percentages of matching controls and a selection of 12 "Treacherous" controls.

Interestingly, except for the TSPC, the results of the selection show a proportionate increase of matching controls for all the frameworks between their older and newer versions. Not only does TSPC 2014 match fewer controls in the CCM than did TSPC 2009, but even the improved TSPC 2016 still has fewer matches than TSPC 2009. The very similar structure of the two most recent versions could be to blame; except for a few adjustments and the addition of the privacy section, the criteria were largely unchanged from 2014 to 2016. TSPC 2009, which was more specific for each principle of the C-I-A, gave more precise directions on the nature of the controls to be implemented. However, we must keep in mind that TSPC is a set of general criteria, and that the SOC reports are based on controls that are built on those criteria but chosen by the CSP and assessed by a CPA. While that may explain the fluctuation and indeed provide partial justification for the omissions, it is a reminder of the importance of choosing adequate controls to fulfill the criteria when they do not fully specify the security measures to be implemented.

A final observation that emerged from the selection of controls concerns the overlap of the three standards. No one control was omitted by all three standards, and that result does not change if we restrict the analysis to the most recent version of each standard. While the presence of ISO/IEC 27001:2013 in the analysis obviously reduces the possibility of overlapping omissions (since only two controls are omitted in ISO), even if we narrow the observation to just TSPC and FedRAMP, we find that only two controls are missing in both. Those two controls are both from the control family that covers virtualization security, and are poorly addressed in TSPC and FedRAMP because they do not clearly require security measures for overseeing the use of virtualization-aware assessment tools or the hardening of operating systems in the virtualization stack.

8.5 The Future

The observation of omitted controls that emerged from our comparison of TSPC, FedRAMP, and ISO/IEC 27001 against CSA CCM answers our questions on the effectiveness and completeness of each standard. Although gaps and weaknesses emerged from the analysis, a more careful observation reveals that they offer a good security baseline when confronted with current issues, threats, and vulnerabilities. We also observed that the three standards show high complementarity and can compensate for each other's omissions. That answers the question of what benefit can be obtained by complying with multiple standards at the same time. Multiple certifications assure more complete protection against issues and vulnerabilities, since each one's focus helps cover for contexts in which the others are less effective. However, a CSP's ability to leverage the complementarity of the standards does not justify their creators' failure to make a single standard that offers full protection, as CSPs incur significant costs in obtaining multiple certifications. Partial protection could be both acceptable and justifiable for sectoral standards whose goals are specific enough that completeness is unnecessary, but standards that have a general intent of guaranteeing IT security and privacy cannot overlook baseline controls.

The origin of each standard undoubtedly plays a role in its unique characteristics. FedRAMP, designed by the federal government for the federal government, can rely on provisions in federal laws and regulations to complete the baseline. A CSP, as a contractor to the federal government, might have additional obligations that compensate for omissions, as shown by the case of the omitted controls on interoperability and portability of data and applications. An assessment of these obligations prior to the adoption of a standard helps one understand its limits and supports planning for additional controls where necessary. SOC 2 shows its limits in not being a full certification, but rather a form of guidance for assessing IT security. On the one hand, TSPC's greater flexibility has a negative impact on completeness, and a careful strategy is needed in order to define the objectives and specifications of an SOC 2 assessment. On the other hand, flexibility represents uncharted territory for tenants; they cannot rely on the mere availability of an SOC report from their CSP, but need to do a careful reading of the report to identify the extent to which the assessment it describes actually guarantees cloud security. On the other hand, ISO/IEC 27001 benefits from its mixed approach. As a cross-industry IT certification, it is more general by design; furthermore, as it is more specific than TSPC in dictating the controls to implement during a security assessment, it has the smallest number of omitted controls.

The idea that a good balance between general and specific measures is beneficial to the effectiveness of a standard is reinforced by observing the performance of each standard over time (see Figure 8.1). ISO/IEC 27001, which suffered heavily from not including mobile security controls in its older version and from being outdated by the technology advances of the past 11 years, still shows protection comparable to, if not better than, that offered by the others. Not only is the total number of controls it omits lower than that for TSPC 2014 and FedRAMP rev. 3, but the number of most significant controls omitted after the selection (see Figure 8.2) is equal to that of the most recent release of FedRAMP, and second only to ISO/IEC 27001:2013. In a similar vein, TSPC 2009, which was oriented toward the assessment of cross-industry security but was more specific than its later updates, does not suffer from being 7 years older than its most recent successor, outperforming it in the coverage of the most relevant controls.

Although using a better approach in designing IT security standards is not the only element to consider in order to tackle cloud security, it certainly represents a strong start. Flexibility may be reckoned a weakness at first sight, but can also have value if used correctly, and if strong control measures are implemented to fulfill general criteria. Conversely, less flexibility could hold back a CSP in the implementation of necessary protection measures.

8.5.1 Current Challenges

Consideration of the issues described above opens up uncertainty for both the cloud service provider and the tenants. What certification process could be the most beneficial, and what type of compliance should be attained to assure information in the cloud? Looking at a baseline drawn from a single standard may not be enough, but multiple certifications necessarily entail additional costs for the CSP. External audits are expensive not just because of the involvement of third parties, but also because internal resources will be engaged in the auditing process, raising the cost for the CSP beyond the fees for certification.

The benefit of multiple certifications derives from the different approaches used in the standards and the possibility of leveraging flexible criteria and precise controls together. At the same time, the reciprocal coverage of omitted controls among the standards in our study shows that integrations are possible and that each standard could alone assure better coverage than it currently does. A first step must be to improve current standards with the introduction of measures suitable for covering those areas left unprotected. To ensure the success of this process, the study of threats and vulnerabilities must remain a priority, and the results must be translated into existing standards so that they can be effectively improved and updated.

The importance of maintaining updated standards to cope with current issues and vulnerabilities is apparent when we consider the rapid and constant evolution of technology. The Internet of Things (IoT) is one of the most vivid examples of the innovative forms of data processing that keep springing up, generating unexpected threats. The IoT model implies processing of a massive amount of user data, and is often susceptible to flaws caused by lack of basic attention to patches, updates, or the use of passwords [65]. Connected devices require interaction with cloud systems, since limited storage and processing power could limit their performance; at the same time, they have distinctive characteristics that make them more vulnerable to external attacks [66]. In the IoT scenario, single devices could be the weak link in the chain protecting C-I-A principles, and their violation might have disruptive consequences. Standards must have the resilience to handle the appearance of radical innovations, such as in the interaction of IoT devices and cloud systems. At the same time, the resilience of a standard cannot be determined solely by the frequency of its updates, but rather must be implicit in its design.

An alternative path could be to promote the optimization of the certification processes, and the achievement of better integration of assessment procedures through dialogue with and cooperation of the stakeholders. Choosing this path, and avoiding repetition of controls across standards once a first certification has been obtained, is a necessary step. Initiatives in that direction have been promoted by industry advocacy groups [67] and government agencies [68], but their extent is still unclear and limited to guidance for auditors and CSPs, rather than a full, mutual recognition.

8.5.2 Opportunities

Standards can truly help improve security and privacy. By establishing a security baseline and setting the scene for the implementation of privacy- and security-enhancing controls, standards can optimize the outcomes of CSPs' efforts to ensure information assurance. From that perspective, standards should be seen as a security tool.

In particular, the use of standards can help improve CSPs' efficiency in implementing security and privacy measures. A CSP trying to protect its cloud system without guidance could overprotect certain areas while overlooking the need for other measures. By improving the security controls in relation to the threat landscape, standards can help a CSP achieve the best possible security result while lowering assessment and implementation costs.

Our observations on the impact of the "Treacherous Twelve" on the effectiveness of IT security standards in cloud environments point toward a possible path to the improvement of IT security standards. Observations of threats, issues, and vulnerabilities can help us understand the need for new or different control measures, and their connection to security standards can lead to better effectiveness, completeness, and efficiency.

References

1 Mell, P. and Grance, T. (2011) The NIST definition of cloud computing: recommendations of the National Institute of Standards and Technology, Special Publication 800-145, National Institute of Standards and Technology, U.S. Department of Commerce, September. Available at http://nvlpubs.nist .gov/nistpubs/Legacy/SP/nistspecialpublication800-145.pdf.

2 Fischer, E.A. and Figliola, P.M. (2013) Overview and issues for implementation of the Federal Cloud Computing Initiative: implications for federal information technology reform management. *Journal of Current Issues in Media and Telecommunications*, **5** (1), 1–27.

3 Harms, R. and Yamartino, M. (2010) Economics of the cloud, Microsoft, Nov. 11. Available at https://blogs.technet.microsoft.com/microsoft_on_the_issues/ 2010/11/11/economics-of-the-cloud/.

4 Office of Management and Budget (2017) Analytical Perspectives. Budget of the U.S. Government. Available at Retrieved from https://obamawhitehouse .archives.gov/omb/budget/Analytical_Perspectives.

5 Kundra, V. (2011) Federal cloud computing strategy. Available at https://www.dhs.gov/sites/default/files/publications/digital-strategy/federal-cloud-computing-strategy.pdf (accessed May 27, 2016).

6 Gartner (2016) Gartner says worldwide public cloud services market to grow 17 percent in 2016, Sep. 15. Available at http://www.gartner.com/newsroom/id/3443517 (accessed Nov. 20, 2016).

7 Inci, M.S., Gulmezoglu, B., Irazoqui, G., Eisenbarth, T., and Sunar, B. (2015) Seriously, get off my cloud! Cross-VM RSA key recovery in a public cloud, Cryptology ePrint Archive, Report 2015/898, Sep. 22. Available at http://ia.cr/2015/898 (accessed Oct. 9, 2016).

8 Liu, F., Yarom, Y., Ge, Q., Heiser, G., and Lee, R.B. (2015) Last-level cache side-channel attacks are practical, in Proceedings of the IEEE Symposium Security and Privacy, pp. 605–622.

9 Zhang, Y., Juels, A., Reiter, M.K., and Ristenpart, T. (2012) Cross-VM side channels and their use to extract private keys, in Proceedings of the ACM Conference on Computer and Communications Security, pp. 305–316.

10 Zhang, Y., Juels, A., Reiter, M.K., and Ristenpart, T. Cross-tenant side-channel attacks in PaaS clouds, in (2014) Proceedings of the ACM SIGSAC Conference on Computer and Communications Security, pp. 990–1003.

11 Verizon (2016) 2016 data breach investigations report. Available at http://www.verizonenterprise.com/verizon-insights-lab/dbir/2016/.

12 Timmermans, S. and Epstein, S. (2010) A world of standards but not a standard world: toward a sociology of standards and standardization. *Annual Review of Sociology*, **36**, 69–89.

13 Brunsson, N. and Jacobsson, B. (2000) The contemporary expansion of standardization, in *A World of Standards* (eds. N. Brunsson and B. Jacobsson), Oxford University Press, pp. 1–18.

14 International Organization for Standardization (ISO) About ISO, ISO. Available at http://www.iso.org/iso/home/about.htm (accessed Jun. 1, 2016).

15 National Institute of Standards and Technology (NIST) (2016) About NIST, Aug. 25. Available at https://www.nist.gov/about-nist (accessed Nov. 30, 2016).

16 Circular No. A-119, Revised: Federal Participation in the Development and Use of Voluntary Consensus Standards and in Conformity Assessment Activities, Circular No. A-119, revised, National Institute of Standards and Technology (NIST), U.S. Department of Commerce, Jan. 27. (2016) Available at https://www.nist.gov/standardsgov/what-we-do/federal-policy-standards/key-federal-directives (accessed May 17, 2017).

17 Schroeder, F.E.H. (1986) More "small things forgotten": domestic electrical plugs and receptacles, 1881–1931. *Technology and Culture*, **27** (3), 525–543.

18 Drake J., (2015) Stagefright: scary code in the heart of Android: researching Android multimedia framework security, Black Hat, USA, Aug. 5. Available at

https://www.blackhat.com/docs/us-15/materials/us-15-Drake-Stagefright-Scary-Code-In-The-Heart-Of-Android.pdf (accessed Oct. 15, 2016).

19 Goodin, D. (2013) Google confirms critical Android crypto flaw used in $5,700 Bitcoin heist, Ars Technica, Aug. 14. Available at http://arstechnica.com/security/2013/08/google-confirms-critical-android-crypto-flaw-used-in-5700-bitcoin-heist/ (accessed Jun. 8, 2016).

20 Ormandy, T. (2007) An empirical study into the security exposure to host of hostile virtualized environments. Available at http://taviso.decsystem.org/virtsec.pdf.

21 Defense Information Systems Agency (2017) Department of Defense cloud computing security requirements guide. Available at http://iasecontent.disa.mil/cloud/SRG/ (accessed May 17, 2017).

22 DigitalGov (2016) FedRAMP high baseline requirements for federal agencies, YouTube, Jun. 30. Available at https://www.youtube.com/watch?v=CSRH5wjlcEk (accessed Nov. 20, 2016).

23 U.S. Department of State (2016) 12 FAM 540: sensitive but unclassified information (SBU), CT:DS-256, Foreign Affairs Manual, Apr. 7. Available at https://fam.state.gov/FAM/12FAM/12FAM0540.html (accessed Nov. 20, 2016).

24 Kundra, V. (2010) 25 Point implementation plan to reform federal information technology management, The White House, Washington, DC, Dec. 9. Available at https://www.dhs.gov/sites/default/files/publications/digital-strategy/25-point-implementation-plan-to-reform-federal-it.pdf (accessed May 23, 2016).

25 American Institute of Certified Public Accountants (AICPA) (2014) Service Organization Control reports. Available at http://www.aicpa.org/InterestAreas/FRC/AssuranceAdvisoryServices/DownloadableDocuments/SOC_Reports_Flyer_FINAL.pdf (accessed May 16, 2017).

26 McClure, D. (2010) Statement of Dr. David McClure, Associate Administrator, Office of Citizen Services and Innovative Technologies, General Services Administration, before the House Committee on Oversight and Government Reform Subcommittee on Government Management, Organization, and Procurement, U.S. General Services Administration (GSA), Jul. 1. Available at http://www.gsa.gov/portal/content/159101 (accessed Nov. 24, 2016).

27 VanRoekel, S. (2011) Security authorization of information systems in cloud computing environments. Available at https://s3.amazonaws.com/sitesusa/wp-content/uploads/sites/482/2015/03/fedrampmemo.pdf (accessed May 17, 2017).

28 FedRAMP (2014) Guide to understanding FedRAMP, version 2.0, Jun. 6. Available at https://s3.amazonaws.com/sitesusa/wp-content/uploads/sites/482/2015/03/Guide-to-Understanding-FedRAMP-v2.0-4.docx (accessed June 6, 2014).

29 European Telecommunications Standards Institute (ETSI) (2013) Cloud standards coordination: final report, version 1.0, November. Available at http://csc.etsi.org/resources/CSC-Phase-1/CSC-Deliverable-008-Final_Report-V1_0.pdf (accessed May 25, 2016).

30 Joint Task Force Transformation Initiative (2010) Guide for Applying the Risk Management Framework to Federal Information Systems: A Security Life Cycle Approach, NIST Special Publication 800-37, Revision 1, National Institute of Standards and Technology (NIST), U.S. Department of Commerce, February. (with updates as of Jun. 5, 2014). Available at http://nvlpubs.nist.gov/nistpubs/SpecialPublications/NIST.SP.800-37r1.pdf. (accessed Aug. 2, 2017).

31 International Organization for Standardization (ISO) (2012) ISO/IEC 17020:2012: Conformity Assessment – Requirements for the Operation of Various Types of Bodies Performing Inspection, March. Available at http://www.iso.org/iso/catalogue_detail?csnumber=52994 (accessed Nov. 24, 2016).

32 FedRAMP (2016) Program overview. Available at https://www.fedramp.gov/about-us/about/ (accessed May 26, 2016).

33 Joint Task Force Transformation Initiative (2013) Security and privacy controls for federal information systems and organizations, NIST Special Publication 800-53, Revision 4, National Institute of Standards and Technology (NIST), U.S. Department of Commerce, April. Available at http://nvlpubs.nist.gov/nistpubs/SpecialPublications/NIST.SP.800-53r4.pdf (accessed May 31, 2016).

34 Computer Security Division, Information Technology Laboratory, National Institute of Standards and Technology (NIST) (2004) Standards for security categorization of federal information and information systems, Federal Information Processing Standards Publication FIPS PUB 199, February. Available at http://nvlpubs.nist.gov/nistpubs/FIPS/NIST.FIPS.199.pdf (accessed May 31, 2016).

35 FedRAMP High Baseline Release, Jun. 17. (2016) Available at https://www.fedramp.gov/fedramp-releases-high-baseline/ (accessed Nov. 24, 2016).

36 FedRAMP (2016) Marketplace. Available at https://marketplace.fedramp.gov/index.html#/products?sort=productName (accessed Nov. 21, 2016).

37 American Institute of Certified Public Accountants (AICPA) (2016) About the AICPA. Available at http://www.aicpa.org/About/Pages/default.aspx (accessed Nov. 26, 2016).

38 Gartner (2010) Gartner says SAS 70 is not proof of security, continuity or privacy compliance, Jul. 14. Available at http://www.gartner.com/newsroom/id/1400813 (accessed Nov. 25, 2016).

39 Nickell, C.G. and Denyer, C. (2007) An introduction to SAS 70 audits. *Benefits Law Journal*, **20** (1), 58–68.

40 American Institute of Certified Public Accountants (AICPA) (2011) New SOC reports for service organizations replace SAS 70 reports, Feb. 7. Available at

https://www.aicpastore.com/Content/media/PRODUCER_CONTENT/
Newsletters/Articles_2011/CPA/Feb/SOCReplaceSAS70Reports.jsp (accessed
Nov. 24, 2016).

41 International Organization for Standardization (ISO) ISO/IEC JTC 1 –
Information Technology. Available at http://www.iso.org/iso/jtc1_home.html
(accessed Jun. 1, 2016).

42 International Organization for Standardization (ISO) ISO/IEC 27000 family:
information security management systems, ISO Available at https://www.iso
.org/isoiec-27001-information-security.html (accessed Jun. 1, 2016).

43 Charlet, L. (2015) ISO Survey, International Organization for Standardization
(ISO). Available at http://www.iso.org/iso/iso-survey (accessed Nov. 24, 2016).

44 Gantz, S.D. (2014) *The Basics of IT Audit: Purposes, Processes, and Practical
Information*, Syngress, Waltham, MA.

45 Bird, K. (2013) New version of ISO/IEC 27001 to better tackle IT security
risks, International Organization for Standardization (ISO), Aug. 14.
Available at http://www.iso.org/iso/news.htm?refid=Ref1767 (accessed
May 29, 2016).

46 Watkins, S.G. (2013) *An Introduction to Information Security and
ISO27001:2013: A Pocket Guide*, 2nd edn, IT Governance Publishing, Ely, UK.

47 PCI Security Standards Council, LLC PCI Security. Available at https://www
.pcisecuritystandards.org/pci_security/ (accessed Nov. 30, 2016).

48 Montanari, M., Huh, J.H., Dagit, D., Bobba, R.B., and Campbell, R.H. (2012)
Evidence of log integrity in policy-based security monitoring, in 2012 IEEE/
IFIP 42nd International Conference on Dependable Systems and Networks
Workshops, Article No. 6264693.

49 Wright, S. (2011) *PCI DSS: A Practical Guide to Implementing and
Maintaining Compliance*, 3rd edn, IT Governance Publishing, Ely, UK.

50 Bundesamt für Sicherheit in der Informationstechnik (BSI) (2016)
Anforderungskatalog Cloud Computing (C5). Available at https://www.bsi
.bund.de/C5 (accessed Nov. 29, 2016).

51 Cloud Security Alliance (CSA) (2016) About. Available at https://
cloudsecurityalliance.org/about/ (accessed May 24, 2016).

52 Cloud Security Alliance (CSA) CSA STAR: the future of cloud trust and
assurance. Available at https://cloudsecurityalliance.org/star/ (accessed May
31, 2016).

53 Cloud Security Alliance (CSA) Introduction to the Cloud Control Matrix
Working Group. Available at https://cloudsecurityalliance.org/group/cloud-
controls-matrix/ (accessed May 21, 2016).

54 Cloud Security Alliance (CSA) (2016) 'The Treacherous Twelve' cloud
computing top threats in 2016. Available at https://cloudsecurityalliance.org/
download/the-treacherous-twelve-cloud-computing-top-threats-in-2016/
(accessed May 23, 2016).

55 Bayuk, J.L. (2011) Alternative security metrics, in Proceedings of the 8th International Conference on Information Technology: New Generations, pp. 943–946.

56 Adobe (2015) Adobe security and privacy certifications, Adobe Systems Incorporated. Available at http://www.adobe.com/content/dam/Adobe/en/security/pdfs/adobe-ccf-012015.pdf (accessed Jun. 1, 2016).

57 Microsoft Simplify compliance with the Microsoft Common Controls Hub, Microsoft Trust Center. Available at https://www.microsoft.com/en-us/trustcenter/Common-Controls-Hub (accessed Nov. 30, 2016).

58 Ardagna, C.A., Asal, R., Damiani, E., and Vu, Q.H. (2015) From security to assurance in the cloud: a survey. ACM Computing Surveys, 48 (1), Article 2.

59 Cloud Security Alliance (CSA) (2010) Top threats to cloud computing V1.0. Available at https://cloudsecurityalliance.org/topthreats/csathreats.v1.0.pdf (accessed May 23, 2016).

60 Cloud Security Alliance (CSA) (2013) The Notorious Nine: Cloud Computing Top Threats in 2013. Available at https://downloads.cloudsecurityalliance.org/initiatives/top_threats/The_Notorious_Nine_Cloud_Computing_Top_Threats_in_2013.pdf (accessed May 24, 2016).

61 Di Giulio, C., Kamhoua, C., Campbell, R.H., Sprabery, R., Kwiat, K., and Bashir, M.N. (2017) Cloud standards in comparison: are new security frameworks improving cloud security? in Proceedings of the IEEE 10th International Conference on Cloud Computing (CLOUD), pp. 50–57.

62 Di Giulio, C., Kamhoua, C., Campbell, R.H., Sprabery, R., Kwiat, K., and Bashir, M.N. (2017) IT security and privacy standards in comparison: improving FedRAMP authorization for cloud service providers, in Proceedings of the 17th IEEE/ACM International Symposium on Cluster, Cloud and Grid Computing (CCGRID), pp. 1090–1099.

63 Di Giulio, C., Sprabery, R., Kamhoua, C.A., Kwiat, K., Campbell, R.H., and Bashir, M.N. (2017) Cloud security certifications: a comparison to improve cloud service provider security, in Proceedings of the 2nd International Conference on Internet of Things, Data and Cloud Computing (ICC), ACM, New York, NY.

64 Cloud Security Alliance (CSA) FedRAMP cloud controls matrix v3.0.1 candidate mapping. Available at https://cloudsecurityalliance.org/download/fedramp-cloud-controls-matrix-v3-0-1-candidate-mapping/ (accessed Jun. 1, 2016).

65 Roos, G. (2015) How cloud, IoT are altering the security landscape, Channel Insider, Sep. 13. Available at http://www.channelinsider.com/security/slideshows/how-cloud-iot-are-altering-the-security-landscape.html (accessed May 17), 2017.

66 Fraga-Lamas, P., Fernández-Caramés, T.M., Suárez-Albela, M., Castedo, L., and González-López, M. (2016) A review on Internet of Things for defense and public safety. Sensors, 16 (10), Article 1644.

67 FedRAMP Fast Forward Industry Advocacy Group Fix FedRAMP: a 6-point plan. Available at https://www.meritalk.com/study/fix-fedramp/.

68 U.S. Government Accountability Office (GAO) (2010) Report to the Chairwoman, Subcommittee on Government Management, Organization, and Procurement, Committee on Oversight and Government Reform, House of Representatives: Information Security: Progress Made on Harmonizing Policies and Guidance for National Security and Non-National Security Systems, GAO-10-916, Washington, DC, September. Available at http://www.gao.gov/assets/310/309573.pdf.

9

Summary and Future Work

Roy H. Campbell

Department of Computer Science, University of Illinois at Urbana-Champaign, Urbana, IL, USA

The adoption of cloud computing by the U.S. government, including the Department of Defense, is proceeding quickly [1–4] and is likely to become widespread [5]. As government becomes more comfortable with the technology, mission-oriented cloud computing seems inevitable. However, security remains a top concern in the use of clouds for dependable and trustworthy computing [6], even as FedRAMP [7] and other standards converge to a common set of requirements, as discussed in Chapter 8. The cloud computing environment is maturing, but we are observing the rise of new aspects of cloud computing – such as mobiles interconnected into clouds, real-time concerns, edge computing, and machine learning – that are challenging the existing techniques for testing, validation, verification, robustness, and resistance to attack. As reflected in this book, academia and industry are attempting to respond quickly to rapidly changing cloud technologies, as driven by the value of these technologies in today's society.

The preceding chapters of this book have touched on many of the concerns arising from cloud technology: survivability, risks, benefits, detection, security, scalability, workloads, performance, resource management, validation and verification, theoretical problems, and certification. In this final chapter, we will consider what has been learned since 2007 and what issues and obstacles remain in any mission-critical system deployment on cloud computing.

9.1 Survivability

Cloud computing systems as a cyberinfrastructure supporting mission-oriented tasks must survive long enough for the missions to accomplish their goals. Any number of challenges face cloud computing survivability, including the unpredictability of technological advances. In Chapter 2, we focused on design, formal modeling, and validation, giving cloud storage systems as an example. Without excellent requirements and design, survivability is problematic. (The security and dependability aspects of survivability were addressed in separate

Assured Cloud Computing, First Edition. Edited by Roy H. Campbell, Charles A. Kamhoua, and Kevin A. Kwiat.

chapters.) Key to cloud computing survivability is the reality that both the infrastructure and applications are built on distributed computing. Survivability for the cloud requires correct requirements and design of systems in which the fundamental concerns include parallelism, communications, distributed algorithms, and undecidability.

Maude and its real-time extensions were used formally to specify and analyze the correctness and performance of Apache Cassandra, Megastore, and ZooKeeper (all industrial systems) as well as RAMP (an academic system). The approach was also used to design and formalize significant extensions of these systems (a variant of Cassandra, Megastore-CGC, a key management system on top of ZooKeeper, and variations of RAMP), building confidence that there is an approach to formalize and design clouds with assurance that they satisfy desired correctness properties and performance. Furthermore, in the case of Cassandra, we compared the performance estimates provided with the performance actually observed when running the real Cassandra code on representative workloads; they differed by only 10–15%.

The abovementioned findings represent the first published work on the use of formal methods to model and analyze such a wide swathe of industrial cloud storage systems and demonstrate that these distributed, real-time cloud systems are amenable to formal description and analysis. Since many of the faults, failures, and security breaches of systems are caused by human error in requirements and design, the work paves the way for cloud computing assurance by showing that it is feasible to verify and validate cloud computing support for missions. Such formal studies should be made of cloud computing environments and their applications to help assure the missions using them.

9.2   Risks and Benefits

Cloud computing infrastructure for mission-oriented tasks allows many new avenues of attack, and some of the opportunities for launching an attack remain obscure until an attack occurs. Cloud service providers may mitigate vulnerabilities and repel attempted attacks, but providing an assured cloud computing environment continues to be a struggle. Cloud users may build applications with the goal of ensuring secure services, while others may have different objectives requiring less security. Risk assessment for an application depends on many factors, like the security of the provider's services, user applications, and the values and costs associated with security breaches. In a multitenancy environment, risk is also associated with issues of externality: Can an attacker find a way to compromise a cloud application by using some less secure component running on that cloud, from which an attack could be more easily launched and would be more likely to succeed against the target cloud application? Clearly, when left to choose a cloud, cloud users will move their applications to clouds that reduce their risk, even while they improve their applications' security provisions. However, a cloud that attracts many applications with high-risk assets, even with improved application security, would become a high-value target for attack. Chapter 3 presented one of the first theoretical ways to use game theory to evaluate risk externality in terms of the benefit versus risk of using a cloud computing environment (the cloud environment, other cloud users, and risks of successful attacks).

As cloud providers add functionality to their services, cloud users can create systems with higher value and benefit, including applications that implement mission-oriented tasks. Understanding how to assess the risk posed by the externalities of the interactions within multitenancy clouds becomes increasingly more critical. Hence, we believe that applying game theory to the problems of risk, attack, and security in cloud systems is an increasingly important concern. Mission-oriented tasks also raise many new game-theoretic problems in clouds, including ones related to geo-distributed clouds, applications that integrate mobile devices with clouds, mobile cloud computing, large-scale sensor systems generating data for clouds, edge computing for clouds, transport and transmission of encrypted data, use of blue (military) or gray (commercial) networks with clouds, active security, and security response systems. Game theory may also be a useful tool in understanding various trade-offs among authorization, authentication, security policies, and insider attacks.

9.3 Detection and Security

Many of the key topics in security provisions, detection, monitoring, and access control for assured cloud computing are addressed in Chapter 4. Although the problems here are open-ended – because of both the nature of the technologies and the increasing sophistication of attacks – clear themes are emerging in our own research and that of others. While it is not possible to prove that a system is secure, layered security provisions that force attacks to have multiple stages improve the likelihood of success and help guarantee detection, response, and recovery. We discussed examples that highlighted active and passive monitoring as a way to provide situational awareness about a system and users' state and behavior; automated reasoning about system/application state based on observations from monitoring tools; coordination of monitoring and system activities to provide a robust response to accidental failures and malicious attacks; and use of smart access control methods to reduce the attack surface and limit the likelihood of an unauthorized access to the system.

Scalable cloud resources allow more flexible cloud computing. However, as attack and failure modes have increased in their complexity and impact, the effort to protect flexible cloud computing has also increased. In Chapter 4, we explained how virtual machine monitoring plays an essential role in achieving resiliency. Virtual monitoring can be integrated with Trusted Platform Module hardware [8] to build resilient and resistant monitoring solutions. However, existing virtual monitoring systems are not a panacea, as multiple operating system versions and requirements have added complexity.

Recognizing common operating system design patterns is suggested as a way to infer monitoring parameters from a guest operating system. The patterns and parameters might be extracted directly from a cloud user's guest operating system. However, monitoring can create overhead in performance. Further, virtual machine monitoring systems may require setup and configuration as part of the boot process, or modification of guest operating system internals. In our approach, monitoring probes using low-level hardware traps (hprobes) may be inserted dynamically according to perceptions about the dynamic nature of possible attacks. The hprobe framework is characterized by its simplicity, dynamism, and ability to perform application-level monitoring. The prototype for this framework uses hardware-assisted virtualization and satisfies protection requirements presented in the literature. Compared to past work, the simplicity with which the detectors can be implemented and inserted/removed at runtime allows monitoring solutions to be developed quickly.

As a proof of concept, some virtual machine monitors have been built through application of this technique, including a return-to-user attack detector and a process-based keylogger detector. Extending this approach to other behaviors that might be used in attacks will be an interesting direction for future research. Best practices suggest that robust and efficient monitoring and protection require formal and experimental validation. Further research is needed on validation frameworks that integrate the use of tools such as model checkers (for formal analysis and symbolic execution of software) and fault/attack injectors (for experimental assessment).

An additional security concern arises because of multitenancy concerns and the use of hypervisor-supported virtual machines in cloud architectures. Side-channel attacks, in which cache behavior is used to determine information about the nature of the processing within a virtual machine, has been shown to reduce the time needed to deduce encryption keys or identify the processing of algorithms of interest to an attacker, like monitoring.

Chapter 4 discussed hypervisor introspection as a technique to determine the presence of and evade a passive virtual machine introspection monitoring system through a timing side-channel. Through hypervisor introspection, hypervisor activity was shown to be not perfectly isolated from the guest virtual machine. In addition, an example of an insider threat attack model was shown that utilizes hypervisor introspection to hide malicious activity from a realistic, passive virtual machine introspection system. Some inherent weaknesses of passive virtual machine introspection monitoring can be avoided by using active virtual machine introspection monitoring.

A Bayesian network modeling approach was described and used to detect compromised users in a shared computing infrastructure. The approach was validated using real incident data collected over 3 years at the National Center for Supercomputing Applications (NCSA). The results demonstrate that the Bayesian network approach is a valuable strategy for driving the investigative efforts of security personnel. Furthermore, it was able to significantly reduce the number of false positives (by 80%, with respect to the analyzed data). However, the deficiencies of the underlying monitoring tools could affect the effectiveness of this network approach.

Access control plays an essential role in preventing potentially malicious actors from entering a system. RBAC is a popular access scheme but has weaknesses: a pure RBAC system lacks flexibility to adapt efficiently to changing users, objects, and security policies. In particular, it is time-consuming to make and maintain manual user-to-role assignments and role-to-permission assignments in the context of a cloud that might have a large number of users and/or security objects. One solution to this problem is to combine ABAC with RBAC, bringing together the advantages of both models. We developed our model in two levels: aboveground and underground. A simple and standard RBAC model is extended with environment constraints, which retains the simplicity of RBAC and supports straightforward security administration and review. Attribute-based policies are used to create the simple RBAC behavior automatically. The attribute-based policies bring the advantages of ABAC: They are easy to build and easy to change for a dynamic application. The approach can be applied to RBAC system design for large-scale Internet cloud system applications.

Clearly, much work remains to be done in detection and security in cloud computing. However, recent work shows progress toward making intrusions in cloud infrastructure more difficult for the attacker. Virtual machines provide some measure of isolation, and implementations of the hypervisors that enable virtual machines are becoming more secure. Container technologies [8] are now often used in cloud computing. Isolation and security for container technologies are a little more difficult than for virtual machines. However, use of cache partitioning in recent Intel products [16] coupled with compiler and other techniques may lead to isolation and assurance for container-based cloud applications.

9.4 Scalability, Workloads, and Performance

The work discussed in Chapter 5 was aimed at the scalability, performance, algorithms, and application workloads of assured cloud computing and addressed issues of network performance, geographic distribution, and stream processing. In cloud computing, scalability is a key parameter underlying performance. As a consequence, this chapter focused on issues related to change in size, volume, and geographical distribution. It considered the evaluation of scaling solutions by using traces and synthetic workload generators.

A key practical solution to many of these issues is to provide appropriate availability to cloud resources, including redundant resources, services, networks, file systems, and nodes. However, this raises a corresponding problem of how to offer cloud applications appropriate consistency. The analysis of the issues and results of the research have been encapsulated in prototypes or systems that exemplify many assured cloud computing problems, solutions, and tools.

Specifically, in Chapter 5 we described DARE, a distributed data replication and placement algorithm that adapts to workload, synthetic workload generation, and clustered renewal processes; Ambry, a geographically distributed blob store; and, briefly, Samza, a stream processing system.

The growth of data analytics for big data encourages the design of next-generation storage systems to handle peta- and exascale storage requirements. DARE demonstrated that a better understanding of the workloads for big data is becoming critical for proper design and tuning. Specifically, popularity, temporal locality, and arrival patterns were studied for a 6-month period from file access patterns of two multipetabyte Hadoop clusters at Yahoo! across several dimensions. Data *popularity* measures accesses: both the number of and intensity of those accesses. The workloads were dominated by high file churn in which most files were accessed fewer than 10 times. A small percentage of files were highly popular. Young files accounted for a high percentage of accesses but a small percentage of bytes stored. The observed request interarrivals (opens, creates, and deletes) were bursty and exhibited self-similar behavior. The files were very short-lived. From the point of view of an individual data node, the DARE algorithm quickly identifies the most popular set of data and creates replicas for this set.

Data analytics of the behavior of big data systems suggested new algorithms and data to improve the performance of applications using Hadoop and HDFS. In general, it is difficult to obtain real traces of systems. Often, when data are available, the traces must be de-identified to be used for research. However, workload generation can often be used in simulations and real experiments to help reveal how a system reacts to variations in the load. Such experiments can be used to validate new designs, find potential bottlenecks, evaluate performance, and do capacity planning based on observed or predicted workloads.

Two important characteristics of object request streams are popularity (access counts) and temporal reference locality. For the purpose of synthetic workload generation, it is desirable to simultaneously reproduce the access counts and the request interarrivals of each individual object, as both of these dimensions can affect system performance. In Chapter 5, single-distribution approaches – which summarize the behavior of different types of objects with a single distribution per dimension – were shown not to reproduce both behaviors accurately at the same time. In particular, the common practice of collapsing the per-object interarrival distributions into a single system-wide distribution (instead of individual per-object distributions) obscures the identity of the object being accessed, thus homogenizing the otherwise distinct per-object behavior. Further, as big data applications lead to emerging workloads and these workloads keep growing in scale, the need for workload generators that can scale up the workload and/or facilitate its modification based on predicted behavior is increasingly urgent. Chapter 5 described a lightweight model that used unsupervised statistical clustering to identify groups of objects with similar behavior, and this significantly reduced the model space by modeling "types of objects" instead of individual objects. As a result, the clustered model can be suitable for synthetic generation and scaled as needed. The synthetic trace generator used this approach, which was evaluated across several dimensions. Using a big data storage workload from Yahoo!, we validated the approach by demonstrating its ability to approximate the original request interarrivals and popularity distributions.

New applications for clouds, such as social networks and file sharing (e.g., LinkedIn, Facebook, and YouTube), demonstrate a need for low-latency worldwide services. High fault tolerance and reliability for these systems necessitate geo-distributed storage with replicas in multiple locations. Chapter 5 discusses the design trade-offs of Ambry, a scalable geo-distributed object store. For several years, Ambry has been the mainstream storage for all of LinkedIn's media objects across all four of its data centers, serving more than 450 million users. The experimental results show that Ambry reaches high throughput and low latency, works efficiently across multiple geo-distributed data centers, and improves the imbalance among disks, while moving minimal data. The chapter also mentioned a collaboration with LinkedIn to develop Samza, a scalable large-state stream-processing system. The system is designed to handle very large state in stream-processing systems, which enables large joins and aggregations over streams. The large-state handling on Samza can reach two orders of magnitude better performance than the traditional way of handling state. Further, the failure recovery can be in parallel and almost constant irrespective of the number of failures. Overhead for failure recovery is reduced by preventing state rebuild as much as possible. Overall, the mechanism reaches low latency, high throughput, and almost constant failure recovery time. Samza is a currently running system of LinkedIn, and the code is open-source.

Future research is needed to address the coming data analytics and machine learning innovations now occurring. The reliability, robustness, accuracy, and performance of cloud-based large machine learning computations is likely to become a major issue in cloud computing. In the few months prior to the time of this writing, production implementations of TensorFlow and other deep-learning systems have come online at Google, Amazon, and Microsoft. The models from these systems are used for inferencing in both clouds and local devices like cell phones. These systems, when coupled with Edge learning systems or smartphones, form complex distributed learning systems that require performance analysis and evaluation. Ubiquitous sensors, autonomous vehicles that exchange state information about traffic conditions, and a host of close-to-real-time and health applications continue to expand the boundaries of cloud computing. The techniques discussed in this chapter, including measurement, modeling, and optimization based on performance, will govern the design of such systems and contribute to assured cloud computing.

9.5 Resource Management

Building assured cloud computing applications and services that perform predictably remains one of the biggest challenges today, both in mission-critical scenarios and in nonreal-time scenarios. The work outlined in Chapter 6 has made deep inroads toward solving key issues in this area. Specifically, the work described constituted the starting steps toward realization of a truly autonomous and self-aware cloud system for which the mission team merely needs to specify SLAs/SLOs (service-level agreements and objectives), and the system then reconfigures itself automatically and continuously over the lifetime of the mission to ensure that these requirements are always met. This chapter described some key design techniques and algorithms, outlined designs and implementation details, and touched on key experimental results. The experimental results were obtained by deploying both the original system(s) and modified systems on real clusters and subjecting them to real workloads.

We overviewed five systems that are oriented toward offering performance assuredness in cloud computing frameworks, even while the system is under change. They include Morphus, which supports reconfigurations in sharded distributed NoSQL databases/storage systems; Parqua, which supports reconfigurations in distributed ring-based key-value stores; Stela, which supports scale-out/in in distributed stream processing systems; an unnamed system to support scale-out/in in distributed graph processing systems; and Natjam, which supports priorities and deadlines for jobs in batch processing systems.

For each system, the motivations, design, implementation, and experimental results were presented. Our systems have been implemented in popular open-source cloud computing frameworks, including MongoDB (Morphus), Cassandra (Parqua), Storm (Stela), LFGraph, and Hadoop (Natjam). We described multiple new approaches to attacking different pieces of the broad problem and incorporated performance assuredness into cloud storage/database systems, stream processing systems, and batch processing systems.

Specifically, the Morphus system supports reconfigurations in NoSQL distributed storage/database systems such as shard key change in sharded databases such as MongoDB. Optimal decisions place new chunks by using bipartite graph matching (which minimizes network volume transferred). Morphus supports concurrent migration of data, processing of foreground queries, and replay of logged writes received while reconfiguration was in progress. The Parqua system extended the Morphus design to ring-based key-value stores; the implementation is in Apache Cassandra, the most popular key-value store in industry today.

The design of NoSQL storage, database systems, and key value stores will continue to evolve as more performance studies characterize new and different applications and specific storage device concerns. The design of such stores must balance competing performance goals, including low latency for searches, high write throughput, high read throughput, concurrency, low write amplification (in solid-state drives), and reconfiguration overhead. Devices such as phase-change memory and memristors offer new opportunities for research and study.

The Stela system supports automated scale-out/in of distributed stream processing systems. The implementation uses Apache Storm, the most popular stream-processing system in industry today. The congestion levels of operators in the stream-processing job (which is a DAG of operators) are identified. For scale-out, more resources are provided to the most congested operators, while for scale-in, resources are removed from the least congested operators. The changes occur in the background without affecting ongoing processing at other operators in the job.

Trends in data analytics show increasing adoption of low-latency, high-throughput stream processing [9–11]. Stream processing is mainly based either on a record-at-a-time or by bulk synchronous processing using batched records, and many optimizations apply to record-at-a-time and batched records [12]. Providing mechanisms to scale-out/in the building blocks of such stream processing in systems that support SLAs/SLOs for multitenant clusters will increase the rapidity of their adoption and make them more readily available for assured cloud computing.

To demonstrate graph processing elasticity, scale-out/in facilities were designed into LFGraph. The approach repartitions the vertices of the graphs (e.g., a Web graph or Facebook-style social network) among the remaining servers, so as to minimize the amount of data moved and thus the time to reconfigure. Migration occurs in the background, while the iterative computation proceeds normally. The Natjam system incorporates support for job priorities and deadlines (i.e., SLAs/SLOs) in Apache YARN. The implementation is in Apache Hadoop, the most popular distributed batch processing system in industry today. Individual tasks are checkpointed so that if one is preempted by a task of a higher-priority (or lower-deadline) job, then it is resumed from where it left off (thus avoiding wasted work). Scheduling policies for both job-level and task-level eviction decide which running components are victimized when a more important job arrives.

Assured cloud systems would benefit from using declarative ways of specifying requirements from users and developers. SLAs/SLOs should be standardized. Further work is needed on (i) extending the richness of these SLAs/SLOs while still keeping them user-facing and away from the innards of the system, and (ii) extending the notion of such requirements to other emerging areas of distributed systems, such as distributed machine learning, for example, through tools such as TensorFlow, PyTorch, or Caffe [13].

9.6 Theoretical Considerations: Inferring and Enforcing Use Patterns for Mobile Cloud Assurance

Enormous numbers of mobile devices and smart sensors are incrementally relatively inexpensive to scale up and have limited computational resources (memory, processing capability, and energy). At the same time, cloud computing provides elastic on-demand access to virtually unlimited resources at an affordable price. Integrating mobile devices, smart sensors, and elastic on-demand clouds provides new functionality and quality of service for mobile users. Such an integrated system is referred to here as a *mobile cloud*.

Chapter 7 considered these hybrid mobile clouds using an actor-based approach to programming. By using actors, we explored functionality and efficiency while enforcing security and privacy. Two key ideas were studied. First, the fine granularity of actor units of computation allows agility in a mobile cloud by facilitating migration of components. Migration occurs in response to a system context (including dynamic variables such as available bandwidth, processing power, and energy) while respecting constraints on information containment boundaries. Second, information flow between actors can be observed and suspicious activity flagged or prevented by specifying constraints on actor interaction patterns. Our approach facilitates a "holistic" form of assured cloud computing, which we have realized in a prototype mobile hybrid cloud platform; as we discussed in the chapter, suitable formalisms can be used to capture interaction patterns to improve the safety and security of computation in the mobile cloud.

The actor mobile cloud framework can enable effective balancing of resource use with performance while preserving security. There is a particular concern, however, with energy management in mobile devices. Further research is needed in techniques to infer the energy consumed by a specific application on a mobile device. Monitoring techniques that infer coordination constraints and session types are another area in which more research is needed in order to allow patterns of interaction in a running system to be inferred. System assurance will be enhanced by flagging suspicious deviations as well as preventing harmful actions. Moreover, the current notations for representing session types are formal in nature and not suitable for use by system developers. A friendly interface could help programmers visualize the interaction behaviors as well as reduce errors in the system. Moreover, session types may help detect information leaks, as certain sequences may reveal more information than would a single message interaction. In addition, statistical sampling can help detect violations of quantitative coordination constraints when not just information sequences, but also the sum total of information revealed from a very large number of sources, need to be constrained.

The challenge of addressing the impact of failures and faults on interaction patterns needs to be studied further. Such failures are not explicitly incorporated either in the language of synchronizers or in session types. Adding support for dynamic process creation will be an important direction for future work in session types for actor systems. In its current form, System-A cannot express actor creation as a behavior, and global types assume that all participants already exist. Matching a created actor with its subsequent use in a type requires an extra step that is not obvious. Furthermore, System-A omits support for session delegation, and does not deal with issues of progress. Finally, it does not consider overlapping nested indexed names nested and included in multiple operators. That omission disallows, for example, all-to-all communication.

9.7 Certifications

Security standards are tools that help improve security and privacy by establishing a security baseline and supporting the implementation of privacy- and security-enhancing measures. Such standards are revised and improved over time in an effort to keep pace with technological change and the emergence of new threats, and security certifications are a widely used mechanism for demonstrating compliance with standards. In Chapter 8, we looked at how three of the most popular and highly regarded certifications used for cloud security – ISO/IEC 27001, SOC 2, and FedRAMP – have evolved over time, how they stack up in comparison to each other, and what the implications are of having multiple certification options instead of a single universally used standard.

We concluded that the three standards have important similarities and similar goals; all three are based on a set list of criteria and controls that must be assessed by independent, accredited third parties. We found a number of key differences as well. For example, as discussed in the chapter, while FedRAMP focuses specifically on cloud services offered by private outsourcers to the federal government [14, 15], ISO/IEC 27001 and SOC 2 have a more general scope; their security assessment tools are available for use by any kind of service organization. At the same time, FedRAMP requires more than twice as many controls as the other standards, reflecting FedRAMP's more specific and detailed controls, while those in ISO/IEC 27001 and SOC/TSPC are more general.

Likewise, SOC 2 and ISO/IEC 27001 are more flexible than FedRAMP in specifying how to implement the criteria and controls that they require.

Naturally, the value of a security standard lies in how well the approaches that it enforces actually protect systems against threats. Therefore, the existence of multiple standards with notable similarities raises the question of why multiple standards are needed, if everyone is trying to achieve the same goal of having the best possible security?

Upon close examination, we found that the three standards are not in fact redundant; rather, they show high complementarity and compensate for each other's weaknesses and omissions. There is good reason for a cloud service provider to invest in complying with multiple standards instead of only one. Even so, given that obtaining certifications is costly, it would still be desirable for the cloud computing community to develop a single standard that offers all the protections that are currently articulated piecemeal across multiple standards.

However, another challenge to standardization is the reality that new vulnerabilities and threats are continually appearing, and new defensive countermeasures are being continually developed in response. The impact of the "Treacherous Twelve" on the effectiveness of IT security standards in cloud environments points toward a possible path to the improvement of IT security standards. Observations of threats, issues, and vulnerabilities can help cloud providers and users understand the need for new or different control measures, and their connection to security standards can lead to better effectiveness, completeness, and efficiency.

The goal of locking down standardized protection methods will for the foreseeable future be in tension with the constant evolution of the threat landscape that those methods must handle. Worse, to date, the academic and industrial stakeholders have tended to study new threats more or less in isolation, and coordination with standards developers has been limited. To optimize the responsiveness of standards, the study of vulnerabilities should be more actively and closely connected to the maintenance of security standards.

References

1 U.S. Department of Defense (2014) The DoD Cloud Way Forward, version 1.0, Jul. 23. Available at http://iase.disa.mil/Documents/dodciomemo_w-attachment_cloudwayforwardreport-20141106.pdf.

2 Defense Information Systems Agency (DISA) (2015) Best Practices Guide for Department of Defense Cloud Mission Owners, version 1.0, Aug. 6. Available at http://iasecontent.disa.mil/stigs/pdf/unclass-best_practices_guide_for_dod_cloud_mission_owners_FINAL.pdf.

3 Owens, K. (2017) MilCloud 2.0 upgraded with commercial cloud infrastructure, Defense Systems, Jun. 12. Available at https://defensesystems .com/articles/2017/06/12/milcloud.aspx.

4 International Organization for Standardization (ISO) (2009) ISO/IEC 11889-1:2009: Information Technology – Trusted Platform Module – Part 1: Overview. Available at https://www.iso.org/standard/50970.html (accessed Nov. 29, 2013).

5 Holgate, R. and Cannon, N. (2017) Get ready for the inflection point in U.S. federal government cloud adoption, Gartner, Inc. Jun. 9. Available at https:// www.gartner.com/doc/3187120/ready-inflection-point-federal-government.

6 Top Threats Working Group, Cloud Security Alliance (CSA) (2016) The treacherous 12: cloud computing top threats in 2016. Available at https:// downloads.cloudsecurityalliance.org/assets/research/top-threats/Treacherous-12_Cloud-Computing_Top-Threats.pdf.

7 FedRAMP. Program overview. Available at https://www.fedramp.gov/about-us/about/ (accessed May 26, 2016).

8 Bernstein, D. (2014) Containers and cloud: from LXC to Docker to Kubernetes. *IEEE Cloud Computing*, **1** (3), 81–84.

9 Murray, D.G., McSherry, F., Isaacs, R., Isard, M., Barham, P., and Abadi, M. (2013) Naiad: a timely dataflow system, in Proceedings of the 24th ACM Symposium on Operating Systems Principles (SOSP), pp. 439–455.

10 Ousterhout, K., Wendell, P., Zaharia, M., and Stoica, I. (2013) Sparrow: distributed, low latency scheduling, in Proceedings of the 24th ACM Symposium on Operating Systems Principles (SOSP), pp. 69–84.

11 Noghabi, S.A., Paramasivam, K., Pan, Y., Ramesh, N., Bringhurst, J., Gupta, I., and Campbell, R.H. (2017) Samza: stateful scalable stream processing at LinkedIn. Proceedings of the VLDB Endowment, **10** (12), 1634–1645.

12 Venkataraman, S., Panda, A., Ousterhout, K., Ghodsi, A., Armbrust, M., Recht, B., Franklin, M.J., and Stoica, I. (2017) Drizzle: fast and adaptable stream processing at scale, in Proceedings of the 26th Symposium on Operating Systems Principles (SOSP), pp. 374–389.

13 Wikipedia. (2017) Comparison of deep learning software. Available at https:// en.wikipedia.org/wiki/Comparison_of_deep_learning_software (accessed Sep. 2, 2017).

14 Halvorsen, T.A., (2014) Updated Guidance on the Acquisition and Use of Commercial Cloud Computing Services, Chief Information Officer, U.S. Department of Defense, Dec. 15. Available at http://iase.disa.mil/Documents/commercial_cloud_computing_services.pdf.

15 Information Assurance Support Environment (IASE) (2017) DoD Cloud Computing Security, May 17, Defense Information Systems Agency (DISA). Available at http://iase.disa.mil/cloud_security/Pages/index.aspx.

16 Intel (2015) Improving Real-Time Performance by Utilizing Cache Allocation Technology: Enhancing Performance via Allocation of the Processor's Cache: White Paper, Document No. 331843-001US. Available at http://www .intel.com/content/dam/www/public/us/en/documents/white-papers/cache-allocation-technology-white-paper.pdf. (accessed Aug. 23, 2017).

Index

Assured Cloud Computing, First Edition. Edited by Roy H. Campbell, Charles A. Kamhoua, and Kevin A. Kwiat.